# Theoretical Chemistry Accounts / New Century Issue

Special reprint edition of Vol. 103  Number 3–4

Springer-Verlag Berlin Heidelberg GmbH

C. J. Cramer    D. G. Truhlar (Eds.)

# Theoretical Chemistry Accounts

## New Century Issue

With contributions by numerous experts
Special reprint edition of Vol. 103  Number 3–4

With 28 Figures and 2 Tables

Springer

PROFESSOR CHRISTOPHER J. CRAMER
PROFESSOR DONALD G. TRUHLAR

University of Minnesota
Department of Chemistry
207 Pleasant Street SE
55455-0431 Minneapolis, MN
USA

Special reprint edition of Theoretical Chemistry Accounts, Volume 103, Number 3–4, February 2000,
C. J. Cramer, D. G. Truhlar (Eds.)
With contributions by numerous experts

ISSN 1432-881X
ISBN 978-3-540-67867-0

CIP data applied for
Die Deutsche Bibliothek – CIP-Einheitsaufnahme
Theoretical chemistry accounts: new century issue / ed.: Christopher J. Cramer; Donald G. Truhlar.

ISBN 978-3-540-67867-0        ISBN 978-3-662-10421-7 (eBook)
DOI 10.1007/978-3-662-10421-7

© Springer-Verlag Berlin Heidelberg 2001
Originally published by Springer-Verlag Berlin Heidelberg New York in 2001

Printed on acid-free paper    SPIN 10778582    51/3130/as  5 4 3 2 1 0

*Preface*
# A new century of theoretical chemistry

The end of the century provides a time to look back, and the start of a new century inspires one to look ahead. For the New Century issue of *Theoretical Chemistry Accounts,* we asked the advisory editors of the journal to identify a paper from the first century of theoretical chemistry and discuss its importance for the twentieth century with an eye towards the twenty-first century. Although we collected 66 such perspectives, they are far from representing a full set of the papers deserving our reflection. At one point we thought of inviting additional perspectives to fill some obvious gaps or provide better balance, but we decided that any such attempt would spoil the spontaneity of the collection, which is part of its interest.

We hope the *Theoretical Chemistry Accounts* New Century Issue will be entertaining, but even more we hope it will help younger scientists to see the role that selected creative efforts of the past have played in getting us to the present. Further, to the extent that this issue inspires members of the theoretical chemistry community to reread the papers that led to the present state of the art, that is all to the good, because there is still a lot to be learned from the last century's literature. Finally, taking this issue as prolog, we can all try to make our upcoming efforts as influential for the next century as these papers were for the last.

C. J. Cramer, D. G. Truhlar
Minneapolis, USA
December 31, 1999

# Theoretical Chemistry Accounts

Theory, Computation, and Modeling
Theoretica Chimica Acta

Springer

## New Century Issue

# Theoretical Chemistry Accounts

Theory, Computation, and Modeling
Theoretica Chimica Acta

Springer

Table of contents continued

# Theoretical Chemistry Accounts

Theory, Computation, and Modeling
Theoretica Chimica Acta

Online edition in LINK - Chemical Sciences
Online Library http://link.springer.de

Indexed in *Current Contents*

Theor Chem Acc (2000) 103:168–170
DOI 10.1007/s002149900081

Theoretical
Chemistry Accounts
© Springer-Verlag 2000

*Perspective*

# Perspective on "Zur Quantentheorie der Spektrallinien"

## Sommerfeld A (1916) Ann Phys (Leipzig) 51:1–94, 125–167

**Bernd Artur Hess**

Lehrstuhl für Theoretische Chemie, Friedrich-Alexander-Universität Erlangen-Nürnberg, Egerlandstrasse 3, D-91058 Erlangen, Germany

Received: 4 March 1999 / Accepted: 29 July 1999 / Published online: 4 October 1999

**Abstract.** Among many other results, Arnold Sommerfeld gave in his article the correct expression for the relativistic bound-state energy levels of the hydrogen atom, well before the development of wave mechanics, clear ideas about the electron spin, and Dirac's relativistic wave equation. He correctly attributed the fine structure of atomic spectra to relativistic effects, and thus published the first paper giving a quantitative perspective on relativistic quantum chemistry.

**Key words:** Relativistic quantum chemistry – Sommerfeld transformation

Sommerfeld's paper features three major sections, each presenting new ideas for the explanation of the phenomenology of the spectra of hydrogen and hydrogen-like atoms. The first section develops the theory of the Balmer series in the spectrum of the hydrogen atom. The theoretical method employed is the application of Bohr–Wilson–Sommerfeld quantization rules to the nonrelativistic bound-state Kepler problem, i.e., the physics of a charged body moving in an attractive central potential. The extension of previous ad hoc quantization rules to phase integrals, enabling the application of the "old quantum theory" to nonspherical orbits, is introduced in this section. Sommerfeld uses these new rules to derive the expressions for the elliptic (bound-state) orbits, in particular the ones with nonspherical symmetry, i.e., nonvanishing eccentricity, $\varepsilon$.

The second section deals with the fine structure of the hydrogen spectral lines. The discussion of the bound-state Kepler problem is extended to the relativistic case. Following an idea of Bohr [1], who had already conjectured that the fine structure of the hydrogen spectrum could be a relativistic effect proportional to $e^4/\hbar^2 c^2$, the fine structure of the hydrogen spectrum is

thus explained using the "quantized" result of this calculation. The fine-structure constant is introduced as a measure of the size of the relativistic effects. It is subsequently used as an expansion parameter for various quantities, thus defining a "nonrelativistic limit". The separation of kinematical relativistic effects and fine-structure splitting is discussed considering the spherical orbits.

In the third section, X-ray spectra (Röntgenspektren) of the hydrogen atom are discussed. The principles developed in the first two parts of the paper are applied to inner-shell spectroscopy. The paper is characterized by a quick adaptation and rapid development of the ideas of the "old quantum theory" introduced by Bohr, which was vividly discussed in the contemporary literature [2–4]. The simultaneous treatment of relativity and quantum effects is the first of its kind.

Sommerfeld's quantization rules

$$\int p_i \, dq_i = 2\pi n_i \hbar, \quad i = 1, \dots, D$$

for a system with $D$ degrees of freedom are not independent of the coordinate system, and he had to carry out an explicit separation of variables in order to apply them. Einstein [5] reinvestigated the problem and, besides giving a coordinate-invariant formulation of the quantum conditions, pointed out that a quantization by means of classical action integrals is possible only for integrable systems. These quantization rules for integrable systems were later refined by Keller [6] and are known as Einstein–Brillouin–Keller rules. The method of semiclassical quantization was revived in the early 1980s, when Gutzwiller [7] found a way to apply semiclassical quantization also to nonintegrable systems. His method is based on the Feynman path integral and its expansion around closed classical paths in phase space, the "periodic orbits", and is nowadays instrumental in the context of strongly chaotic systems. For an account of these developments see the recent monograph by Grosche and Steiner [8].

In the following I shall focus on the second section of the paper, dealing with fine structure, since to some extent it survived the quantum revolutions of the 1920s, and until the end of the century still fostered interesting insight, in particular into symmetry aspects of the problem.

The results presented in Sommerfeld's papers are remarkable for three reasons. Sommerfelds tackled (and successfully solved) a problem for which the appropriate theoretical tools were only available a decade later. This fact has been called the "Sommerfeld puzzle", and a solution to the puzzle was given in a beautiful paper by Biedenharn [9], addressing this issue in the following words:

"Clearly Sommerfeld's methods were heuristic (Bohr quantization rules), out-dated by *two* revolutions (Heisenberg–Schrödinger nonrelativistic quantum mechanics and Dirac's relativistic quantum mechanics) and his methods obviously had no place at all for the electron spin, let alone the four-components of the Dirac electron. So Sommerfeld's correct answer could only be a lucky accident, a sort of cosmic joke at the expense of serious minded physicists."

Biedenharn analyzed Sommerfeld's method and gives a surprising explanation for Sommerfeld's success, thus showing that Sommerfeld did indeed obtain the right answer for the right reason.

Biedenharn first explains the agreement of Sommerfeld's nonrelativistic quantum numbers with the exact answer. This agreement is by no means trivial, since usually Bohr–Sommerfeld quantization rules yield quantum number which are shifted by an unknown numerical constant from the exact ones. In the nonrelativistic Kepler problem there is, however, a quantum-mechanical operator corresponding to the classical eccentricity. This makes it possible to define the "spherical" orbits (i.e., those with vanishing eccentricity) in an unambiguous manner, which gives an absolute frame of reference for the Bohr–Sommerfeld quantum numbers.

Second, there is a special reason that Sommerfeld's procedure works at all in the relativistic case. In a space-fixed frame of reference, the relativistic Kepler orbit is not closed. Rather, the perihel advances in each revolution, which leads to a rosettelike orbit. Sommerfeld uses a rotating frame of reference, which effects that the relativistic orbit is again of the form a conic section (i.e., an ellipse for the bound states), albeit with an angle variable different from the nonrelativistic analogue. In this frame of reference, the phase integral

$$\int_{\phi=0}^{\phi=2\pi} p_\phi \, d\phi = 2\pi n_\phi \hbar \; ,$$

with the classical angular momentum $p_\phi = mr^2\dot{\phi}$, can be used to quantize the angular motion also in the relativistic case. Biedenharn shows that there is a quantum-mechanical counterpart to the (classical) "Sommerfeld transformation" to the moving frame, such that Sommerfeld's solution carries over to the exact treatment.

A third condition is required to explain Sommerfeld's success. Biedenharn shows that, surprisingly, the nonrelativistic problem solved by Sommerfeld is that of a nonrelativistic particle with (dynamically independent) spin, rather than a spinless nonrelativistic particle. If the spinless Schrödinger equation is used, the operator

analogue for the eccentricity is given by the length of the Runge–Lenz vector

$$\mathbf{A} = \hat{r} + \frac{\hbar}{2Ze^2m}(\mathbf{L} \times \mathbf{p} - \mathbf{p} \times \mathbf{L}) \; ,$$

where $\hat{r}$ is a unit vector in the direction of the radius vector, and the other symbols have obvious meanings. From the commutation relations of $\mathbf{A}$ with $\mathbf{L}$, the relation $L^2 + n^2A^2 + 1 = n^2$ obtains, where for simplicity we work in a basis of energy eigenfunctions, with $n$ denoting the principal quantum number. If we solve for $\varepsilon \equiv (\mathbf{A} \cdot \mathbf{A})^{1/2}$, and replace $L^2$ by its eigenvalue, we get

$$\varepsilon = \left(1 - \frac{l(l+1)+1}{n^2}\right)^{\frac{1}{2}} \; .$$

From this relation we see that (except for $n = 1$) the eccentricity does not vanish for orbitals with nodeless radial probability density, which are characterized by the condition $l = n - 1$ and correspond to the classical circular orbits.

If, on the other hand, a dynamically independent spin is introduced, the appropriate orbital angular momentum operator is

$$K = \beta\left(\boldsymbol{\sigma}\mathbf{L} + \tfrac{1}{2}\right)$$

rather than $\mathbf{L}$ itself. As is well known from the Dirac treatment of the relativistic electron, this operator has eigenvalues $\kappa = \pm 1, \pm 2, \ldots,$. If the commutation relations of $\mathbf{A}$ with $K$ are employed, one arrives at the relation

$$n^2(\boldsymbol{\sigma} \cdot \mathbf{A})^2 + K^2 = n^2 \; ,$$

and defining the operator analogue of the eccentricity as

$$(\boldsymbol{\sigma}\mathbf{A}\boldsymbol{\sigma}\mathbf{A})^{\frac{1}{2}} \equiv \varepsilon = \left(1 - \frac{\kappa^2}{n^2}\right)^{\frac{1}{2}} \; ,$$

again replacing $K^2$ by its eigenvalue $\kappa$, we arrive at vanishing eccentricity for orbits, which correspond to the condition $l = n - 1$ for nodeless radial probability density. Moreover, the problem of "Pendelbahnen", i.e., orbits passing through the nucleus, which had to be excluded heuristically, does not occur since $\kappa$ cannot be zero. Sommerfeld himself remarked in his paper (page 21) that "at this point already a relativistic generalization is required …". In fact, the generalization required is not a relativistic one, but rather the inclusion of spin (which was unknown at the time Sommerfeld wrote his paper). Consideration of spin in the quantum-mechanical treatment resolves this problem and gives the right expectation value for the above-mentioned operator for the eccentricity of spherical orbits.

The same is true for two more special cases, namely the relativistic particle with spin (yielding the correct Dirac energy levels), and a relativistic particle without fine structure. In these cases, the Runge–Lenz vector is no longer a constant of motion, and the $O_4$ symmetry of the nonrelativistic problem is broken. It is, however, not "seriously" broken, and an analogue of the Runge–Lenz vector, the so-called Johnson–Lippman operator [10] can be used in lieu of **A** defined above [11]. This has the effect that a residual (super) symmetry is left in the energy levels of Dirac–Kepler problem, which has been analyzed and explained only recently [12].

The previous discussion shows that neither the spin nor the operator $K$ are related to relativistic effects (as is often claimed), but rather they are compatible with nonrelativistic motion (Galilei group relativity) as well as relativistic motion (Poincaré group relativity). This point was also made in several of the papers by Lévy-Leblond [13].

Besides making implicit use of these really puzzling properties of the relativistic Kepler problem, the second major impact of Sommerfeld's article lies in several notions introduced which lie at the foundation of relativistic quantum chemistry and have since been instrumental in the field: the notion of scalar (kinematical) relativistic effects versus fine-structure effects, the introduction of the fine-structure constant, $\alpha = e^2/\hbar c$, and the expansion of the relativistic expressions in powers of the square of this constant. The idea that relativistic effects decisively influence the structure of the outer electrons of the atoms is at the root of relativistic quantum chemistry.

Last but not least, in Sommerfeld's article, a spirit of theoretical work is developed which is, on the one hand, deeply rooted in the experimental observations (making ample use of the spectroscopic results to derive heuristic concepts), and on the other hand, is led by the belief that there is a microscopic explanation for the experimentally observed phenomena. This paradigm is the foundation for the physics of the whole of the twentieth century, including theoretical chemistry. It is interesting to observe that it was again (this time molecular) spec-troscopy which turned out to be the field of the first successes of quantum chemistry, by means of guidelines very similar to those behind Sommerfeld's work. The first one is the quest for the "right answers for the right reasons", even under the conditions of inappropriate theoretical methods and the need for heuristic concepts, and the danger that the right reasons will be fully known only decades later. The second one is the emphasis on the treatment of "real systems", i.e., systems which are of current interest for experimentalists. I am personally convinced that successful work in theoretical chemistry will continue to build on these guidelines for quite some time in the century to come.

Invoking those analogies to the development of theoretical chemistry, Sommerfeld's paper could be termed a twentieth century theoretical chemist's paper, published long before this branch of science was called into existence.

# References

1. Bohr N (1915) Philos Mag 174: 332–335
2. Wilson W (1915) Philos Mag 174: 795–802
3. Planck M (1916) Ann Phys (Leipzig) 50: 385–418
4. Schwarzschild K (1916) Berl Akad Wiss 548
5. Einstein A (1917) Verh Deutsch Phys Ges 19: 82–92
6. Keller JB (1958) Ann Phys (NY) 4: 180–188
7. Gutzwiller MC (1980) Phys Rev Lett 45: 150–153
8. Grosche C, Steiner F (1998) Handbook of Feynman path integrals. Springer, Berlin Heidelberg New York
9. Biedenharn LC (1983) Found Phys 13: 13–34
10. Johnson MH, Lippmann BA (1950) Phys Rev 78: 329
11. Biedenharn LC (1962) Phys Rev 126: 845–851
12. Dahl JP, Jørgensen T (1995) Int J Quantum Chem 53: 161–181
13. Lévy-Leblond JM (1974) Riv Nuovo Cimento 4: 99

Theor Chem Acc (2000) 103:171–172
DOI 10.1007/s002149900036

Theoretical
Chemistry Accounts
© Springer-Verlag 2000

## Perspective

# A Perspective on "Volume and heat of hydration of ions"

## Born M (1920) Z Phys 1: 45

**B.M. Pettitt**

Department of Chemistry, University of Houston, Houston, TX 77204-5641, USA

Received: 9 March 1999 / Accepted: 30 March 1999 / Published online: 14 July 1999

**Abstract.** Born's simple derivation of the free energy of hydration of ions is classic. It connects the microscopic atomic properties with the macroscopic thermodynamics in a transparent fashion.

**Key words:** Free energy – Poisson–Boltzmann – Solvation

Born's paper [1] in the first volume of *Zeitschrift für Physik* holds importance in many regards for my thoughts on the subject of hydration and the way in which many solvation calculations, new and routine, are now performed. The subject of calculating the thermodynamics, and in particular the free energies of hydration for species in aqueous solution, owes much to this rather brief (only three pages) work of Born. Whether by thermodynamic integration, thermodynamic perturbation or by a dressed quantum mechanical calculation, modern attempts to calculate equilibrium solvation effects are common in the literature [2–4]. Solvation effects are now often calculated for molecules as complicated as peptides and oligonucleotides in efforts to correlate experimental observations with our current understanding of the various phenomenological components of hydration [5].

With this work Born took the step of using atomic concepts and parameters, mixing them with continuum ideas (implicit solvent models) to make correlations with bulk thermodynamics. Not only was this a successful calculation but it remains a common theme in much current work on the subject some 80 years later. Born's theoretical understanding still underpins the field today, whether approached by many-body approximations [6] or by computer simulations [7].

Born's work on this paper evidently began when he read a pair of articles in which Fajans [8] had attempted to calculate the Born-cycle component of the work or free energy (there is some confusion in the works of the time using the words energy and free energy somewhat indiscriminately as there is even today on occasion) for taking ions from an aqueous salt solution to a vacuum by using lattice energies calculated earlier by Born [9]. The calculations were regarded with some doubt by Born because of the need for ionization energies and electron affinities which were not available at the time. This produced "heats" with the wrong sign. Indeed Fajans had used hydrogen at a platinum electrode as a reference and in his first paper had omitted the "heat of evaporation of the electrons from the platinum" or the ionization energy. Even with an ad hoc correction, the calculation's accuracy still did not impress Born.

Born was, however, taken with the fundamental idea Fajans had about the process of hydration. Fajans recognized that the polarization of water in the presence of an ion and not the formation of stoichiometric hydrates was the dominating characteristic of ionic hydration. The idea that the dipole moment of water in proximity to the ions would be partially aligned was noted in the writings of Fajans [8] and Born [1]. Born utilized the concept of Nernst, by then familiar, that the dissolution of salts is correlated with solvent dielectric or solvent polarity. Born sought to quantify the idea.

Born then decided to neglect the explicitly detailed structure of water molecules and replace them with a continuous electrically polarizable medium. This approximation is the same as that made in any Poisson–Boltzmann calculation, but unlike the Debye–Hückel approximation, the ions in Born's calculation retained finite size. Thus, the Born treatment had the possibility of seeing chemically relevant differences due to ionic size. It should be remarked that Fajans was also looking for the chemically interesting dependence on ionic size in his less successful calculations.

Utilizing the relation between the integral of the field strength squared and the energy in the field, Born simply set the field contribution on the interior of ions to zero by integrating from the ionic radius to infinity. Subtracting the result in water (or any dielectric $\varepsilon$) from the result in air ($\varepsilon = 1$) gave his famous equation for the work of charging an ion in a dielectric continuum.

$$W = \frac{1}{2}\left(1 - \frac{1}{\varepsilon}\right)\frac{z^2 e^2}{r_i}$$

From this, Born fit $r_i$ to the experimental numbers for W. He was able to deduce that reasonable numbers for the ionic radii could be found. In addition, he found that the numbers for the positive ions were universally smaller than the then accepted atomic values (from crystal densities) and those for the anions were bigger. This he correctly interpreted in terms of the size change of the atom with the state of ionization long before the quantum mechanics of such systems was worked out.

This simple calculation, of which Born was quite skeptical, was the beginning of a quantitative understanding of ionic solutions in terms of atomic parameters. This would not be significantly improved in a systematic way until the work of Mayer in 1950 [10]. Building upon Mayer, Freidman [11] in the 1960s followed by a flood of researchers succeeded in increasing the level of correlations in the solution and the details of molecular structure in the solvent. With the rediscovery of finite-difference calculation methods for differential equations, the popularity of Born's method utilizing the Poisson–Boltzmann equation for irregularly shaped objects has allowed the extension of the method to large proteins and nucleic acids [12] in dilute saline solution. Theories to bring such macromolecular system calculations up to the level of Mayer have been in progress for the last decade or more.

Born's short paper brought the best ideas of the time together. It produced a work of lasting significance in terms of the ideas and concepts. Born first took the step of connecting atomic ideas with simple solvent field models to calculate bulk thermodynamics. It is still often cited as a fundamental intellectual source for ionic hydration theories and therefore must rank as one of the seminal contributions to theoretical chemistry in the twentieth century.

*Acknowledgements.* I wish to thank Michael Feig for help with the original German text. I thank A.D.J. Haymet for helpful comments on the manuscript.

# References

1. Born M (1920) Z Phys 1: 45
2. Levy RM, Gallicchio E (1998) Annu Rev Phys Chem 49: 531
3. Beveridge DL, DiCapua FM (1989) Annu Rev Biophys Chem 18: 431
4. Kollman PA (1996) Acc Chem Res 29: 461
5. Marlow GE, Perkyns JS, Pettitt BM (1993) Chem Rev 93: 2503
6. Perkyns J, Pettitt BM (1994) Biophys Chem 51: 129
7. Smith PE, Pettitt BM (1994) J phys Chem 39: 9700
8. (a) Fajans K (1919) Verh Dtsch Phys Ges 21: 549; (b) Fajans K(1919) Verh Dtsch Phys Ges: 21: 709
9. Born M (1919) Verh Dtsch Phys Ges 21: 13
10. Mayer JE (1950) J Chem Phys 18: 1426
11. Friedman HL (1962) Ionic solution theory Wiley New york
12. Nicholls A, Honig B (1995) Science 268: 1144

Theor Chem Acc (2000) 103:173–176
DOI 10.1007/s002149900049

Theoretical
Chemistry Accounts
© Springer-Verlag 2000

*Perspective*

# Perspective on "Zur Quantentheorie der Molekeln"

## Born M, Oppenheimer R (1927) Ann Phys 84: 457

**John C. Tully**

Department of Chemistry, Yale University, New Haven, CT 06520-8107, USA

Received: 2 March 1999 / Accepted: 13 April 1999 / Published online: 14 July 1999

**Abstract.** The Born–Oppenheimer approximation, introduced in the 1927 paper "On the quantum theory of molecules", provides the foundation for virtually all subsequent theoretical and computational studies of chemical binding and reactivity, as well as the justification for the universal "ball and stick" picture of molecules as atomic centers attached at fixed distances by electronic glue.

**Key words:** Adiabatic – Nonadiabatic – Potential-energy surface – Born-Oppenheimer approximation

Chemistry is about structure and reactivity. Modern discussions of both these subjects center on the concept of the "potential-energy surface", $\mathscr{E}(\mathbf{R})$. As illustrated in Fig. 1, $\mathscr{E}(\mathbf{R})$ is the energy of a molecular system or, more generally, of any collection of interacting atoms when the nuclei are fixed at position $\mathbf{R} = \{\mathbf{R}_1, \mathbf{R}_2, ...\}$. For a diatomic molecule, $\mathscr{E}(\mathbf{R})$ is a one-dimensional "potential-energy curve" representing the energy of the molecule as a function of the internuclear distance. The location, $R_0$, of the minimum energy along the curve is the bond length of the molecule. The energy difference, $\mathscr{E}(\infty) - \mathscr{E}(R_0)$, is the energy required to break the bond, i.e., the bond strength. The curvature at the bottom of the potential well is the force constant that determines the vibrational frequency of the molecule. The bond length, $R_0$, determines the moment of inertia of the molecule, i.e., its rotational motion. Similarly, for a polyatomic system, stable conformations correspond to local minima of $\mathscr{E}(\mathbf{R})$, for example, points A and B in Fig. 1. The properties of $\mathscr{E}(\mathbf{R})$ in the vicinity of each local minimum govern the vibrational and rotational spectrum and the energetic stability of the conformation. Reactivity is determined by the pathways that lead from one stable minimum to another, or between a minimum and a valley corresponding to separated reactants or products, for example, regions C and D in Fig. 1. The minimum energy path connecting two stable conformers

is often identified as the "reaction coordinate". The saddle point $\mathbf{R}_{ts}$, or position of maximum energy along the minimum energy path, is the "transition state" (point E in Fig. 1.). The properties of $\mathscr{E}(\mathbf{R})$ in the vicinity of $\mathbf{R}_{ts}$ are the input for the widely used "transition-state-theory" or "activated-complex theory" of chemical reaction rates. The actual time-dependent trajectory that the system follows as a reaction progresses (e.g., the solid curve with directional arrows in Fig. 1) is the focus of the field of chemical dynamics. Old chemical bonds may be broken and new ones formed as the system evolves along the trajectory, and dynamical questions can be addressed such as is a long-lived intermediate involved or how is the energy of reaction deposited among the degrees of freedom of the products? (The oscillatory motion of the trajectory in Fig. 1 illustrates vibrational excitation of products).

The above discussion of chemical structure, properties, stability, reactivity and dynamics does not mention the word "electron". Indeed, when we picture a molecule in our minds, on paper or on the computer screen, we assign positions for each of the nuclei but rarely designate the positions of electrons. The justification for this is the Born-Oppenheimer approximation [1]. In the 1920s the new quantum theory was able to demystify the electronic structure of atoms, even quantitatively for hydrogen; however, molecules exhibit all the electronic complexity of atoms and, in addition, comparably complex interactions among the nuclei. Born and Oppenheimer recognized that a great simplification results because the mass of the electron is much less than that of any nucleon. (The mass of the lightest atom, hydrogen, is 1836 times that of the electron.) To a good approximation (with exceptions, as discussed later) electrons respond instantaneously to the much slower motions of the nuclei. As the nuclei move through a position $\mathbf{R}$, the electrons readjust to the same optimum configuration that they would have if the nuclei were stationary at position $\mathbf{R}$. The energy of this optimal electronic configuration is $\mathscr{E}(\mathbf{R})$, a point on the potential-energy surface. Knowledge of $\mathscr{E}(\mathbf{R})$ is sufficient to determine structure and reactivity; explicit knowledge of electronic motion is not required once $\mathscr{E}(\mathbf{R})$ has been determined.

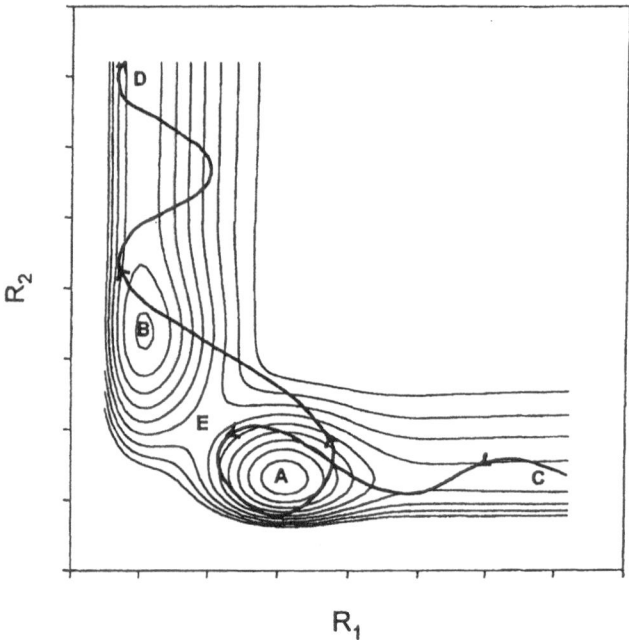

$R_2$

$R_1$

**Fig. 1.** A schematic two-dimensional representation of a multidimensional potential-energy surface. $R_1$ and $R_2$ represent bond distances. The *solid curves* are contours of equal electronic energy, $\mathscr{E}(\mathbf{R})$

This prompted Eyring and Polanyi [2], in 1931, to present the first formulation of chemical reaction dynamics in terms of a multidimensional potential-energy surface, using the hydrogen-exchange reaction as an example. The electrons have not been removed from the problem, of course. For any nuclear configuration $\mathbf{R}$, the energy $\mathscr{E}(\mathbf{R})$ is determined by the full, many-electron Schrödinger equation for fixed nuclear positions $\mathbf{R}$. Thus, the Born-Oppenheimer approximation separates the problem of chemical structure and reactivity into two parts: the electronic structure part and the nuclear motion part.

The Born–Oppenheimer separation is derived as follows. The total nonrelativistic quantum mechanical Hamiltonian for a system of interacting atoms is

$$\mathscr{H} = -\frac{\hbar^2}{2}\sum_\alpha M_\alpha^{-1}\nabla^2_{\mathbf{R}_\alpha} - \frac{\hbar^2}{2}\sum_i m_e^{-1}\nabla^2_r + V(\mathbf{r},\mathbf{R})$$
$$= -\frac{\hbar^2}{2}\sum_\alpha M_\alpha^{-1}\nabla^2_{\mathbf{R}_\alpha} + \mathscr{H}_{el}(\mathbf{r};\mathbf{R}) \ , \qquad (1)$$

where $\mathbf{r}$ and $\mathbf{R}$ denote the positions of the electrons and nuclei, respectively, $M_\alpha$ is the mass of nucleus $\alpha$ and $m_e$ is the electron mass. $V(\mathbf{r},\mathbf{R})$ includes all interparticle interactions: electron–electron repulsions, electron–nuclear attractions, and nuclear–nuclear repulsions. Thus $\mathscr{H}_{el}(\mathbf{r};\mathbf{R})$ is the entire Hamiltonian of the system with the exception of the kinetic-energy operator for the slow particles. $\mathscr{H}_{el}(\mathbf{r};\mathbf{R})$ can be viewed as the Hamiltonian that governs the electrons when the nuclei are fixed at position $\mathbf{R}$. We now define the adiabatic (Born–Oppenheimer) electronic wave functions $\Phi_j(\mathbf{r};\mathbf{R})$ to be the eigenfunctions of $\mathscr{H}_{el}(\mathbf{r};\mathbf{R})$ for a fixed $\mathbf{R}$:

$$\mathscr{H}_{el}(\mathbf{r};\mathbf{R})\Phi_j(\mathbf{r};\mathbf{R}) = \mathscr{E}_j(\mathbf{R})\Phi_j(\mathbf{r};\mathbf{R}) \ . \qquad (2)$$

$\mathscr{E}_j(\mathbf{R})$ is the adiabatic or Born-Oppenheimer potential-energy surface corresponding to electronic state $j$. The ground-state ($j = 0$) potential-energy surface $\mathscr{E}_0(\mathbf{R})$ is the same as that referred to in qualitative terms in the first paragraph. The $\mathscr{E}_j(\mathbf{R})$ for $j \neq 0$ are the potential-energy surfaces corresponding to the excited electronic states. The $\Phi_j(\mathbf{r};\mathbf{R})$ depend only parametrically on the nuclear positions $\mathbf{R}$; thus the semicolon in $\Phi_j(\mathbf{r};\mathbf{R})$. The ground- and excited-state electronic wave functions $\Phi_j(\mathbf{r};\mathbf{R})$, for any fixed $\mathbf{R}$, constitute a complete set that spans the space of the electrons; thus, we can express the exact molecular wave function $\Psi(\mathbf{r},\mathbf{R})$ as

$$\Psi(\mathbf{r},\mathbf{R}) = \sum_i \Phi_i(\mathbf{r};\mathbf{R})\Omega_i(\mathbf{R}) \ . \qquad (3)$$

Substituting Eq. (3) into the Schrödinger equation using the Hamiltonian of Eq. (1), multiplying from the left by $\Phi_j^*(\mathbf{r};\mathbf{R})$, and integrating over electronic coordinates $\mathbf{r}$, we obtain a set of coupled Schrödinger equations for the wave functions $\Omega_j(\mathbf{R})$ describing nuclear motion on each potential-energy surface, $\mathscr{E}_j(\mathbf{R})$:

$$-\frac{\hbar^2}{2}\sum_\alpha M_\alpha^{-1}\nabla^2_{\mathbf{R}_\alpha}\Omega_j(\mathbf{R}) + \mathscr{E}_j(\mathbf{R})\Omega_j(\mathbf{R}) - E\Omega_j(\mathbf{R})$$
$$= -\frac{\hbar^2}{2}\sum_i D_{ji}(\mathbf{R})\Omega_i(\mathbf{R}) + \hbar^2\sum_{i\neq j}\mathbf{d}_{ji}(\mathbf{R})\cdot\nabla_{\mathbf{R}_\alpha}\Omega_i(\mathbf{R}) \ , \quad (4)$$

where the first and second derivative matrix elements are defined as

$$\mathbf{d}_{ij}(\mathbf{R}) = -\sum_\alpha M_\alpha^{-1}\int\left\{\Phi_i^*(\mathbf{r},\mathbf{R})\left[\nabla_{\mathbf{R}_\alpha}\Phi_j(\mathbf{r},\mathbf{R})\right]\right\}d\mathbf{r} \ , \quad (5)$$

$$D_{ij}(\mathbf{R}) = -\sum_\alpha M_\alpha^{-1}\int\left\{\Phi_i^*(\mathbf{r},\mathbf{R})\left[\nabla^2_{\mathbf{R}_\alpha}\Phi_j(\mathbf{r},\mathbf{R})\right]\right\}d\mathbf{r} \ . \quad (6)$$

By neglecting the right-hand side of Eq. (4), we obtain

$$\left[-\frac{\hbar^2}{2}\sum_\alpha M_\alpha^{-1}\nabla^2_{\mathbf{R}_\alpha} + \mathscr{E}_j(\mathbf{R}) - E\right]\Omega_j(\mathbf{R},t) = 0 \ . \qquad (7)$$

Note that the first term on the right-hand side of Eq. (4) contains a diagonal term involving $D_{jj}(\mathbf{R})$. This term is frequently included in Eq. (7) as a correction to $\mathscr{E}_j(\mathbf{R})$. We omit it here for simplicity; it is of the same order of magnitude as the neglected off-diagonal $D_{ij}(\mathbf{R})$ terms. The final result, Eq. (7), is that nuclear motion is governed by a Schrödinger equation, with the potential-energy function given by $\mathscr{E}_j(\mathbf{R})$. $\mathscr{E}_j(\mathbf{R})$, in turn, is obtained from Eq. (2) for each required nuclear geometry $\mathbf{R}$. This is the Born-Oppenheimer approximation.

Born and Oppenheimer presented the theoretical justification for neglecting the two terms on the right-hand side of Eq. (4) using perturbation theory, with the small parameter $\xi$ chosen to be

$$\xi = (m_e/M)^{1/4} \ . \qquad (8)$$

$M$ is a typical nuclear mass and $m_e$ is the electron mass, taken to be unity. For a diatomic molecule vibrating near its minimum, the internuclear separation is expressed as

$$R = R_0 + \xi u . \tag{9}$$

where the reduced distance $u$ was argued to be of order unity, i.e., of the same order as $R_0$. From Eqs. (8) and (9), $M^{-1} \propto \xi^4$ and $\partial/\partial R \propto \xi^{-1}$. This gives the following scaling relationships:

Typical vibrational energy $\propto 1/\sqrt{M} \propto \xi^2$ ,

Typical rotational energy $\propto 1/MR_0^2 \propto \xi^4$ ,

First derivative coupling $\propto \mathbf{d}_{ij}(\mathbf{R})\partial/\partial R \propto M^{-1}\partial/\partial R \propto \xi^3$ ,

Second derivative coupling $\propto D_{ij}(\mathbf{R}) \propto M^{-1} \propto \xi^4$ .

For $M$ of order $10^4 m_e$, $\xi$ is of order 0.1. Thus, typical vibrational and rotational energies are of order $10^{-2}$ and $10^{-4}$, respectively, compared to electronic energies. The $D_{jj}(\mathbf{R})$ correction to the potential-energy surface $\mathscr{E}_j(\mathbf{R})$ is of order $10^{-4}$, similar to a rotational spacing. The mixing of electronic states $i \neq j$ is proportional to the squares of the first and second derivative couplings, i.e., $\xi^6 \approx 10^{-6}$ and $\xi^8 \approx 10^{-8}$, respectively. This provides the justification for the Born-Oppenheimer approximation.

The impact of the Born-Oppenheimer approximation in chemistry is pervasive. It is not an exaggeration to say that the single activity that has demanded the most effort (both intellectual and computational) from theoretical chemists in the twentieth century is the electronic structure problem, i.e., the calculation of $\mathscr{E}_j(\mathbf{R})$ for fixed $\mathbf{R}$ by ab initio, semiempirical, and empirical approaches. The tremendous improvements in the accuracy and practicality of ab initio methods that have been achieved were recognized by the award of the 1998 Nobel Prize in Chemistry to Walter Kohn and John Pople. Electronic structure computer codes have now become an invaluable tool of the modern theoretical and experimental chemist. This tool is applied routinely to elucidate structure and reactivity in a myriad of ways. With the help of methods for computing analytical derivatives, local minima of $\mathscr{E}_j(\mathbf{R})$ can be found and characterized. This directly gives the structure and stability of bound molecular species. In addition, analytical second derivatives of $\mathscr{E}_j(\mathbf{R})$ provide the force constants needed to compute vibrational spectra. Electronic spectra can be computed from the energy spacings of $\mathscr{E}_j(\mathbf{R})$ for different electronic states $j$, combined with the overlap integrals between vibrational wave functions computed on the two potential surfaces. The latter, called "Franck-Condon factors" were described in the original Born-Oppenheimer paper. Many other properties can be computed from the fixed-nuclei electronic wave functions, including spin-orbit interactions, magnetic resonance chemical shifts, molecular polarizabilities, dipole moments and transition dipoles, ionization potentials and electron affinities, etc. Analytical first and second derivatives of $\mathscr{E}_j(\mathbf{R})$ also facilitate the location and characterization of transition states, allowing ab initio determination of chemical reaction rates using transition-state theory. Monte Carlo sampling of the potential-energy surface has become a practical method in statistical mechanics for obtaining thermodynamic properties of molecules, fluids, adsorbates on surfaces, etc.

The Born-Oppenheimer approximation is also the cornerstone of the field of chemical dynamics. "Molecular dynamics", the simulation of the classical mechanical motion of interacting atoms, has important applications in fields ranging from biology to materials engineering. Molecular dynamics is based on, first, the Born–Oppenheimer approximation, and second, the classical limit of Eq. (7). The latter requires calculation of the classical forces, i.e., derivatives of $\mathscr{E}_j(\mathbf{R})$. This can be done either in advance or point-by-point along the trajectory, employing ab initio, semiempirical, or empirical methods. For all these alternatives, the nuclei evolve via classical mechanical equations of motion on an approximation to the Born-Oppenheimer potential-energy surface $\mathscr{E}_j(\mathbf{R})$, of Eq. (2). Thus, the terminology "quantum molecular dynamics" that is sometimes applied when ab initio forces are used is unfortunate. This classification should be reserved for situations where the heavy-particle motion is treated by quantum mechanics, i.e., Eq. (7).

Central to the argument of Born and Oppenheimer is the assumption that electronic wave functions vary on a spatial scale comparable to the reduced distance $u$ of Eq. (9), i.e.,

$$\int \{\Phi_i^*(\mathbf{r}, u)[\partial\Phi_j(\mathbf{r}, u)/\partial u]\}d\mathbf{r} \approx 1 . \tag{10}$$

This condition is not always satisfied. For example, in regions where two electronic states approach very closely in energy, the adiabatic electronic wave functions can change significantly in character in response to a very small change internuclear distance. For such "avoided-crossing" situations, the Born-Oppenheimer approximation may be invalid [3]. There are many such situations in chemistry. Examples include nonradiative transitions in molecules and solids, electron transfer, quenching of excited electronic states, collisional electronic excitation, and inelastic electron scattering. To describe such processes, "nonadiabatic transitions" among different potential-energy surfaces must be accounted for; however, this language is still based on the Born–Oppenheimer concept of potential-energy surfaces. Thus, the Born-Oppenheimer separation of electronic and nuclear motion remains central to the description, but the theory must be extended to multiple electronic states. This requires calculation of the derivative coupling matrix elements of Eqs. (5) and (6), an area where further progress is required.

Another area that will draw attention in the future is "true" quantum molecular dynamics, in which Eq. (7) or its time-dependent form describing the nuclear motion is solved quantum mechanically, without the classical approximation. This represents the ultimate limit of the Born-Oppenheimer approximation. In some cases it may be advantageous to make an additional, second-level Born-Oppenheimer separation of fast and slow nuclear motions, with inclusion of nonadiabatic transitions among nuclear quantum states.

While quantum mechanical simulation of nuclear motion will become more practical in the future, classical mechanical molecular dynamics will remain an important tool for simulating large molecular systems for many years to come. Ab initio determination of forces will play an increasingly large role. But a system of $N$ atoms requires at least $10^{3N}$ points to completely map out $\mathscr{E}_j(\mathbf{R})$ (ten points along each degree of freedom). For $N$ of order 100, it is clearly prohibitive to comprehensively tabulate $\mathscr{E}_j(\mathbf{R})$ in advance (in the absence of simplifications such as pairwise additivity). By contrast, a 1-ns trajectory with 1-fs time steps requires $10^6$ evaluations of $\mathscr{E}_j(\mathbf{R})$ and its derivatives, a very formidable task but far more accessible than the alternative. Thus, it will be essential in the future to develop "on-the-fly" methods for ab initio calculation of forces [4].

The Born-Oppenheimer approximation separates the theoretical study of molecules into two parts, the electronic structure part, Eq. (2), and the nuclear motion part, Eq. (7). This has produced a separation among theoretical chemists themselves: the electronic structure theorists and the statistical mechanics/dynamics theorists, i.e., those that compute $\mathscr{E}_j(\mathbf{R})$ and those that use it. While this is an overstatement and there have been many theorists with a foot in each camp, this bifurcation has been unhealthy for chemistry. Theoretical chemists of the future will need to be expert in both electronic structure and statistical mechanics/dynamics (perhaps as well as in biology, condensed matter physics, materials science, environmental science, computer science, ...?).

*Acknowledgement.* This work was supported by the U.S. National Science Foundation, Grant CHE-9707798.

## References

1. Born M, Oppenheimer R (1927) Ann Phys 84: 457
2. Eyring H, Polanyi M (1931) Z Phys Chem B 12: 279
3. (a) Landau LD (1932) Phys Z Sowjetunion 2: 46; (b) Zener C (1932) Proc R Soc Lond Ser A 137: 696
4. Car R, Parrinello M (1985) Phys Rev Lett 55: 2471

Theor Chem Acc (2000) 103:177–179
DOI 10.1007/s002149900040

**Theoretical
Chemistry Accounts**
© Springer-Verlag 2000

*Perspective*

# Perspective on "Wechselwirkung neutraler Atome und homöopolare Bindung nach der Quantenmechanik"

## Heitler W, London F (1927) Z Phys 44: 455–472

**Gernot Frenking**

Fachbereich Chemie, Philipps-Universität Marburg, Hans-Meerwein-Strasse, D-35037 Marburg, Germany

Received: 8 March 1999 / Accepted: 29 March 1999 / Published online: 28 June 1999

**Abstract.** The paper of Heitler and London was the first quantum theoretical study which explains the nature of the covalent bond. It also contains the noncrossing rule.

**Key words:** Covalent bond – Hydrogen molecule – Helium atom – Noncrossing rule

One of the greatest puzzles in natural sciences before the paper of Heitler and London (HL) was published in 1927 was the nature of the strong attractive interactions between neutral atoms which lead to the formation of a covalent chemical bond. It was clear that only electrostatic forces could be responsible for the interatomic attraction, but the application of classical laws of electrostatic interactions gave bond energies which were much too low. The explanation of a chemical bond in terms of classical electrostatic forces also violated the law of Earnshaw, which states that a static system held together by charge attractions should not be stable.

HL showed that strong attractive interactions between two hydrogen atoms in $H_2$ are predicted for certain interatomic distances when the then newly developed quantum theory of Schrödinger and Heisenberg is applied to a molecule in a similar way as was previously done only for atoms. The crucial point was that first the wavefunctions of the two atoms should be used to construct the wavefunction of the molecule which then gives the energy and charge distribution of $H_2$, rather than using the atomic charges in order to build up the charge distribution of the hydrogen molecule. The formation of the chemical bond was shown to be a typical quantum phenomenon and, thus, could not be understood before quantum theory was developed. The paper of HL is the birth of quantum chemistry.

The mathematical part of the paper belongs now to the introductory parts of many textbooks of quantum chemistry and is easy to read. More interesting are the comments and arguments which are given by the authors. The paper is divided into a short introduction and five chapters. The authors say in the introduction that the interaction between neutral atoms has been a difficult subject for theoretical treatments, and that the developement of quantum mechanics opens new avenues to deal with the problem. Two aspects would be new: the $e^{-r}$ behaviour of the charge distribution, which introduces a different interplay of forces between the atoms, and more importantly, "a characteristic quantum mechanical resonance phenomenon that is closely related to the resonance vibrations that were found by Heisenberg." HL consider not only the H—H interactions in $H_2$ but also the forces between two helium atoms. They anticipate in the introduction that there will be two solutions for the interaction energy between two hydrogen atoms. One solution gives attractive interactions at medium interatomic distances, which is suited for the formation of covalent bonds ("homeopolar molecules"). This solution would not be allowed for two helium atoms because of the Pauli principle. The second solution is allowed for hydrogen and helium, and it gives repulsive forces at all distances.

Chapter 1 describes the ansatz, which is now known as the valence bond method, for the interactions between two hydrogen atoms. The Schrödinger equation for the hydrogen molecule is given, and the eigenfunction for $H_2$ is given as the product of the eigenfunctions, $\psi$ and $\varphi$, of the hydrogen atoms. The authors point out that there are two equivalent molecular eigenfunctions, $\psi_1\varphi_2$ and $\psi_2\varphi_1$, which have the same energy, and that all pairs of orthogonalized and normalized linear combinations

$$\alpha = a\psi_1\varphi_2 + b\psi_2\varphi_1 \tag{1a}$$

$$\beta = c\psi_1\varphi_2 + d\psi_2\varphi_1 \tag{1b}$$

can be considered as undisturbed eigenfunctions of $H_2$. The correct linear combinations, for which the proof would be given later, would be

$$\alpha = 1/(2 + 2S)^{0.5}(\psi_1\varphi_2 + \psi_2\varphi_1) \tag{2a}$$

$$\beta = 1/(2 - 2S)^{0.5}(\psi_1\varphi_2 - \psi_2\varphi_1) \ . \tag{2b}$$

The disturbed eigenfunctions are then given by

$$\chi_\alpha = \alpha + v_\alpha \tag{3a}$$

$$\chi_\beta = \beta + v_\beta \tag{3b}$$

The ansatz (Eq. 3) for the disturbed wavefunctions is then used in the Schrödinger equation for $H_2$, which leads in a straightforward way to the energies of the perturbation

$$E_\alpha = E_{11} - (E_{11}S - E_{12})/(1 + S) \tag{4a}$$

$$E_\beta = E_{11} + (E_{11}S - E_{12})/(1 - S) , \tag{4b}$$

where $E_{11}$ gives the Coulombic interactions and $E_{12}$ is now known as the resonance integral, but the latter term is not used in the paper. It is only said that the $E_{12}$ terms are not easy to interpret.

Chapter 2 discusses the results (Eq. 4). The integrals for $E_{11}$ can be solved analytically, but for $E_{12}$ only an upper limit can be given because it involves an integral over $1/r_{12}$ which cannot be solved exactly. Using the upper-limit expression for $E_{12}$ leads to the curves for $E_\alpha$ and $E_\beta$ that are shown in Fig. 1, which also shows the energy curve given by the pure Coulombic interactions $E_{11}$.

The authors say: "The observed nonpolar attraction appears as a characteristical quantum mechanical effect. It arises already without considering perturbation by polarization. It is also remarkable that the repulsion $E_\beta$ is visibly a quantum mechanical effect, too, which is not based on Coulombic interactions ($E_{11}$)." It is interesting that the authors, while saying that Fig. 1 gives approximate values for the dissociation energy and equilibrium distance, do not give numbers, although these can be estimated from the scale that is used in the figure. They say that it is not their goal to calculate as accurate values as possible, but rather to give insight into the nature of the covalent ("homeopolar") bond. The calculated values for the equilibrium distance ($\sim$1.5 a.u.) and the binding energy ($\sim$2.5 eV) are remarkably close to the

exact values (1.401 a.u. and 4.7466 eV) considering the crude approximation.

Chapter 3 is devoted to the interaction of two helium atoms. Again, the Schrödinger equation for $He_2$ is given first, and the solution of the fourth-order secular equations leads to three eigenvalues (one of them doubly degenerate) and four zeroth-order wavefunctions

$$\alpha = \psi_{12}\varphi_{34} + \psi_{34}\varphi_{12} + \psi_{14}\varphi_{23} + \psi_{23}\varphi_{14} \tag{5a}$$

$$\beta = \psi_{12}\varphi_{34} + \psi_{34}\varphi_{12} - \psi_{14}\varphi_{23} - \psi_{23}\varphi_{14} \tag{5b}$$

$$\gamma = \psi_{12}\varphi_{34} - \psi_{34}\varphi_{12} \tag{5c}$$

$$\delta = \psi_{14}\varphi_{23} - \psi_{23}\varphi_{14} , \tag{5d}$$

where $\psi_{ab}$ and $\varphi_{cd}$ are atomic two-electron functions.

Chapter 4 is entitled "Pauli principle and molecule formation". HL expand the Pauli principle, which was hitherto applied only to atoms, to systems with two interacting atoms. Because the Pauli principle demands that the sign of the eigenfunction changes when two electrons become exchanged, it follows that only Eq. (5b) ($\beta$) is a possible solution for $He_2$. This leads to the high-energy eigenvalue $E_\beta = H_{11} - H_{12}$. The authors do not give mathematical expressions for the energy terms and say that, in analogy to the $H_2$ case, one should expect the same curve for $He_2$ as for $E_\beta$ ($H_2$) shown in Fig. 1.

The final part of this chapter is a discussion about the principal difference between systems which can form chemical bonds and those which cannot. The authors try to generalize their results for $H_2$ and $He_2$. HL point out that two helium atoms in their ground state cannot have different electron spins, while two hydrogen atoms can. The authors conclude that this is a general condition for two interacting systems to form a covalent bond, and that only repulsive forces yielding energy curves like $E_\beta$

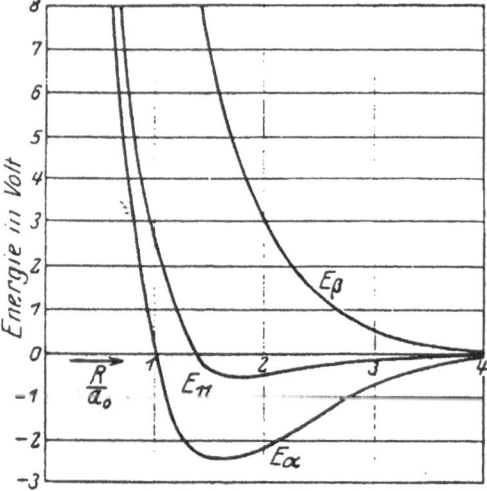

**Fig. 1.** Translation of the original figure caption: Potential of two neutral H atoms. ($E_\alpha$ = homeopolar attraction, $E_\beta$ = elastic reflection.)

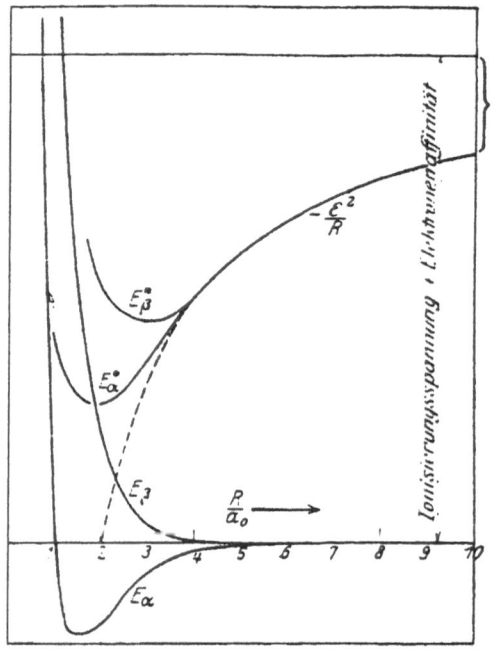

**Fig. 2.** Translation of the original figure caption: Ion potential ($E_\alpha^*$, $E_\beta^*$), compared with the potential of neutral atoms ($E_\alpha$, $E_\beta$)

in Fig. 1 will be found when one of the two systems has a closed-shell configuration like $H_2 + H$, $He + H$, $H_2 + H_2$, etc.

The final chapter discusses the case when $H_2$ dissociates into $H^+$ and $H^-$, i.e. the next higher-lying dissociation limit of $H_2$. The undisturbed eigenfunction for the two-electron system ($H^-$) is now $\psi_{12}$ or, equivalently, $\varphi_{12}$. HL discuss qualitatively the energy curves of the eigenvalues $E_\alpha^*$ and $E_\beta^*$ which belong to the zeroth-order eigenfunctions

$$\alpha^* = \psi_{12} + \varphi_{12} \tag{6a}$$

$$\beta^* = \psi_{12} - \varphi_{12} \ . \tag{6b}$$

Figure 2 is a reproduction of the original figure. Note that there is no scale for the energy values. The most important result from this section, which may surprise the reader, is that HL concluded that the energy curves of $E_\alpha$ and $E_\alpha^*$ cannot cross; likewise, the curves for $E_\beta$ and $E_\beta^*$ do not cross. Thus, the paper of HL is not only the first quantum chemical study which explains the nature of the covalent bond; it also contains the non-crossing rule. The authors clearly say that the two solutions $\alpha$ and $\alpha^*$ (likewise $\beta$ and $\beta^*$) can be combined linearly; however, the exact form of the combination would not be predictable without further investigation.

Theor Chem Acc (2000) 103:180–181
DOI 10.1007/s002149900051

Theoretical
Chemistry Accounts
© Springer-Verlag 2000

*Perspective*

# Perspective on "Neue Berechnung der Energie des Heliums im Grundzustande, sowie des tiefsten Terms von Ortho-Helium"

## Hylleraas EA (1929) Z Phys 54: 347–366

**Trygve Helgaker**[1], **Wim Klopper**[2]

[1] Department of Chemistry, University of Oslo, P.O. Box 1033 Blindern, N-0315 Oslo, Norway
[2] Theoretical Chemistry Group, Debye Institute, University of Utrecht, Paddualaan 14, NL-3584 CH Utrecht, The Netherlands

Received: 24 March 1999 / Accepted: 7 April 1999 / Published online: 14 July 1999

**Abstract.** Hylleraas' paper on the ground state of the helium atom represents the first ab initio calculation on a two-electron system in quantitative agreement with experiment. As such, it served not only as an important numerical verification of wave mechanics, it also exposed the shortcomings of determinantal expansions for accurate calculations of molecular and atomic electronic structure. By the simple addition of terms linear in the interelectronic distance, dramatic improvements in the description of the electronic wave function were achieved. The generalization to larger systems of chemical interest has proven difficult but interest in the techniques pioneered by Hylleraas has grown in recent years, prompted by progress in computer technology.

**Key words:** Helium atom – Electron correlation – Explicitly correlated wave functions – Hylleraas expansion

When Egil Andersen Hylleraas came to Göttingen to work with Max Born in 1926, he arrived with a background in crystallography and began work on the optical properties of quartz crystals. However, this was in the early days of wave mechanics and Born was more eager to work on the helium atom. The one-electron hydrogen atom had, of course, been worked out by then, but the study of many-electron atoms had barely begun. Heisenberg [1] had formulated the helium problem quantum mechanically in 1926, but a simple first-order perturbational treatment by Unsöld [2] had yielded an ionization potential of only 20.41 eV[1]. Compared with the experimental measurement of 24.59 eV, this was not better than the numerical value of 28.29 eV, which followed from the earlier Bohr theory. Born, therefore, considered it crucial to have a much better agreement with experiment to confirm the correctness of wave mechanics.

Born already had a student working on this problem, but this student fell sick and Hylleraas was given the task to carry on the urgent work. Hylleraas modified the original attempt in two ways. First, in the expansion of the wave function, he replaced the incomplete set of the bound-state hydrogenic functions by the complete set of Laguerre functions. Second, he simplified the treatment of the ground state by reducing the number of coordinates from 6 to 3: the distances $r_1$ and $r_2$ of the two electrons from the nucleus and the angle $\theta$ between the two position vectors of the electrons. Then, taking advantage of a newly installed desk calculator, he obtained an ionization potential of 24.47 eV [3] – merely 0.12 eV below the experimental number.

In the meantime, while Hylleraas was carrying out his work, other papers on the same problem appeared during 1927. By essentially the same variational method – today known as the configuration-interaction (CI) method – Kellner [4] in Berlin calculated the ionization potential of helium but obtained a result less conclusive than that of Hylleraas since a shorter expansion was used. Furthermore, using a different, nonvariational method, Slater [5] in Cambridge, Massachusetts and Sugiura [6] in Copenhagen both obtained good agreement with experiment – in fact, as good as Hylleraas. So, by 1928, several independent calculations on the helium atom appeared to have confirmed the validity of wave mechanics.

Still, there was a discrepancy of 0.12 eV that continued to bother Hylleraas when he returned to Oslo in 1928. Hylleraas considered this discrepancy a serious problem. On the one hand, attaching great importance to the variational principle, he was happy to see that his calculations approached the experimental number from the right side. On the other hand, he could not see how further extensions of the CI expansion could improve the situation much, hinting at a limit of 24.49 eV [3] – still

---

[1] The unit of 1 eV in the old literature is different from today's value (used in this article), the relationship being 1 eV (old) = 1.005 eV (new).

*Correspondence to*: T. Helgaker

0.10 eV below the experimental value. Nevertheless, he stated cautiously that it would be premature to conclude that wave mechanics is in some manner deficient [3]. Rather, he kept on trying to improve the helium wave function.

Later in 1928, a breakthrough was reached when he noted a peculiarity of the CI wave function. Expressing $\cos\theta$ in terms of $r_1$, $r_2$, and $r_{12}$, he realized that the wave function contains arbitrary powers of $r_1$ and $r_2$ but only even powers of $r_{12}$. This seemed to him unsatisfactory since the radial Hamiltonian treats these three variables in a more symmetrical fashion. He therefore decided to treat the three variables on the same footing, adding to the CI expansion terms of odd powers in $r_{12}$. With this inclusion, everything fell into place; to Hylleraas himself, it had almost the effect of a miracle. Thus, with only three terms in the first *Hylleraas expansion*, he obtained 24.56 eV; with six terms, he obtained 24.58 eV. The error had been reduced to 0.01 eV – i.e. to the same order of magnitude as the corrections arising from relativity and nuclear motion. Hylleraas was aware of these corrections and their approximate magnitudes and concluded that there were no remaining discrepancies between wave mechanics and experiment for the helium atom. His results were published in *Zeitschrift für Physik* in early 1929 under the title "Neue Berechnung der Energie des Heliums im Grundzustande, sowie des tiefsten Terms von Ortho-Helium" [7]. As the title indicates, he also considered the singlet–triplet splitting of the helium atom, which he reproduced to within 0.01 eV of the experimental number.

Hylleraas was thus not only the first person to experience the slow convergence of the CI expansion. He also solved the problem – at least for the two-electron system. However, Hylleraas had not arrived at his wave function from a consideration of the singularities of the Hamiltonian. In fact, in 1928, Slater had analyzed the properties of the helium wave function and had found that the Coulomb singularity in the Hamiltonian imposes a certain behaviour on the wave function when the electrons coincide and had suggested that the wave function be multiplied by a factor of $\exp(r_{12}/2)$ in order to model this behaviour [8]; however, less interested in numerical solutions and applied mathematics than Hylleraas, he did not attempt to include this factor in the wave function to resolve the discrepancy with experiment for helium. In retrospect, therefore, it appears that the explicitly correlated techniques of quantum chemistry were pioneered by both Hylleraas and Slater.

Hylleraas' work on helium was quickly applied to other two-electron atoms and in 1933 generalized to diatoms by James and Coolidge [9], who were able to compute the energy of the hydrogen molecule with an error of less than 0.03 eV. Clearly, Hylleraas' method had the potential for high accuracy for molecules as well. Nevertheless, the immediate practical impact of Hylleraas' work on chemistry was limited since it appeared difficult if not impractical to apply his ansatz to polyatomic systems. For many years, the most fruitful non-empirical approach to the many-electron problem in chemistry was firmly based on the ideas of orbitals and determinants. Certainly, the determinantal approach has been a highly successful one and still continues to be the workhorse of computational chemistry; however, with the steady improvements in computational techniques and computer technology, the slow convergence of the determinantal expansion, first realized by Hylleraas for the helium atom, has come back to frustrate progress towards highly accurate solutions for chemical systems. Like Hylleraas for the helium atom, chemists must now find a way to solve this problem by including in their ansatz for the wave function some explicit dependence on the interelectronic distance.

*Acknowledgement.* In preparing this article, we made use of Hylleraas' own recollections "Reminiscences from early quantum mechanics for two-electron atoms" [10].

# References

1. Heisenberg W (1926) Z Phys 39: 499
2. Unsöld A (1927) Ann Phys 82: 355
3. Hylleraas EA (1928) Z Phys 48: 469
4. Kellner GW (1927) Z Phys 44: 91
5. Slater JC (1927) Proc Natl Acad Sci USA 13: 423
6. Sugiura Y (1927) Z Phys 44: 190
7. Hylleraas EA (1929) Z Phys 54: 347
8. Slater JC (1928) Phys Rev 31: 333
9. James HM, Coolidge AS (1933) J Chem Phys 1: 825
10. Hylleraas EA (1963) Rev Mod Phys 35: 421

Theor Chem Acc (2000) 103:182–186
DOI 10.1007/s002149900029

**Theoretical
Chemistry Accounts**
© Springer-Verlag 2000

*Perspective*

# Perspective on "Quantum mechanics of many-electron systems"

## Dirac PAM (1929) Proc R Soc Lond Ser A 123: 714

**Werner Kutzelnigg**

Lehrstuhl für Theoretische Chemie, Ruhr-Universität Bochum, D-44780 Bochum, Germany

Received: 17 February 1999 / Accepted: 29 March 1999 / Published online: 21 June 1999

**Abstract.** Four prophetic statements in the introductory paragraph of Dirac's probably most cited paper are analyzed. Not only has his claim been disproved that the quantum mechanical equations needed to solve chemical problems are too complicated to ever be solved, even the reduction of chemistry from quantum mechanics is a tricky epistemological problem. Most surprising is that Dirac believed that relativistic effects are unimportant for chemistry.

**Key words:** Chemical bond – Chemical concepts – Many – electron systems – Reductionism – Relativistic effects

## 1 Introduction

P.A.M. Dirac, who shared the 1933 Nobel prize for physics with Schrödinger (the 1932 price went to Heisenberg), was one of the greatest pioneers of quantum mechanics. Most of his achievements entered textbooks so fast that his original papers are hardly cited. Nobody, who uses Dirac's bra–ket notation or his "$\delta$ function" would cite the original references [1].[1] The same is true of Dirac's time-dependent perturbation theory [2] or of the Dirac equation [3], the basis of relativistic quantum mechanics or of his subsequent work on positrons and holes [4].

There is, however, one paper of Dirac [5] that keeps being cited, namely the one at which we want to have a look now. Most people who cite this paper hardly know that its title is "*Quantum mechanics of many-electron systems*" and are unaware of its scientific context. It deals mainly with the relation between permutation symmetry and spin and contains a formula which relates the expectation value of the operator of electron exchange to the total spin of the state.

$$\langle S^2 \rangle = -\left\langle \sum_{i<j} P_{ij} \right\rangle - \frac{1}{4} n(n-4) \ , \tag{1}$$

where $n$ is the number of electrons and where the sum goes over all distinct exchanges of the spatial coordinates of two electrons.

The popularity of this paper [5] has, however, nothing to do with its scientific content. It is entirely based on the introductory paragraph to be quoted now in full length.

*The general theory of quantum mechanics is now almost complete, the imperfections that still remain being in connection with the exact fitting in of the theory with relativity ideas. These give rise to difficulties only when high-speed particles are involved, and are therefore of no importance in the consideration of atomic and molecular structure and ordinary chemical reactions, in which it is, indeed, usually sufficiently accurate if one neglects relativity variation of mass with velocity and assumes only Coulomb forces between the various electrons and atomic nuclei. The underlying physical laws necessary for the mathematical theory of a large part of physics and the whole of chemistry are thus completely known, and the difficulty is only that the exact application of these laws leads to equations much too complicated to be soluble. It therefore becomes desirable that approximate practical methods of applying quantum mechanics should be developed, which can lead to an explanation of the main features of complex atomic systems without too much computation.*

This introductory paragraph is essentially prophetic, and it is not cited so much because the prophecy has been fulfilled, but rather because it was strongly disproved. The quoted text contains four statements:

1. Relativistic effects are not relevant for chemistry. This claim occupies roughly one half of the introductory paragraph, and apparently the author regarded it as important.
2. A large part of physics and the whole of chemistry can be deduced from quantum mechanics.
3. The equations to be solved for chemical problems are too complicated to be solvable.
4. Approximate practical methods should be developed in order to explain atomic (and probably molecular) structure without too much computation.

We note, by the way, that usually (like recently in the press release of the Nobel committee concerning the 1998 Nobel prize for chemistry) only one sentence of

---

[1] These concepts are outlined in Ref. [1], but they may go back to earlier work. In the book no references to original papers are given

the introductory paragraph is quoted, namely the one which contains essentially statements 2 and 3.

Which, if any, of these statements can still be regarded as valid, and what can we learn from their analysis? Let us comment on them in reverse order.

## 2 Need for simplified methods

Statement 4 was an important guiding principle in the early days of the theory of atomic and molecular structure. Results based entirely on group theory were very helpful. It also turned out to be necessary to formulate approximative theories in terms of empirically adjustable parameters. A nice example of a theory in line with Dirac's suggestion is Slater's theory of complex atoms [6] in which certain parameters $F_k$ or $G_k$, appeared, that could, in principle, be calculated, but which were rather evaluated from spectral data to get better agreement with experiment. Semiempirical adjustments of theoretical parameters was also essential in Hückel's molecular orbital theory of $\pi$-electron systems [7] in order for it to be practically useful. Although the recent developments of ab initio theory make semiempirical parameters obsolete, the trend to oversimplified theories that need to be calibrated by experiment or by benchmark calculations is tending to return, as is demonstrated by the apparent success of modern density functional methods [8, 9].

## 3 Equations of quantum mechanics are too complicated

When Dirac made statement 3 he could hardly anticipate the spectacular progress in computer technology, without which the verification of statement 2 would hardly have been possible.

Nevertheless the equations to be solved to treat chemical problems by means of the Schrödinger equation are extremely complicated. After all, for an $n$-electron system one has to solve an eigenvalue problem for a $3n$-dimensional partial differential equation. I still believe that if one asked a competent mathematician, who does not know of the achievements of ab initio quantum chemistry, he would hardly see a chance that these equations could ever be solved, even with modern computers. One should not forget that the very best quantum chemical methods contain serious approximations, often guided by physical or chemical intuition, and that good agreement with experiment is, to a larger extent than is generally admitted, based on a – controlled or fortunate – cancellation of errors. One is also lucky that to answer many chemically relevant questions moderately accurate calculations are sufficient.

While semiempirical methods (including to some extent density functional approaches) rely on adjustable parameters, the quality of the ab initio method could, in principle, be judged by intrinsic criteria. These are, unfortunately, only useful for extremely sophisticated methods applied to very small systems. For the majority of ab initio methods one first checks from the application to known systems how reliable predictions can be expected for unknown ones.

The importance of Boys' ingenious idea [10] to expand wave functions in a Gaussian basis can hardly be overestimated. Gaussian basis functions behave incorrectly near the nuclei and very far from them, but they allow an extremely efficient integral implementation. Only recently has a formal proof been given [11] that an expansion in a Gaussian basis has a satisfactory convergence behavior. Much poorer is the convergence of configuration-interaction-like approaches to treat electron correlation effects [12, 13], but even for this problem progress has been possible recently [14].

## 4 Can chemistry be derived from quantum mechanics?

When Dirac wrote that "the underlying physical laws are for the whole of chemistry are thus completely" known, this was certainly revolutionary, since it implied that the only forces responsible for chemical phenomena are Coulombic, and that there is no genuine chemical force, as was still widely believed. A few years later Hellmann [15] formulated the program of quantum chemistry "which claims nothing less than to predict all chemical and physical properties of matter purely theoretically based on a simple mathematical law".[2] Some 20 years ago the present author tried to formulate a consistent and rigorous theory of the chemical bond [16] and found that "there is a long way from Dirac's statement (2) to an explicit theory of chemistry on the basis of quantum mechanics". In view of the spectacular success of numerical quantum chemistry that culminated in the Nobel prize for chemistry 1998, there are hardly doubts that at least with statement 2 Dirac was right.

Nevertheless, in his stimulating review, Primas [17] criticizes Dirac as a naive reductionist. According to Primas, Dirac was wrong because his "postulate of reductionism" was based on what Primas calls the "pioneer quantum mechanics" which he contrasted with "modern nonrelativistic quantum mechanics", and that Dirac did not consider the complicated epistemological problems related to the reduction of chemistry from quantum mechanics. It is not the scope of our perspective to comment on this criticism. Note, however, that the two "generations" of quantum theory differ more in the interpretation than in the operative formalism, that was, in fact, fully formulated in 1929. As to the relevance of problems of interpretation for the application of quantum mechanics, the reader is referred to a refreshing paper by Lévy-Leblond [18].

Theory reduction is, in a philosophical sense, certainly not a trivial problem; already the definition of how one understands reduction is crucial. Conclusions on the possibility of theory reduction depend much on this definition. Some scenarios are possible.

---

[2] The original quotation is "*Die Quantenchemie maßt sich nicht weniger an als sämtliche chemische und physikalische Materialeigenschaften rein theoretisch verausberechnen zu können, nur auf Grund eines einzigen mathematischen Gesetzes [... und] die chemische Welt theoretisch nachzukonstruieren*"

1. It may be necessary to amend the underlying theory before the reduction process. In order to derive thermodynamics from classical statistical mechanics, Boltzmann was, for example, obliged to introduce the quasiergodic hypothesis. There are vague speculations that quantum theory may have to be amended for systems with a very large number of particles [17].
2. On the way of reduction, by some limiting process, quantities and concepts may arise that have no place in the underlying theory, such as temperature or entropy or phase transitions as derived from statistical mechanics. The appearance of macroscopic irreversibility from microscopic irreversibility is still a controversial topic.
3. Finally reductionism may fail, i.e. it may not be possible to derive essential concepts of a "subtheory" from a supertheory. There is no evidence that this might be the case for the reduction of chemistry from quantum mechanics. The most challenging candidate for such a new concept is "life", which arises when one goes one step further down in Comte's [19] hierarchy of sciences, namely to biology.

The best understood examples of theory reduction are those of classical mechanics from quantum mechanics and thermodynamics from quantum statistics. Both classical mechanics and statistical mechanics can be either formulated by theory reduction or directly in terms of intrinsic axiomatics without reference to the underlying higher level theory. They have a different logical structure than quantum mechanics, for example, all variables commute.

In addition to quantum chemistry, which is a theory at a microscopic scale, there is place for macroscopic chemical theories such as chemical thermodynamics. In this context the reformulation of electrochemistry by Hertz [20] also deserves attention. On the other hand, many "prequantum" theories of the chemical bond had to be abandoned, because they turned out to be inconsistent with quantum chemistry. The "theory of mesomerism" survived for a while because it was regarded as a mapping to a simplified quantum chemical model, that of the semiempirical valence-bond theory, until both the former and the latter became obsolete.

There is certainly a challenge for genuine chemical theories, but there appears to be agreement that they have to conform with molecular quantum mechanics.

Typical chemical concepts are not as sharp as typical physical concepts. A nice example is that of atomic charge densities which were analyzed [21] in terms of "factor analysis" and were found to be "scalar" quantities (different possible definitions do not lead to the same numerical values, but these correlate satisfactorily), at variance with "aromaticity" [22, 23] which turned out to be a "multidimensional half-ordered" concept [23]. Primas [17] suggested that "molecular structure" is a genuine chemical concept, which has no counterpart in rigorous quantum mechanics. On the other hand, situations where molecular structures become undefined do show up in chemistry, or at least in molecular spectroscopy. Quantum phenomena apparently have a stronger tendency to survive in chemistry than in macroscopic physics. After all, the chemical bond can only be understood in terms of quantum mechanics [16, 24–26].

A useful theory reduction generates convenient intermediate levels. In order to understand the three-dimensional structure of proteins one will certainly not go back to the Schrödinger equation for the protein, but to a force-field model, that by itself is derivable – at least in principle – from quantum mechanics.

Whether a mathematical formulation of chemistry (as is now familiar to us) is possible or even desirable has been a controversial subject over the centuries. In 1786 Kant wrote [27] that chemistry will never be a genuine science because it cannot be formulated in mathematical terms.[3] While Gay-Lussac [28] believed in 1808 that in the near future the majority of chemical phenomena will be calculable, Comte [19] in 1830 even went beyond Kant and claimed that mathematics and chemistry are mutually exclusive, and that a – fortunately unlikely – mathematical access to chemistry would imply its decline.[4]

More recently Wigner argued that even if we were able to solve the Schrödinger equation numerically to any desired accuracy, this would not provide physical insight [29]. Wigner's view of physical insight is very puristic. What one cannot verify on the back of an envelope is, in his view, no acceptable theory. This opinion is to some extent shared by Longuet-Higgins [30], who has proposed to divide chemists into three classes: experimentalists, computationalists and theorists. In his opinion – and to some extent that of Primas [17] – computational chemistry is not theory. Future generations will probably have difficulties to understand the somewhat irrational aversion of some scientists of the twentieth century to insight obtained by means of computer application, although of course use of computers should not replace thinking.

---

[3] The original quotation is

*Solange als noch für die chymischen Wirkungen der Materien aufeinander kein Begriff ausgefunden wird, der sich construieren läßt, d. i. kein Gesetz der Annäherung oder Entfernung der Theile angeben läßt, nach welchem etwa in Proportionen ihrer Dichtigkeiten u. d. g. ihre Bewegungen samt ihren Folgen sich im Raume a priori anschaulich machen und darstellen lassen (eine Forderung, die schwerlich jemals erfüllt werden wird), so kann Chymie nichts mehr als systematische Kunst, oder Experimentallehre, niemals aber eigentliche Wissenschaft werden, weil die Principien derselben blos empirisch sind und keine Darstellung a priori in der Anschauung erlauben, folglich die Grundsätze chymischer Erscheinungen ihrer Möglichkeit nach nicht im mindesten begreiflich machen, weil sie der Anwendung der Mathematik unfähig wird*

[4] The original quotation is

*Toute tentative de faire rentrer les questions chimiques dans le domaine des doctrines mathémathiques doit être réputée jusqu'ici, et sans doute à jamais, profondément irrationnelle, comme étant antipathique à la nature des phénomènes... si, par une aberration heureusement presque impossible, l'emploi de l'analyse mathématique acquérait jamais, en chimie, une semblable prépondérance, il déterminerait inévitablement, et sans aucune compensation, dans l'économie entière de cette science, une immense et rapide rétrogradation, en substituant l'empire des conceptions vagues à celui des notions positives, et un facile verbiage algébrique à une laborieuse exploration des faits*

## 5 Are relativistic effects unimportant for chemistry?

Dirac's statement 1 is most surprising, if one realizes that he was the very one who formulated in an ingenious way the relativistic theory of an electron [3]. He did not say explicitly why he thought that relativistic effects were negligible in atomic or molecular theory, but apparently he had an argument in mind that was still popular some 50 years later. It is undeniable that relativity affects the motion of electrons if their speed is of the order of magnitude of the velocity of light, and this is the case for the K-shell electrons in heavy atoms. It was argued that chemical bonding only involves the valence electrons, which "move slowly", and that the inner shells are not affected by bonding, such that their relativistic effects cancel in binding energies or other valence-shell problems. It took a long time until it was found that this argument is invalid. The inner shells do affect the valence shell in two ways:

1. By the requirement of orthogonality, which affects mainly valence orbitals with low angular momentum quantum number (in a classical picture the corresponding orbits penetrate into the core region).
2. By a screening of the nuclear charge which is different from that in the nonrelativistic situation.

On the whole valence orbitals of $s$- and $p$-type are shrunk, $d$- and $f$-type orbitals expanded. Although we know that electron spin is not a relativistic effect, since a formulation of nonrelativistic quantum mechanics is possible in which the electron spin appears naturally [31], spin–orbit interaction definitely has a relativistic origin, and is actually the most important relativistic effect. It leads, for example, to a splitting of $p$ atomic orbitals into $p_{1/2}$ and $p_{3/2}$ orbitals, which not only have different energy, but also different radial extension. Relativistic effects on the energy are generally of the order $O(Z^2\alpha^2)$ relative to the nonrelativistic energies, where $Z$ is the nuclear charge and $\alpha$ is about 1/137, the fine-structure constant. The fact that relativity mainly affects the inner electrons, was for a long time interpreted in the sense that relativistic effects for different shells in an atom scale as approximately $Z_{\text{eff}}^2\alpha^2$, with the effective charge $Z_{\text{eff}}$ for the valence shell differing rather little in one column of the periodic table. In reality there is a dependence of approximately $Z^2\alpha^2$ with the full $Z$ even for valence electrons.

The importance of relativistic effects on chemistry has been reviewed, for example, by Pyykkö [32]. We just mention that gold is regarded as the most relativistic element and that its chemistry is largely determined by relativistic effects, even its colour, as well as the low boiling point of mercury. Many examples on relativistic effects on bond length are known [32], but relativity also affects the overall structure of molecules [33]. Relativistic effects are not limited to heavy atoms, even the reaction $H_2 + F \rightarrow HF + H$ cannot be described quantitatively if one ignores spin–orbit coupling [34]. Let us mention finally that, in spite of its uncontested validity, there are mathematical problems with the Dirac equation [36]. Moreover it is a one-electron equation, the generaliza-tion of which to $n$-electron systems is by no means trivial, but available approximate relativistic Hamilto-nians appear to work.

## 6 Conclusions

1. Controversial papers are cited more frequently than uncontested ones, especially so long after their publication [35].
2. Scientific progress consists largely of disproving authoritative statements. There is hardly a better challenge to do something than a claim by someone like Dirac that it is impossible.
3. The question to what extent chemistry is reducible from quantum mechanics is still a nontrivial philosophical problem.
4. Applied quantum mechanics will continue to be an important tool in chemistry. There is still need for methodological advances.
5. Dirac cannot be blamed for his belief that the practical use of quantum mechanics for chemistry will never come; however, he should have realized that relativistic effects are important in chemistry.

*Acknowledgements.* The author thanks C. van Wüllen for critically reading this manuscript. He is indebted to S.D. Peyerimhoff for the references to Comte and Gay-Lussac and to W.H.E. Schwarz for a copy of the Staatsexamensarbeit by Neus [23]. Kant [27] was – together with Dirac [5] – previously quoted in a textbook by Hartmann [37].

## References

1. Dirac PAM (1930) The Principles of quantum mechanics, 1st edn. Clarendon, Oxford
2. Dirac PAM (1927) Proc R Soc Lond Ser A 114: 243
3. (a) Dirac PAM (1928) Proc R Soc Lond Ser A 117: 610; (b) Dirac PAM (1928) Proc R Soc Lond Ser A 118: 351 (c) Dirac PAM (1932) Proc R Soc Lond Ser A 136: 453
4. (a) Dirac PAM (1929) Proc R Soc Lond 126: 360; (b) Dirac PAM (1931) Proc R Soc Lond 133: 60
5. Dirac PAM (1929) Proc R Soc Lond Ser A 123: 714
6. Slater JC (1929) Phys Rev 34: 1293
7. (a) Hückel E (1931) Z Phys 70: 204; (b) Hückel E (1931) Z Phys 72: 310; (c) Hückel E (1932) Z Phys 76: 628
8. Becke AD (1993) J Chem Phys 98: 5648
9. Becke AD (1999) J Comput Chem 20: 63
10. Boys SF (1950) Proc R Soc Lond Ser A 200: 542
11. Kutzelnigg W (1994) Int J Quantum Chem 51: 447
12. Schwartz C (1962) Phys Rev 126: 1015
13. Hill RN (1985) J Chem Phys 83: 1973
14. Kutzelnigg W, Herigonte Pv (1999) Adv Quantum Chem (in press)
15. (a) Hellmann H (1936) Front Nauki Tekh 6: 34; (b) Hellmann H (1936) Front Nauki Tekh 7: 39 (Russian translation of a German original manuscript)
16. Kutzelnigg W (1978/1994) Einführung in die Theoretische Chemie, Bd 2: Die Chemische Bindung. VCH, Weinheim. See especially the preface to the 1st edition
17. Primas H (1981) Chemistry, quantum mechanics, and reductionism. Lecture Notes in Chemistry, vol 24. Springer, Berlin Heidelberg New York
18. Levy-Leblond JM (1977) In: Lopes JL, Paty M (ed) Quantum mechanics, a half century later. Reidel, Dordrecht, pp 1–206

19. Comte A (1830) Cours de philosophie positive. Schleicher, Paris
20. Hertz HG (1980) Lecture notes in chemistry, vol 17. Electro-chemistry – A reformulation of the basic principles. Springer, Berlin Heidelberg New York
21. Meister J, Schwarz WHE (1994) J Phys Chem 98: 8245
22. Schleyer PvR (1996) Pure Appl Chem 68: 209
23. Neus J (1998) Staatsexamensarbeit. Siegen, Germany
24. Ruedenberg K (1962) Rev Mod Phys 34: 326
25. (a) Kutzelnigg W (1973) Angew Chem 85: 551; (b) Kutzelnigg W (1973) Angew Chem Int Ed Engl 12: 546
26. Kutzelnigg W (1990) In: Maksić ZB (ed) Theoretical models of chemical bonding, part 2: The concept of the chemical bond. Springer, Berlin Heidelberg New York, pp 1–44
27. Kant I (1786) Metaphysische Anfangsgründe der Naturwissenschaft. Hartknoch, Leipzig
28. Gay-Lussac JL (1808) Mem Soc d'Arcueil 2: 207
29. Wigner EP, Seitz F (1955) In: Seitz F, Turnbull D (eds) Solid state physics, vol 1. Academic Press, New York, p 97
30. Longuet-Higgins HC (1977) Discuss Faraday Soc 62: 347
31. Levy-Leblond JM (1967) Commun Math Phys 6: 286
32. Pyykkö P (1988) Chem Rev 88: 563
33. Kaupp M, van Wüllen Ch, Franke R, Schmitz F, Kutzelnigg W (1996) J Am Chem Soc 118: 11939
34. Stark K, Werner H-J (1996) J Chem Phys 104: 6515
35. Thaller B (1992) The Dirac equation. Springer, Berlin Heidelberg New York
36. Kutzelnigg W (1998) Nachr Chem Tech 46: 826
37. Hartmann H (1954) Theorie der chemischen Bindung. Springer, Berlin Heidelberg New York

Theor Chem Acc (2000) 103:187–189
DOI 10.1007/s002149900023

Theoretical
Chemistry Accounts
© Springer-Verlag 2000

*Perspective*

# Perspective on "Quantentheoretische Beiträge zum Benzolproblem. I. Die Elektronenkonfiguration des Benzols und verwandter Beziehungen"

## Hückel E (1931) Z Phys 70: 204–286

**Gernot Frenking**

Fachbereich Chemie, Philipps-Universität Marburg, Hans-Meerwein-Strasse, D-35037 Marburg, Germany

Received: 24 February 1999 / Accepted: 25 March 1999 / Published online: 21 June 1999

**Abstract.** Guided by an intuitive choice of approximations which shows remarkable chemical insight into the topic of aromaticity, Hückel mastered the difficult mathematical treatment of a complex molecule like benzene at a very early stage of quantum theory using method 1 (now valence bond theory) and method 2 (now molecular orbital theory). He concluded that method 2 is clearly superior to method 1 because the results of this method explain directly the peculiar behaviour of planar molecules with 6 π electrons.

**Key words:** Aromaticity – Molecular orbital theory – Valence bond theory

The year of the birth of quantum chemistry is 1927, when Heitler and London [1] showed for the first time that the chemical bonding between two neutral atoms can be understood in terms of fundamental laws of physics if the newly developed quantum theoretical principles are applied to the interactions between the two hydrogen atoms in $H_2$ [2]. It was only 4 years later that the epochal quantum theoretical study by Erich Hückel about the electronic structure of benzene, which has become the theoretical basis for our present understanding of aromaticity [3], was published.[1] It is remarkable

that a much more complicated molecule than $H_2$ such as benzene could be treated in the infant days of quantum chemistry in a meaningful way.

Hückel's paper on benzene was his second quantum theoretical investigation of chemical bonding. His first topic was the nature of the double bond, which was published in two papers [7]. The study of ethylene was inspired by a molecular orbital (MO) analysis of $O_2$ by Lennard-Jones [8], which focussed on the question of why the ground state of dioxygen is a triplet. It was an early triumph of MO theory (the name was not used at that time though) that the $^3\sum_q^-$ ground state of $O_2$ could be easily explained when the symmetric and antisymmetric combinations of the atomic eigenfunctions are used as molecular eigenfunctions, which are then occupied according to the aufbau principle and Hund's rule. Hückel recognized in the work of Lennard-Jones that there are two different types of O—O bonds, which were at that time already labeled σ and π. By subsequent replacement of O with isoelectronic $CH_2$, he developed a qualitative model for the bonding situation, first in $CH_2$=O and then in $H_2C$=$CH_2$. This work was important for his study of benzene because it led him to conclude that the two C—C bonds in ethylene are different and not equivalent as was generally believed and taught at that time. It was common to use van't Hoff's prequantum theoretical model of two tetrahedra sharing one edge for a discussion of the double bond [9]. Only 1 year after Hückel published his two papers on ethylene, Pauling [10] and Slater [11] independently developed a quantum theoretical description of ethylene with two equivalent C—C bonds, which are made up by overlapping two pairs of $sp^3$ orbitals. Pauling strongly opposed the idea that the C—C bonds in olefins are not equivalent.

Hückel began his quantum theoretical studies of chemical bonding following postdoctoral work at several locations, the most important one being Zürich, where he developed together with his Ph.D. advisor Peter Debye the well-known Debye–Hückel theory of electrolytic

---

[1] The paper about benzene is part 1 of a series with the translated title "Quantum chemical contributions to the benzene problem. I. The electron configuration of benzene and related compounds." Three other papers followed. The second paper [4] has the same series title and the translated subtitle "II. Quantum theory of induced polarizabilities." The study is an attempt to correlate chemical behaviour of substituted benzenes with the charge distribution in the ring which becomes disturbed by the substituents. The third paper [5] has a slightly different series title with no further subtitle: "Quantum theoretical contributions to the problem of aromatic and unsaturated compounds. III." The final paper in the series [6] has the subtitle "Free radicals in organic chemistry"

solutions [12]. The work concerning the double bond was suggested to him by Nils Bohr, whom Hückel visited in Copenhagen in 1929 when he was in Leipzig as a stipendiary [13]. The benzene results were published in the famous landmark paper which has 83 printed pages. The publication was also his habilitation thesis to become a docent at the Technical High School (now University) Stuttgart, to where he moved in 1930.

The paper is divided into six parts, an abstract, and an appendix with mathematical details. Part 1 gives an introduction and outline of the problem, which shows that the physicist Erich Hückel had a pronounced knowledge about the chemical and physical properties of aromatic compounds. The chemical knowledge was certainly aquired from discussions with his brother Walter Hückel, who was a chemistry professor and textbook author [14]. Hückel discusses benzene, pyridine, pyrrol, furan, thiophene, isomeric forms of dihydrobenzene and chinone, cyclobutadiene, cyclooctatetraene, and cyclopentadiene. He points out that there is no satisfactory explanation for the observation why cyclobutadiene does not (at that time) exist, and why benzene and cyclooctatetraene are stable but chemically very different in their reactivity. Hückel emphasizes that the number 6 appears to play a particular role for aromatic compounds, which had been pointed out for the first time by Bamberger in 1890 [15].

The second part of the paper is entitled "General remarks about quantum theoretical methods for treatment of unsaturated ring systems." This part describes his fundamental approach of constructing the electronic structure of benzene in terms of the electronic state of the carbon atoms. Hückel argues that in a planar regular ring system, $C_nH_n$, each carbon atom interacts primarily with the three neighbouring atoms (two carbons, one hydrogen). The perturbation of the electronic states of the carbon atom by the nine valence electrons of the neighbouring atoms leads to energetically different atomic states with the order $(s) < (p) < (p)_v < (p)_h$, where v means vertical and h horizontal with regard to the ring plane.[2] Hückel uses symmetry arguments for deriving the energy order, which is determined by the number of neighbouring atoms that are found in the nodal planes. Thus, $(p)_h$ is the energetically highest-lying atomic state because all three neighbouring atoms lie in the nodal plane of this state (orbital) [16]. The atomic electronic states lead to four different arrangements for benzene:

$(s)^2, \quad (p)^2, \quad (p)_v^2, \quad (p)_h$
$(s)^2, \quad (p)^2, \quad (p)_h^2, \quad (p)_v$
$(s)^2, \quad (p)_v^2, \quad (p)_h^2, \quad (p)$
$(p)^2, \quad (p)_v^2, \quad (p)_h^2, \quad (s)$ .

Hückel argues that the first of these terms, where the $(p)_h$ state of carbon [$p(\pi)$ orbital in modern notation] is singly occupied, should be the energetically lowest-lying term for benzene; therefore, only this term is considered

in the mathematical treatments of the electronic structure. In the final section of part 2 Hückel says that he is going to use two different mathematical treatments for the benzene problem. He calls them "first method" and "second method", which are now known as Valence bond (VB) and MO methods, respectively. Hückel cites papers by Heitler and London [1], Heisenberg [16], Slater [17], and Bloch [18] as fundamental for the first (VB) method and says that he is going to use Bloch's version for his work. For the second (MO) method he cites another paper by Bloch [19] and the famous publication of Hartree [20], but not Fock [21]. Hückel writes that Hartree's approximation neglects exchange interactions, but he notes that this can be accounted for later in the calculations. Before coming to the two central mathematical parts of the paper, Hückel notes that "...it will be seen that the second method is better suited for the real behaviour than the first method."

In the third part of the paper Hückel derives energy expressions for the different electronic terms of benzene using method 1. The wavefunction is given as a Slater determinant, and the solution of the Schrödinger equation is expressed in analogy to the Heitler–London paper [1] as

$$W = n \times w + \Delta W \tag{1}$$

$$\psi = \sum a_\mu \varphi \mu \ , \tag{2}$$

where $\Delta W$ is the interaction energy between the $n$ $\pi$ electrons which is determined by the coulomb integrals $J_0$ and the exchange integrals $J$.[3] The equations for determining $\Delta W$ and $a$ are then given as

$$[\Delta W - nJ_0 + (n-t)\,J]a_\mu + J\sum a_v = 0 \tag{3}$$

Hückel uses symmetry and group theoretical arguments to finally derive (in chapter 5) the following energy expressions for $n$ electrons in $n$-cyclic CH compounds given by method 1:

$$n = 3: \quad \Delta W_3 = 3J_0 \tag{4}$$

$$n = 4: \quad \Delta W_4 = 4J_0 + 2J \tag{5}$$

$$n = 5: \quad \Delta W_5 = 5J_0 + 1.24J \tag{6}$$

$$n = 6: \quad \Delta W_6 = 6J_0 + 2.6J \tag{7}$$

Chapter 4 describes the energy expressions for $\pi$ electrons in cyclic conjugated systems using the familiar integrals $\alpha$, for the energy of an unperturbed electron, and $\beta$, for the interaction between electrons at neighbouring atoms. Hückel notes that $\beta > 0$ if the wavefunction is nodeless, while $\beta < 0$ when the wavefunction has a node (Hückel defines $\alpha$ and $\beta$ in such a way that positive values are stabilizing and negative values destabilizing).

The important part of this chapter concerns the wavefunction. Hückel cites Bloch [19] and notes that, for cyclic groups, each wavefunction, $\chi$, may be expressed as

$$\chi^k(r, z, \varphi) = \exp(ik\varphi)\, u^k(r, z, \varphi) \ , \tag{8}$$

---

[2] Hückel uses the words "state" (Zustand) and "term" in a different meaning than is used now

[3] This is the notation which is used by Hückel

where $k$ is either a positive or negative integer or zero. The energy eigenvalues are then given by

$$W^k = W_0 - \alpha - 2\beta \cos(2\pi k/n) \quad (k = 0, 1, \ldots, n-1) \ . \tag{9}$$

Hückel notes that the eigenfunctions and eigenvalues are determined by the values for $k$

$$k = 0, \ \pm 1, \ \pm 2, \ \pm 0.5(n-1) \quad \text{if } n \text{ is odd} \tag{10}$$

$$k = 0, \ \pm 1, \ \pm 2, \ \pm(0.5n - 1), \ +0.5n \quad \text{if } n \text{ is even} \ . \tag{11}$$

Another important result of this part of the paper is the number of different terms which arise when the lowest-lying eigenfunctions become occupied. The results are graphically shown in part 5, where the energy expressions which arise from method 1 (Eqs. 4–7) and method 2 (Eqs. 9–11) are compared. Hückel displays the qualitative ordering of the energies of the different terms which are given by the two methods for $n = 3$–6 in four figures. It becomes obvious that the energies given by method 1 (Eqs. 4–7) do not reveal the particular stability of the ring system with $n = 6$. Hückel notes that the only conclusion which can be drawn from Eqs. (4)–(7) is that cyclic compounds with an odd number of ring atoms should be higher in energy than those with an even number. He mentions that the underlying $\sigma$ frame also influences the reactivity of the cyclic molecules, and that it is known from saturated compounds that small ring compounds with $n = 3, 4$ are higher in energy than compounds with larger rings. Similar behaviour should be expected for unsaturated compounds. He points out that the stability of a molecule is not only determined by the total energy of the compound. The chemical reactivity may also depend on the way the energy of the molecule changes when it is subject to external perturbation. Hückel says that there is substantial chemical evidence for a correlation between the constitution and the reactivity of organic molecules, and that only modest theoretical explanations are found in the literature. Here he cites the two-volume textbook of his brother Walter Hückel [14]. Then he switches to the discussion of the results of method 2 with the sentence: "We now believe that we can offer a new perspective for the ring systems considered."

The rest of chapter 5 is a far-sighted discussion of the conclusions which can be drawn from the results of method 2. Hückel points out that the occupation of the lowest-lying eigenfunctions leads to only one electronic state for $n = 6$, while there are four states for $n = 4$ and 8. This would indicate a closed-shell structure for benzene, while cyclobutadiene and cyclooctatetraene are not only high-energy molecules, but the open-shell structure should make them highly reactive. Hückel assumed planar structures for all the cyclic molecules considered (the nonplanar structure of $C_8H_8$ was not known at that time), but he was aware of possibly strong influences of nonplanar geometries on the stability and reactivity of the compounds. He says explicitly that this may be the case for $n = 8$ and 10. One formula which is frequently associated with the name Hückel is not found in the publication: $4n + 2$. This now familiar way to explain the aromaticity of cyclic conjugated compounds with the number of $\pi$ electrons (Hückel's rule) was introduced much later by Doering [22].

Hückel discusses the experimental observation that cyclopentadiene reacts easily with potassium, while cycloheptatriene does not. He says that the formation of $C_5H_5^-$ could explain the results because it would have the same electron configuration as benzene. Further examples which are discussed in chapter 5 are pyridine, pyrrol, furan, and thiophene. The final chapter is devoted to hydrobenzenes. Hückel shows that the energy ordering of 1,2-dihydrobenzene and 1,4-dihydrobenzene predicted by method 2 is in agreement with experimental findings.

In retrospect, there are two aspects which make this paper a century-contribution. One aspect is the mathematical treatment of the immensely difficult problem at a very early stage of quantum chemistry, which was guided by an intuitive choice of approximations. The choice was made possible because of Hückel's chemical insight into the problem, which is the second remarkable aspect. It is surprising that the theoretical physicist Erich Hückel was able to build a bridge between the mathematical results and chemical observations. It is a pity that it took two decades before chemists started to become interested in Hückel's work.

*Acknowledgement.* This work was supported by the Deutsche Forschungsgemeinschaft and the Fonds der Chemischen Industrie.

## References

1. Heitler W, London F (1927) Z Phys 44: 455
2. Frenking G (1999)
3. Minkin VI, Glukhovtsev MN, Simkin BYa (1994) Aromaticity and antiaromaticity. Wiley, New York
4. Hückel E (1931) Z Phy 72: 310
5. Hückel E (1932) Z Phy 76: 628
6. Hückel E (1933) Z Phy 83: 632
7. (a) Hückel E (1930) Z Phy 60: 423; (b) Hückel E (1930) Z Elektrochem Angew Phys Chem 36: 641
8. Lennard-Jones JE (1929) Trans Faraday Soc 25: 668
9. van't Hoff JH (1898) The arrangement of atoms in space. Longmans, London
10. Pauling L (1931) J Am Chem Soc 53: 1367
11. (a) Slater JC (1931) Phys Rev 37: 481; (b) Slater JC (1931) Phys Rev 38: 1109
12. (a) Debye P, Hückel E (1923) Phys Z 24: 185; (b) Debye P, Hückel E (1923) Phys Z 305
13. Hückel E (1975) Ein Gelehrtenleben. VCH, Weinheim
14. Hückel W (1931) Theoretische Grundlagen der organischen Chemie. VCH, Weinheim. English translation: Theoretical principles of organic chemistry. Elsevier, Amsterdam
15. (a) Bamberger E (1890) Liebigs Ann 257: 47. (b) Bamberger E (1893) Liebigs Ann 273: 373
16. Heisenberg W (1928) Z Phys 49: 619
17. Slater JC (1929) Phys Rev 34: 1293
18. Bloch F (1930) Z Phys 61: 206
19. (a) Bloch F (1928) Z Phys 52: 555; (b) Bloch F (1928) Z Phys 561
20. (a) Hartree DR (1928) Proc Camb Philos Soc 24: 89; (b) Hartree DR (1928) Proc Camb Philos Soc 426; Gaunt JA (1928) Proc Camb Philos Soc 24: 326
21. Fock V (1930) Z Phys 61: 126
22. Doering WvE, Detert FL (1951) J Am Chem Soc 73: 876

Theor Chem Acc (2000) 103:190–195
DOI 10.1007/s002149900102

Theoretical
Chemistry Accounts
© Springer-Verlag 2000

*Perspective*

# Perspective on "The activated complex in chemical reactions"

## Eyring H (1935) J Chem Phys 3: 107

### George A. Petersson

Hall-Atwater Laboratories of Chemistry, Wesleyan University, Middletown, CT 06459-0180, USA

Received: 20 July 1999 / Accepted: 23 August 1999 / Published online: 15 December 1999

**Abstract.** A general theory of the absolute rates of chemical reactions proved to be an elusive goal for nineteenth century chemists. This goal would only be achieved through a combination of statistical mechanics with the new quantum mechanics of the early twentieth century, when the insights of Henry Eyring and his contemporaries lead to the absolute rate equation that we are only now beginning to rigorously evaluate. The conceptual focus of absolute rate theory is the transition state (or activated complex), the window through which the future plunges into the past, and this is still the foundation of our understanding of chemical reaction rates as we enter the new millennium.

**Key words:** Transition state theory – Absolute rate theory – Chemical kinetics

## 1 Introduction

The great achievement of fundamental chemical theory of the nineteenth century was the conceptualization of bulk matter in equilibrium through thermodynamics and statistical mechanics. The early twentieth century brought an understanding of the periodic table and the behavior of individual molecules through quantum mechanics. The transition-state theory (TST) of Henry Eyring [1] combined these two theories to provide the fundamental conceptual framework for our understanding of the rates of chemical reactions. Very few single articles have had such a dramatic impact on chemistry. This paper is the starting point for any qualitative or quantitative discussion of bulk chemical reaction rates. Our concept of the nature of the transition state (TS) has evolved somewhat, but it is still a dividing surface separating products from reactants: "the now and the here – through which all future plunges into the past" [2].

The absolute rate of a chemical reaction is a problem first addressed in the middle of the last century, but it was the insights of Eyring and his contemporaries in the

1930s that lead to the equation that we are only now beginning to rigorously evaluate. I shall give just enough selective history to set the scene for Eyring's work and to present these ideas in the context in which they were introduced. Then I shall give a selective sampling of the path down which Eyring sent us – a path that promises unimagined new discoveries well into the next millennium.

## 2 Background

The temperature dependence of rate constants was examined by Wilhelmy as early as 1850 [3], but it was not until 1889 that the correct equation,

$$k_{\text{rate}}(T) \approx A \exp(-\Delta E^{\ddagger}/kT) \ , \tag{1}$$

for the basic temperature dependence was proposed by Arrhenius [4]. The development of the theory of chemical kinetics then focused on the preexponential factor, $A$, and the activation energy, $\Delta E^{\ddagger}$. The simple interpretation of the Arrhenius equation assigned $A$ to the collision frequency and $\Delta E^{\ddagger}$ to the minimum energy required for a successful (i.e. reactive) collision. It gradually became clear that nature was a good bit more subtle and thus far more interesting than this very simple suggestion.

A macroscopic thermodynamic understanding of the preexponential factor was greatly advanced by the work of Kohnstamm and Scheffer [5], who introduced the concept of the Gibbs free energy of activation. A complementary molecular understanding came from the collision theory approach of McC Lewis [6], which is rather remarkable for its simplicity. If a "collision" between molecules A and B is defined as the two molecules coming within $(r_A + r_B)$, the sum of the radii, then each molecule of A with relative velocity $v_{AB}$ will sweep out a cylinder of length $v_{AB} \, dt$ and cross section $\pi(r_A + r_B)^2$. The mean relative velocity, $v_{AB}$, had been calculated 50 years earlier by Maxwell and Boltzmann [7]:

$$v_{AB} = [8kT/\pi\mu_{AB}]^{1/2} \ , \tag{2}$$

where the reduced mass, $\mu_{AB}$, is $m_A m_B/(m_A + m_B)$. The number of B molecules in the collision cylinder is then

$$dN_B = [N_B/V]\pi(r_A + r_B)^2 \left[\frac{8kT}{\pi}\left(\frac{m_A + m_B}{m_A m_B}\right)\right]^{1/2} dt \quad (3)$$

and so the total rate of collisions per unit volume for all A molecules is

$$\frac{d(N_{coll}/V)}{dt} = [N_A/V][N_B/V](r_A + r_B)^2$$

$$\times \left[8\pi kT\left(\frac{m_A + m_B}{m_A m_B}\right)\right]^{1/2} \quad (4)$$

giving the collision theory rate constant

$$k_{rate}(T) = (r_A + r_B)^2 \left[8\pi kT\left(\frac{m_A + m_B}{m_A m_B}\right)\right]^{1/2}$$

$$\times \exp(-\Delta E^\ddagger/kT) \ . \quad (5)$$

The preexponential factor is often called the collision frequency and is abbreviated $z_{AB}$. In order to achieve even qualitative agreement with experiment, it was necessary to introduce an empirical "steric factor", $\mathscr{P}_{,AB}$, which is generally between 0.01 and 1.0, but occasionally is outside this range:

$$k_{rate}(T) = \mathscr{P}_{AB} z_{AB} \exp(-\Delta E^\ddagger/kT) \ . \quad (6)$$

The mathematical problem was thus reduced to the very physical problem of understanding why most collisions between molecules with sufficient energy to react nevertheless do not lead to chemical reaction.

By 1914 Marcelin [8] had shown that the Arrhenius equation could be derived from statistical mechanics. However, the nature of both the preexponential factor, $A$, and the activation energy, $\Delta E^\ddagger$, remained undetermined until 1919 when Herzfeld [9] derived the correct form of $A$ for the dissociation of a diatomic molecule:

$$k_{rate}(T) = (kT/h)Q_{vib}^{-1} \exp(-\Delta E^\ddagger/kT) \ . \quad (7)$$

The first understanding of the real nature of $\Delta E^\ddagger$ can be traced to Tolman's proof that the activation energy is the difference between the energy of the activated state necessary for reaction and the average energy of the reactant molecules [10, 11].

The formalism for the quasiseparation of nuclear coordinates from electronic coordinates was introduced by Born and Oppenheimer in 1927 [12]. However, it was the 1929 paper by London [13] and the 1931 paper by Eyring and Polanyi [14] that introduced the potential-energy surface (PES) for a chemical reaction and thus created the conceptual framework for modern kinetic theory. The following year Pelzer and Wigner [15] identified the saddle point on the PES as the TS. The stage had been set.

## 3 Henry Eyring's TST

Eyring formulated absolute reaction rates "in terms of quantities which are available from potential energy surfaces" [1]. He defined the "activated complex" (we

now prefer transition state) as "a saddle point with positive curvature in all degrees of freedom except the one which corresponds to crossing the barrier [i.e. the reaction path] for which it is of course negative." He assumed that once the forces between atoms have been treated quantum mechanically, the reaction rate can then be calculated by the methods of statistical mechanics. The fundamental idea was to treat the molecular flux from reactants to products as a one-dimensional problem along the reaction path by taking an ensemble average over all other degrees of freedom.

The partition function for the forward motion along the reaction path (rp) is

$$\frac{dQ}{d(rp)} = (2\pi\mu_{rp}kT)^{1/2}/2h \ . \quad (8)$$

Eyring multiplied this sum of states by the average velocity along the reaction path

$$\langle p/\mu_{rp}\rangle = \int_0^\infty (p/\mu_{rp})\exp[-(p^2/2\mu_{rp}kT)]dp$$

$$\Big/ \int_0^\infty \exp[-(p^2/2\mu_{rp}kT)]dp$$

$$= [2kT/\pi\mu_{rp}]^{1/2} \quad (9)$$

to obtain the flux (frequency factor) along the reaction path

$$\frac{dN_{rp}}{dt} = kT/h \ . \quad (10)$$

He performed an ensemble average over all remaining coordinates to obtain the total flux through the dividing surface:

$$\frac{dN^\ddagger}{dT} = (kT/h)N_A N_B\left(Q^\ddagger V/Q_A V Q_B V\right)\exp(-\Delta E_0^\ddagger/kT) \ ,$$

$$(11)$$

where $Q^\ddagger$, $Q_A$, and $Q_B$ are the partition functions of the TS and the reactants, and $\Delta E_0^\ddagger$ is the energy of the TS elative to the reactants. Eyring's "barrier height", $\Delta E_0^\ddagger$, included the change in quantum mechanical zero-point energy. His TS partition function, $Q^\ddagger$, includes all degrees of freedom except the reaction path, which was treated separately. Eyring's final expression for the absolute rate constant was

$$k_{rate}(T) = (kT/h)\left(Q^\ddagger/Q_A Q_B\right)\kappa(E)\exp(-\Delta E_0^\ddagger/kT) \ , \quad (12)$$

where $\kappa(E)$ is the "transmission coefficient" (vide infra).

The TST rate equation has the appearance of an equilibrium expression; however, all that is really necessary for the validity of Eq. (12) is that the reactants are at thermal equilibrium (thus defining the temperature, $T$) and that a sufficient number of A and B molecules pass through our "window", or "dividing surface", to give a statistically valid sample of the phase space around the TS. It is of no consequence that few of them may return to the reactant side. Eyring allowed for such recrossing through a factor equal to the reciprocal of the average number of crossings (our $\kappa$), but assumed that "the barriers are so flat near the top that tunneling may be neglected without appreciable error." We shall follow Wigner [16] and include tunneling through the energy-dependent transmission coefficient, $\kappa(E)$:

$$k_{\text{rate}}(T) = (kT/h)(Q^{\ddagger}/Q_A Q_B)$$
$$\times \int_0^{\infty} \kappa(E)\exp(-E/kT)\,\mathrm{d}(E/kT) \ . \qquad (13)$$

Note that Eq. (13) reduces to Eq. (12) if we return to the classical approximation for the transmission coefficient, $\kappa(E)$:

$$\kappa_{\text{classical}}(E) = \left\{ \begin{array}{l} 0: E < \Delta E_0^{\ddagger} \\ 1: E > \Delta E_0^{\ddagger} \end{array} \right\} \ . \qquad (14)$$

If we include quantum interference so that $\kappa(E)$ can be less than 1 for $E > \Delta E_0^{\ddagger}$, then we have also included one-dimensional recrossing in $\kappa(E)$. This requires solution of the one-dimensional Schrödinger equation along the reaction path, rather than just integrating $\exp(-2\pi\{2\mu_{\text{rp}}[V(r_{\text{rp}}) - E]\}^{1/2} r_{\text{rp}}/h)$ over the region with $V(r_{\text{rp}}) > E$.

Although in his original paper [1], Eyring only allowed that "tunneling may occasionally play some role", we can more easily describe the pervasive influence of TST if we include quantum mechanical tunneling in our reference equation. Regrettably, Eq. (13) does not provide a clear intuitive picture of the role of the barrier height, but the additional complexity is necessary to describe both the low-temperature rates of hydrogen-transfer reactions (i.e. tunneling) and the general effects of recrossing.

Simultaneous with Eyring's work, Evans and Polanyi [17] developed equivalent rate equations, but Eyring and Wynne-Jones [18] immediately extended TS to solutions by connecting with the macroscopic thermodynamic quantities $\Delta H^{\ddagger}$ and $\Delta S^{\ddagger}$, thus providing the link from molecular collision theory all the way to these phenomenological macroscopic quantities.

## 4 The impact of TST on chemical kinetics

Eyring's TST has provided the basic conceptual framework for the interpretation of the rates of nearly all chemical reactions on a bulk scale. He quickly applied his new theory to homogeneous gas-phase thermochemical reactions, photochemical reactions, heterogeneous catalysis, and reactions in solution [19]. He even considered such topics as viscosity and diffusion [19].

The greatest impact so far has been as a qualitative interpretive tool, rather than the achievement of the original goal of a quantitative theory of absolute rates. In his original paper Eyring stated: "The calculation of the concentration of activated complexes is a straightforward statistical problem, given the moments of inertia of the complex and the vibrational frequencies. This information is given with sufficient accuracy, even by our very approximate potential surfaces, to give good values

for the partition functions." Even Eyring could be completely wrong. The ab initio calculation of the partition functions has proven difficult, but the ab initio calculation of the barrier heights – until very recently – appeared impossible.

In spite of these difficulties in quantitative implementations, TST has nevertheless provided a framework for understanding even the most complicated reactions. TST immediately provided a deeper understanding of collision theory. The collision theory rate in Eq. (5) can be recovered by evaluating Eq. (12) for the reaction of two point particles (i.e. $Q_{\text{rot}} = Q_{\text{vib}} = Q_{\text{elec}} = 1$) to form a linear TS (i.e. $Q_{\text{rot}\|_{\text{rp}}} = Q_{\text{vib}} = Q_{\text{elec}} = 1$). Comparison of Eq. (6) with the general form then reveals the nature of the steric factor:

$$\mathscr{P}_{\text{AB}}(\text{TST}) = \frac{Q_{\text{rot}\|\text{rp}}^{\ddagger} Q_{\text{vib}}^{\ddagger} Q_{\text{elec}}^{\ddagger} \kappa(E)}{(Q_{\text{rot}} Q_{\text{vib}} Q_{\text{elec}})_A (Q_{\text{rot}} Q_{\text{vib}} Q_{\text{elec}})_B} \ , \qquad (15)$$

where $Q_{\text{rot}\|\text{rp}}^{\ddagger}$ is the partition function for rotation about the reaction path coordinate. The portion of the entropy of activation arising from the translational partition functions was included in Eq. (5), so the dominant additional entropy effect in Eq. (15) is the reduced rotational freedom in the TS, i.e. $Q_{\text{rot}\|\text{rp}}^{\ddagger} \ll (Q_{\text{rot}})_A (Q_{\text{rot}})_B$. Hence, the steric factor is indeed an appropriate description of $\mathscr{P}_{,\text{AB}}$.

Until fairly recently, successful applications of TST have treated the barrier height, $\Delta E_0^{\ddagger}$, as an adjustable parameter. Nevertheless, our basic understanding of general homogeneous and heterogeneous catalysis and even enzyme active sites is based on qualitative considerations of the role of the TS. Qualitative estimates of variations in zero-point energies and in vibrational and rotational partition functions can be based on experience and analogy. Our ability to anticipate the qualitative effect of substituting deuterium for hydrogen has led to a whole industry of mechanistic studies through isotopic substitution [20].

We can expect the new millennium to spawn a new age of routine quantitative absolute rate predictions based on the confluence of chemically accurate ab initio electronic structure theory with continued refinement of Eyring's TST.

## 5 The evolution of absolute rate theory

Refinements in Eyring's TST have been driven both by developments in electronic structure theory and by progress in understanding how to apply this emerging computational technology to the prediction of chemical reaction rates. A very thorough recent review of the current status of TST gives the informed perspective of the primary contributor to these modern refinements [21]. Our starting point is the PES, which we can now routinely generate from first principles.

### 5.1 Ab initio PESs

The journey of the PES from the highly speculative semiempirical formulations of London, Eyring, and

Polanyi [13, 14] to the firm reality of modern ab initio electronic structure theory began with the molecular orbital concepts introduced by Mulliken [22] (1966 Nobel Prize in Chemistry). These ideas were developed into a practical computational method on the new digital computers by Roothaan [23]. Early calculations were severely limited by the difficulty in evaluating integrals over Slater-type orbitals (due to the cusp at each nucleus) [24], until Boys demonstrated the advantages of Gaussian basis sets [25]. At this point Pople released the computer program Gaussian 70, which has evolved into a very comprehensive and widely used computer program [26], implementing his model chemistry concept [27], for which he was to share the 1998 Nobel Prize in Chemistry. The model chemistry concept of Pople provided a well-defined broad approach to general classes of problems, thereby removing the arbitrariness of ad hoc decisions for each particular case. Improvements in both computer hardware and computational methodology have each increased the speed of ab initio calculations by a factor of about $10^4$ since those early days (based on the author's timing comparison of 24 h for a minimum basis set self-consistent-field calculation on ethane in 1964 versus 8 ms for the same calculation today). The combined effect has made ab initio theory into a practical tool for chemical predictions.

Among many other developments important in achieving the current state of the art in computational quantum chemistry, we should recognize the transplanting of coupled-cluster methods from nuclear physics by Bartlett and Purvis [28], providing a manageable approximation to full configuration interaction calculations. The development of systematic sets of atomic basis functions by Dunning [29] has provided a tool to probe convergence to the complete one-electron basis-set limit, which can be compared to results obtained from the efficient implementation of explicit functions of interelectronic coordinates by Kutzelnigg and Klopper [30].

In a complementary development, very rapid, if inexact, evaluations of a PES are now possible through the density functional theory methods developed by Hohenberg and Kohn [31], Kohn and Sham [32], Lee et al. [33a], Parr and Yong [33b], and Becke [34], for which Kohn shared in the 1998 Nobel Prize in Chemistry.

### 5.2 Evaluation of $Q^\ddagger$

It is the genius of TST that only the saddle point of the PES need be examined. At low pressure, we can use the ideal gas partition functions for translational motion:

$$Q_{\text{trans}} = V\left(2\pi mkT/h^2\right)^{3/2} , \tag{16}$$

The geometry of the saddle point provides the moments of inertia, $I_j$, required for evaluation of the rigid rotor rotational partition functions

$$Q_{\text{rot}} = 2\pi\left(\frac{4\pi}{\sigma}\right)\prod_j\left(\frac{2\pi I_j kT}{h^2}\right)^{1/2} \tag{17}$$

which are valid except at very low temperatures. The vibrational partition functions are a bit more problematic. The stretching and bending frequencies, $v_j$, are generally large enough that the harmonic oscillator partition functions

$$Q_{\text{vib}} = \prod_j\left[1 - \exp(-hv_j/kT)\right]^{-1} \tag{18}$$

provide an adequate approximation; however, the torsional modes (especially torsion of one reactant with respect to the other) should be treated as free or hindered rotors [35].

The previous discussion presumes that we know the geometry of the TS and the second derivative of the energy with respect to each of the vibrational normal modes. Since the TS is a stationary point the first derivative of the energy with respect to each coordinate must vanish. Hence, both problems require knowledge of the derivatives of the energy with respect to the positions of the nuclei.

### 5.3 Analytical derivatives

The farsighted development of methods to calculate the analytical derivatives of the ab initio energy with respect to the Cartesian coordinates of the nuclei by Pulay [36] played a key role in the practical implementation of TST. Further developments by Handy and Schaefer [37a] Schlegel et al. [37b], and Johnson and Frisch [38], among others greatly expanded the number of ab initio methods for which these derivatives were available. The tools to evaluate the partition functions were at hand, but first we must locate the TS.

### 5.4 Locating TSs

Once these derivatives are available, efficient and reliable algorithms are still required for locating the TS and in the case of variational TST (VTST) for obtaining a portion of the reaction path. The former was often a rather frustrating search requiring a guess for the reaction path along which a maximum energy point was determined. Once the TS was located, "reaction-path following" was more routine, being essentially a matter of starting off in the direction of negative curvature of the PES and following the energy gradient to reactants and products.

Finding the geometry of a stable molecule is quite straightforward, since following the energy gradient is sufficient to guarantee eventual convergence to the nearest local minimum on the PES. Saddle points provide a more interesting challenge, for which the quadratic synchronous transit (QST) algorithms of Peng and Schlegel [39] represent a dramatic improvement over previous methods, and finally give us a tool, making TS geometry optimizations relatively routine.

### 5.5 Variational transition-state theory

We can trace the origins of VTST all the way back to Wigner [40], but this acronym is now generally associated with the modern development that has been dominated by Truhlar [41].

The exact quantum mechanical absolute rate constant can, in principle, be obtained as a Maxwell–Boltzmann average over the state-to-state reaction cross sections [42]. The rate is also given by the net flux through any surface in phase space dividing products from reactants. VTST seeks the best such window for the plunge from reactants into products. This is the surface giving the minimum one-way forward flux, so the two-dimensional recrossing we ignore must also be a minimum, the sum being constant. Recognizing that Eyring's TST includes the effects of both the zero-point energy and the entropy of activation through the partition functions, it is clear that the forward flux depends on the Gibbs free energy, $\Delta G_0^{\ddagger}(T)$, at the TS. We should therefore select the TS as the point along the reaction path at which the Gibbs free energy is a maximum. The tunneling is then determined by the solution of the one-dimensional Schrödinger equation along the reaction path, with a potential-energy function given by $G(T, r_{rp})$. This is what Truhlar describes as canonical zero curvature (ZC) VTST [43]. Note that the geometry of the TS is now a function of both the temperature and the isotopes of the atoms, in contrast to the saddle point on the Born–Oppenheimer PES.

VTST is especially important for reactions without a saddle point on the Born–Oppenheimer PES, such as radical recombination reactions. The free energy of activation for such reactions is often dominated by entropic effects.

## 5.6 Reaction-path curvature

If the reaction path coordinate were truly separable from the other coordinates, then classical recrossing of the barrier would not occur and our $\kappa(E)$ would give the exact quantum mechanical recrossing rate. Unfortunately, this situation is impossible. For example, if our chemical reaction is

$$A - B + C \quad \rightarrow \quad A + B - C \tag{19}$$

then the initial reaction path must be $r_{AB-C}$, the distance from the center of mass of AB to C, while the final reaction path must be $r_{A-BC}$, the distance from A to the center of mass of BC. The resulting curvature of the reaction path is the primary underlying reason why TST cannot be exact for such a reaction. This curvature necessarily couples the reaction path to the other coordinates and thus leads to corner cutting (i.e. finding a shorter path not quite traversing the saddle point) [44], bob sledding (i.e. riding up the sides of the exit channel) [45], and two-dimensional corrections to recrossing. Refined versions of TST designed for both small [46] and large curvature [47] have been developed. The largest curvature in mass-weighted coordinates corresponds to the transfer of a light atom (i.e. hydrogen) between two heavy atoms – precisely the situation where tunneling will be most important. However, canonical ZC-VTST is certainly adequate for reactions not involving hydrogen.

## 5.7 Unimolecular reactions

The rate of a unimolecular reaction can depend upon both the rate at which collisions provide the required activation energy and the rate at which this energy can be funneled into the reaction coordinate and used to cross the dynamical bottleneck. The crossing of the dynamical bottleneck is properly described by Eyring's TST (or generalizations thereof). The intramolecular energy transfer is usually described by the exchange of energy between coupled harmonic oscillators using the unimolecular rate theory developed by Rice, Ramsperger, Kassel, and Marcus [48]. In the high-pressure limit, the reaction is slow compared to the collision rate and so the reactants maintain a Boltzmann energy distribution, but most unimolecular reactions are studied experimentally under conditions where corrections for non-Boltzmann behavior are required.

## 5.8 Recent progress in barrier heights

For a long time, it was not possible to assess the accuracy of TST. The early PESs were more a result of speculation than firm knowledge, so discrepancies with experiment were impossible to interpret. Two developments have dramatically altered this impasse.

The first is the refinement of quantum dynamics to accurately determine the rate constant that results from a given PES [49–51], thereby providing an evaluation of the various levels of TST that does not rely on the availability of an accurate PES.

The second breakthrough is our rapidly developing ability to remove inaccuracies in the ab initio PES as the bottleneck in the predictive ability of absolute rate theory. The benchmark paper by Diedrich and Anderson [52] on $H_3$ set a new standard of accuracy for a PES. Having at least one PES accurate to ±0.01 kcal/mol allows us to begin the task of objectively evaluating the reliability of methods applicable to larger systems – without reference to problems in the interpretation of experimental rates. Klopper et al. [53] have recently reviewed the state of the art for ab initio calculations on systems with up to about five atoms. An accuracy of ±0.2 kcal/mol is now practical. This error in calculated barrier heights corresponds to an error of ±40% in the rate constant at room temperature. Disagreement with experiment by more than a factor of 2 can then be attributed to a deficiency in the flavor of TST that is employed. The recent compilation of articles on computational thermochemistry by Irikura and Frurip [54] gives a very useful overview of the current state of computational methodology for larger systems. We can anticipate very rapid progress in this area over the next few years.

## 6 Summary and conclusions

The 1935 paper by Eyring represents the successful conclusion of a long search for a quantitative theory of chemical kinetics, providing the link from the microscopic world of the quantum mechanics of single molecules to the macroscopic world of the rates of chemical reactions in bulk matter. TST forms the very core of our conceptual framework for understanding chemical reactions – whether in the gas phase, on surfaces, in solution, or even in the active sites of enzymes.

The extreme sensitivity of chemical reaction rates to small changes in the PES confined early applications to semiempirical qualitative interpretations of experimentally measured rates. It is in this interpretive role that TST has had such a pervasive influence in chemistry. As our ability to calculate accurate PESs and to carry out quantum dynamics calculations has improved, the role of TST has evolved. Refinements in TST and in our ability to generate chemically accurate PESs are opening a new era in which we shall predict the rates of chemical reactions from first principles.

We can be confident that various incarnations of Eyring's TST will continue to provide the foundation for our qualitative and quantitative understanding of chemical reaction rates throughout the next century.

## References

1. Eyring H (1935) J Chem Phys 3: 107
2. Joyce J (1922) Ulysses. Shakespeare and Company, Paris, p 196
3. (a) Wilhelmy L (1850) Pogg Ann 81: 422; (b) Wilhelmy L (1850) Pogg Ann 81: 499
4. Arrhenius S (1889) J Phys Chem 4: 226
5. Kohnstamm P, Scheffer FEC (1911) Proc K Ned Akad Wet 13: 789
6. McC Lewis WC (1918) J Chem Soc 113: 47
7. (a) Niven W (ed) (1952) Maxwell JC, Scientific papers, vol 1. Dover, New York, p 380; (b) Boltzmann L (1876) Wien Ber 74: 503
8. Marcelin R (1914) C R Acad Sci 158: 407
9. Herzfeld KF (1919) Ann Phys 59: 635
10. Tolman RC (1920) J Am Chem Soc 42: 2506
11. Truhlar DG (1978) J Chem Ed 55: 309
12. Born M, Oppenheimer JR (1927) Annl Phys 84: 457
13. London F (1929) Z Elektrochem 35: 552
14. Eyring H, Polanyi M (1931) Z Phys Chem Abt B 12: 279
15. Pelzer H, Wigner E (1932) Z Phys Chem Abt B 15: 445
16. Wigner EP (1932) Z Phys Chem B 15: 445
17. Evans MG, Polanyi M (1935) Trans Faraday Soc 31: 875
18. Eyring H, Wynne-Jones WFK (1935) J Chem Phys 3: 492
19. Glasstone S, Laidler KJ, Eyring H (1941) 'The theory of rate processes' McGraw-Hill, New York
20. Wiberg KB (1955) Chem Rev 55: 713
21. Truhlar DG, Garrett BC, Klippenstein SJ (1996) J Phys Chem 100: 12771
22. (a) Mulliken RS (1949) J Chim Phys 46: 497; (b) Mulliken RS (1955) J Chem Phys 23: 1833; (c) Mulliken RS (1955) J Chem Phys 23: 1841; (d) Mulliken RS (1955) J Chem Phys 23: 2338; (e) Mulliken RS (1955) J Chem Phys 23: 2343
23. Roothan CCJ (1951) Rev Mod Phys 23: 69
24. Slater JC (1930) Phys Rev 36: 57
25. Boys SF (1950) Proc R Soc Lond A 200: 542
26. Frisch MJ, Trucks GW, Schlegel HB, Scuseria GE, Robb MA, Cheeseman JR, Zakrzewski VG, Montgomery JA Jr, Stratmann RE, Burant JC, Dapprich S, Millam JM, Daniels AD, Kudin KN, Strain MC, Farkas O, Tomasi J, Barone V, Cossi M, Cammi R, Mennucci B, Pomelli C, Adamo C, Clifford S, Ochterski J, Petersson GA, Ayala PY, Cui Q, Morokuma K, Malick DK, Rabuck AD, Raghavachari K, Foresman JB, Cioslowski J, Ortiz JV, Stefanov BB, Liu G, Liashenko A, Piskorz P, Komaromi I, Gomperts R, Martin RL, Fox DJ, Keith T, Al-Laham MA, Peng CY, Nanayakkara A, Gonzalez C, Challacombe M, Gill PMW, Johnson B, Chen W, Wong MW, Andres JL, Head-Gordon M, Replogle ES, Pople JA (1998) Gaussian 98. Gaussian Pittsburgh Pa
27. (a) Pople JA (1973) In: Smith DW, McRae WB (eds) Energy, structure and reactivity. Wiley, New York, p 51; (b) Hehre WJ, Radom L, Schleyer PvR, Pople JA (1986) Ab initio molecular orbital theory. Wiley, New York
28. Bartlett RJ, Purvis GD (1978) Int J Quantum Chem 14: 516
29. (a) Dunning TH Jr (1989) J Chem Phys 90: 1007; (b) Kendall RA, Dunning TH Jr, Harrison RJ (1992) J Chem Phys 96: 6796; (c) Woon DE, Dunning TH Jr (1993) J Chem Phys 98: 1358; (d) Dunning TH Jr (1994) 100: 2975; (e) Dunning TH Jr (1995) 103: 4572; (f) Wilson AK, van Mourik T, Dunning TH Jr (1996) J Mol Struct (Theochem) 388: 339; (g) Woon DE, Peterson KA, Dunning TH Jr (1998) J Chem Phys 109: 2233
30. (a) Kutzelnigg W (1985) Theor Chim Acta 68: 445; (b) Klopper W, Kutzelnigg W (1987) Chem Phys Lett 134: 17; (c) Klopper W, Kutzelnigg W (1989) Stud Phys Theor Chem 62: 45
31. Hohenberg P, Kohn W (1964) Phys Rev B 136: 864
32. Kohn W, Sham LJ (1965) Phys Rev A 140: 1133
33. (a) Lee C, Yang W, Parr RG (1988) Phys Rev B 37: 785
33. (b) Parr RG, Yang W (1989) Density functional theory of atoms and molecules. Oxford University Press, New York
34. (a) Becke AD (1993) J Chem Phys 98: 1372; (b) Becke AD (1993) J Chem Phys 98: 5648
35. Truhlar DG (1991) J Comput Chem 12: 266
36. Pulay P (1977) In: Schaefer HF III Applications of electronic structure theory. Plenum, New York, p 153
37. (a) Handy NC, Schaefer HF III (1984) J Chem Phys 81: 5031; (b) Schlegel HB, Binkley JS, Pople JA (1984) J Chem Phys 80: 1976
38. Johnson BG, Frisch MJ (1994) J Chem Phys 100: 7429
39. Peng C, Schlegel HB (1994) Isr J Chem 33: 449
40. Wigner E (1937) J Chem Phys 5: 720
41. Truhlar DG, Garrett BC (1984) Annu Rev Phys Chem 35: 159
42. Eliason MA, Hirschfelder JO (1959) J Chem Phys 30: 1426
43. (a) Truhlar DG (1970) J Chem Phys 53: 2041; (b) Garrett BC, Truhlar DG (1979) J Phys Chem 83: 1052; (c) Garrett BC, Truhlar DG (1979) J Phys Chem 83: 1079
44. Skodje RT, Truhlar DG, Garrett BC (1982) J Chem Phys 77: 5955
45. Marcus RA (1966) J Chem Phys 45: 4493
46. Skodje RT, Garrett BC, Truhlar DG (1981) J Phys Chem 85: 3019
47. Bondi DK, Connor JNL, Garrett BC, Truhlar DG (1983) J Chem Phys 78: 5981
48. (a) Rice OK, Ramsperger HC (1928) J Am Chem Soc 50: 617; (b) Kassel LS (1928) J Phys Chem 32: 1065; (c) Marcus RA, Rice OK (1951) J Phys Colloid Chem 55: 894; (d) Marcus RA (1965) J Chem Phys 43: 2658
49. Truhlar DG, Kuppermann A (1972) J Chem Phys 56: 2232
50. Schatz GC, Kuppermann A (1976) J Chem Phys 65: 4668
51. Friedman RS, Truhlar DG (1999) In: Simon B, Truhlar DG (eds) Multiparticle quantum scattering with applications to nuclear, atomic, and molecular physics. Springer, Berlin Heidelberg New York
52. Diedrich DL, Anderson JB (1992) Science 258: 786
53. Klopper W, Bak KL, Jørgensen P, Olsen J, Helgaker T (1999) J Phys B At Mol Opt Phys 32: R103
54. Irikura KK, Frurip DJ (eds) (1998) Computational thermochemistry. ACS Symposium Series 677. American Chemical Society Washington, D.C.

Theor Chem Acc (2000) 103:196–199
DOI 10.1007/s002149900044

Theoretical
Chemistry Accounts
© Springer-Verlag 2000

*Perspective*

# Cavity and reaction field: "robust" concepts.
# Perspective on "Electric moments of molecules in liquids"

## Onsager L (1936) J Am Chem Soc 58: 1486

**Jacopo Tomasi**

Dipartimento di Chimica e Chimica Industriale, Via Risorgimento 35, I-56126 Pisa, Italy
e-mail: tomasi@dcci.unipi.it

Received: 12 January 1999 / Accepted: 19 April 1999 / Published online: 14 July 1999

**Abstract.** Onsager's model to describe the behavior of molecules in liquids is put in the appropriate historical context of the evolution of chemistry. Some key aspects of the model that justify its success in the past decades are discussed, with emphasis on general features shared with many other good models we have in theoretical chemistry and that should be kept in mind for the development of further models: congruence with physical principles, simplicity and robustness. The present and future evolution of this model is briefly considered, with the aim of learning better from this example how to exploit our studies for the advancement of theoretical chemistry.

**Key words:** Onsager model – Solvation methods – Models in chemistry

## 1 The evolution of chemistry in the first decades of the century

The first readers of Lars Onsager's paper "Electric moments of molecules in liquids", published in the August 1936 issue of the *Journal of the American Chemical Society* [1], probably considered it as a further contribution to a well-established line of research developed in the preceding decades.

One of the main steps in the evolution of chemistry took place in a period of approximately 30 years, beginning with the last decade of the past century; it was essentially led by physicists. Sound experimental evidence for the atomistic description of matter and of its organization into molecular assemblies in the gas, solid, and eventually the liquid phase (1905), prompted the development of detailed models and of a conceptual comprehensive framework in which large-scale and submicroscopic aspects were harmonized, with emphasis on the real existence of the elementary building blocks, atoms and molecules, of this theory.

It was an approach to study matter considerably different from that of traditional chemistry, accustomed to using analogous concepts, but as heuristic tools, without resolute attempts to organize them into a comprehensive and "realistic" theory of matter: important exponents of the chemical community strongly sustained the view that attempts to replace the traditional chemical approach with the new one were futile, and probably misleading and dangerous.

We know now that the physical approach, with the support of new theoretical methods such as statistical thermodynamics, and with the contribution of the vigorous expansion or renovation of many experimental techniques (spectroscopies of various kinds, X-ray diffraction, etc.), organized in a new branch of science, chemical physics, "conquered" the whole realm of chemistry within a short time.

There are no objections, I think, to accepting this historical interpretation: the effort of many scientists, including among them eminent leaders such as Boltzmann, Planck, Einstein, Lorentz, and Debye, to name a few, later crowned by the new formulation of quantum mechanics, has provided the basis for a description of material systems which unifies physics and chemistry and constitutes the conceptual world in which we, theoretical chemists, are working.

## 2 The Onsager model

Among the various directions along which research progressed is the investigation of the properties of polar molecules. An outstanding contribution to this field was made by Debye: we quote his 1912 paper [2] and his 1929 book [3] as examples. Onsager's paper is directly related to this line of research. Debye's dipole theory [2] is quoted in the first line of the article, and then analyzed and greatly modified. Typically, no reference is made to quantum models. Everything is presented in a classical formulation, even if references to recent Pauling articles [4] indicate that quantum theory was present in the background.

There are, therefore, reasons to support my guess that among the first readers of the Onsager paper a large number considered it a modification of the Debye formulation of molecular dipole theory for condensed systems (a theory, we remark, based on older studies of physicists, such as Mossotti [5], for example).

More attentive readers noticed other aspects of the Onsager paper. Kirkwood [6] in 1938 remarked that Onsager introduced a real cavity, conceptually quite different from the Lorentz cavity which is just a mathematical device.

The "reality" of the Onsager cavity is one of the reasons which prompted me to select this paper for this New Century Issue: this is one of the further steps in the development of "real" models using the physical approach. This is not the only reason, however, Onsager introduced new concepts, that of the reaction field and that of the cavity field, with a clear and transparent physical basis, and devised a simple model based on a few parameters with physical meaning, and easily managed computationally.

There is no need to summarize Onsager's paper. The essential point is given by his model of a molecule M within a cavity of appropriate shape encircled by the molecules of the polar solvent. If the liquid is not subjected to external fields, there will be a field of local origin, called a reaction field, $R$, which depends on the displacements of the surrounding molecules produced by the permanent multipoles of M and which at the same time modifies the molecular charge distribution via the polarization functions of M. In the presence of an external field, $E$, there will be a cavity field, $G$, related to $E$ and to the shape of the cavity. In conclusion, the field acting on M will be $G + R$, the first term only contributing to the orienting force-couple in the case of the presence of an external field.

This general formulation is reduced to a simple form. The molecule is represented by a permanent point dipole $\mu$, and by a polarizability, $\alpha$, the cavity has a spherical shape with radius $a$, and the surrounding molecules are reduced to a continuum dielectric medium with fixed dielectric constant.

Onsager is well aware of the limitations thus introduced within his model. He discusses problems about the appropriate choice of the cavity radius and of its dependence on thermal volume changes; he also considers the generalization of the model to other cavity shapes: both subjects are treated briefly but with illuminating remarks about reciprocity and symmetry relations holding in the model. In his article Onsager examines other aspects of the physical problem, such as the effects of hydrogen bonds and the changes in the model on passing from pure polar solvents to mixed polar–nonpolar solvents and finally to nonpolar solvents, and the limitations introduced by considering the medium as a linear continuum dielectric, neglecting aspects related to dielectric saturation or compression, as well as aspects related to the discrete structure of the medium.

These remarks reveal that there was an extended and detailed analysis of the properties of liquid systems in general, under the formulation of the model and its applications to demonstrate the nonexistence of spurious Curie points (which was his starting point) and to present an improved theory for the dielectric constants of pure liquids and solutions (which constitutes the main body of the article).

This accurate analysis is accompanied by a rigorous formal elaboration of the model to which I shall return later. A few years before Onsager, Bell [7] presented a model consisting of a dipole, $\mu$, within a spherical cavity immersed in a continuous uniform dielectric medium. This model is quite similar to the Onsager one, but is summarily treated and with some errors; now it is completely forgotten, and is only quoted in extensive reviews of the subject.

## 3 The success of the Onsager model

Onsager model has not been forgotten. It has been widely used, and is still in use. One of the reasons for this popularity is its very simple mathematical expression. The simplicity of the model is another point to which I shall return later. We have to recall that in 1936, and for many years after, the most sophisticated computational tool available in our laboratories was the slide rule: to use Onsager's formulas a slide rule is not necessary: the back of an envelope and a pencil are sufficient.

Easily obtaining the result is not sufficient to assure the popularity of a formulation. In the Onsager case a reason for the success is the physical "robustness" of the model. It may be modified with little effort and adapted to many different problems. Among these I quote its extension to describe solvent shifts in electronic spectra and to estimate dipole moments of molecules in their excited states. The chemical spectroscopic literature is full of applications of the Onsager model, both in the past (I quote here as examples the texts I have in my room, Refs. [8–10]) as well as at present. The popularity of this model has led to a phenomenon which occurs in several similar cases. There are few quotations of the original Onsager paper in this literature: as the model's so often applied, there is no reason to quote its source. As a sort of compensation there is the adjective "onsagerian", which has found use in some specialized literature.

The model is still in use, and I quote here a recent remark as a bridge to pass from the examination of some aspects of the Onsager paper to more general considerations. Lombardi [11] recently remarked that there are important discrepancies between the dipole moments of molecules in their excited state obtained via solvatochromic shifts, as is normal practice, with respect to the more precise values derived from Stark effect measurements. Solvatochromic values are currently obtained by applying Onsager's formula, adapted in the 1950s to the problem of solvatochromic shifts, but without other changes with respect to the original model. This is another point I must stress: the intrinsic, or maximum, accuracy of the results given by a model.

The chemical elaboration of models derived from the "realistic" physical approach has always been accompanied by the search for higher accuracy. Chemistry is the science of subtle differences among similar systems, a problem not so important in physics. I give here an example I use with my students: in physics there is not much difference between methanol and ethanol; in chemistry we rely very much on their differences. These differences are substantial when we are looking, for example, at the very different chemical effects produced by the ingestion of small quantities of the two substances in living bodies.

So, good models for chemical applications must be robust and flexible, allowing more detailed descriptions of the system and of the property under examination. Onsager's model has these positive features. In the last two decades it has been widely used with important modifications permitting more realistic descriptions of the systems, but keeping the essential points: the cavity, the reaction field, and the cavity field.

The methods now in use range from semiclassical to very sophisticated quantum mechanical descriptions of the solute, permitting the study of solutes of very different size, from a single electron to molecules composed of several thousands of atoms. The solvation methods are now applied to a very broad range of properties, from the evaluation of solvation energies to the study of reaction mechanisms to the analysis and prediction of solvent effects on molecular properties of a very complex nature, such as electro-optical and magnetic properties. For many properties the results are within the error bar of the experimental results, for others they are not far out, and complete agreement with experiment will probably be reached by further refinements in the computational formulation of the model. For many others there are no experimental results to compare with, and the model predictions are used as a first, but reliable, guess.

## 4 The "robustness' of the Onsager model

It is instructive to examine some points of the evolution of the Onsager model to show better its robustness.

### 4.1 The cavity shape

There are no intrinsic reasons in the model to keep a spherical shape for the cavity or to only extend the model to other regular cavities, such as the ellipsoids considered by Onsager. Efforts to keep a spherical shape for the cavity may lead to absurd results: see, for example, the remarks expressed by Luzhkov and Warshel [12] about its use for ion pairs. The most convenient shape is based on the van der Waals envelope of the molecule, modified by factors related to the finite size of the solvent molecules: the solvent excluding surface may be a first acceptable description. Today models adopt cavities with these irregular shapes.

### 4.2 The description of the solute

The point-dipole approximation used by Onsager is not essential: higher multipole moments have been progressively included in the models with spherical or ellipsoidal shapes. Important progress has been the introduction of quantum mechanical descriptions of the solute. This change created new ways of using Onsager's model.

The first is the direct calculation of polarization effects without the use of empirical values of $\alpha$. Polarizability alone is not sufficient to achieve chemical accuracy; better results can be obtained by using group polarizabilities (and hyperpolarizabilities) at the cost, however, of a proliferation in the number of parameters of dubious quality. The direct quantum mechanical calculation avoids these problems and introduces a new dimension to the model. The Schrödinger equation is, in fact, no longer linear: this leads to a refinement of the model (We remark that in the Onsager formulation there was no influence on **R** of the polarizability enhancement of the solute dipole. This was introduced in 1938 by Böttcher [13], but was limited to the original dipole-only model.) and opens the way to "robust" extensions of the model to nonequilibrium problems.

Another consequence of the use of a quantum mechanical description of the solute is that we are no longer obliged to rely on multipole expansions of the solute charge distributions. They may still be used, of course, but other options are open, such as to use the quantum mechanical charge distribution directly or to derive from the quantum mechanical charge distributions other parameters, local charges, for example, explicitly defined to reproduce the solute field in the regions of interest and thus reducing computational costs.

A third important consequence is the possibility of extending the field of application of the models to all the properties of the solute, as we have already remarked. In particular, the solvatochromic shifts and the related excited-state dipole moments we mentioned can now be computed at a level attaining chemical accuracy.

### 4.3 Solute-solvent interactions

We have already mentioned Onsager's warnings concerning possible limitations of his model due to an incomplete description of solute–solvent interactions. All the points he considered have now been examined: it would be too long to give a resumé of the conclusions and of the studies still in progress. Suffice it to say that for all the possible limitations of the model he considered there is a positive answer based on a physically reasonable modification of the model. We add that there are solutions for other limitations Onsager has not considered, the first being related to the intrinsic quantum mechanical overlap of the solute and solvent electron distribution, on which much work has been done [14] and the second to the presence of quantum dispersion contributions to the solute–solvent interactions (the latter being inserted in the reaction-field framework by Lindner [15]).

In this short summary of recent extensions of the Onsager model to illustrate its robustness I have not given recent references. It is not my intention to review here modern continuum solvation methods. Interested readers are referred to a rather comprehensive review covering the pertinent literature from the beginning to recent times [16].

This said about the robustness of the model, I now consider the other points I stressed in the preceding discussion.

1. *The simplicity of the mathematical formulation.* This has already been discussed. In passing to the modern formulations of the model this kind of simplicity has been lost. We have to consider, however, what simplicity now means. We are no longer in the time of the slide rule, and computers are pervasive. Simplicity means codes easy to use, well documented, not requiring skilled operators and complex computer structures, and especially models clearly related to the physical features of the problem.

2. *Reality of the model.* This is a different way of formulating the request that the model should be clearly connected with the physical principles and with the specific physical properties of the system. Good models share this characteristic: other models may be used, but their range of application is limited and is subject to the assumption set out in their formulation. They are not robust models.

3. *Rigorous formal elaboration of the model.* In the Onsager case it has paid off. In extensions made by many researchers, the Onsager lesson of rigor has been conserved in many realizations of the model. The rigor does not exclude the development and use of approximate expressions: theoretical and computational chemistry provides many examples of approximate formulations of robust theories; quantum chemistry is based on approximate realizations of a robust model. Approximate formulations must be accompanied by benchmark studies having as a goal the determination of the intrinsic accuracy of the model. This remark is valid for all the models we use in theoretical chemistry, but in particular for continuum solvation methods which introduce a drastic simplification of the real systems they model.

4. *New concepts.* This last point is the most important. The impact of a model is determined by the quality of the new concepts it introduces: points 1–3 are just additional conditions. Onsager's model has introduced new concepts. More than 60 years have shown their validity and their capability of surviving the revolutionary changes in our methods to describe matter. We have paid attention in the preceding discussion to concepts of the cavity and of the reaction field: also the concept of the internal field, essential to connect microscopic to macroscopic behavior of matter has recently been introduced into the accurate quantum mechanical realizations of the model.

## 5 About the future

Onsager's model shall continue to be exploited in the new century, and I am confident that further development of this approach shall result in important contributions to the struggle of theoretical chemistry to provide interpretations to the enthusing progress of chemistry more and more aimed to define and to study complex systems. I am also confident that theoretical and computational chemistry will also be able to provide more concrete contributions to the advancement of our discipline by modeling new systems with specific prop-

erties, by suggesting new experiments, in a continuous joint effort with other specialists.

The status and role of the Onsager model is not unique, of course. Looking at the provisional list of contributions to the New Century Issue, I notice many other robust models which introduced new concepts. This listing is surely not complete; in addition other proposals of new concepts are buried in the literature because of lack in the proper development of the model. However, we have not to congratulate ourselves on the number of models we have been able to develop or to lament for what we have missed: we have to look to the future.

My last remarks concern this point. The new century coincides with 70 years of application of quantum theories to chemistry. Two generations of theoreticians have spent their scientific lives applying quantum mechanics to chemistry. There are no doubts about their successes or about the necessity of continuing along this way. For the future we cannot, however, limit ourselves to a refinement of our methods and codes. We need new concepts, new models, new methods, and new strategies. The new concepts we are searching for are not limited to the realm of the physical approach to molecular systems, but also concern mathematics, information theory, and other scientific disciplines. Methods and strategies should introduce a more integrated use of the considerable body of knowledge we have, and that is continuously growing.

I do not have the space (nor the intellectual strength) to develop these remarks into a more coherent and detailed research programme. Substantial progress in science can be achieved with the aid of general planning, we have had several examples in our past and recent history, but is often derived from an ingenious look at the problems arising in the study of a specific subject. This second way of introducing innovation is apparently more casual and more modest, but is potentially rich with results.

I urge the readers of this issue to develop every opportunity they have to do it. Science strongly needs these efforts.

## References

1. Onsager L (1936) J Am Chem Soc 58: 1486
2. Debye P (1912) Phys Z 13: 97
3. Debye P (1929) Polar molecules. Chemical Catalog Co, New York
4. (a) Pauling L (1935) J Am Chem Soc 57: 2680; (b) Pauling (1936) J Am Chem Soc 58: 94
5. Mossotti OF (1850) Memorie di matem e fisica; Soc Ital Sci, (Modena) 24: 49
6. Kirkwood JG (1938) J Chem Phys 7: 911
7. Bell RP (1931) Trans Faraday Soc 47: 1143
8. Basu S (1964) Adv Quantum Chem 1: 145
9. Mataga L, Kubota T (1970) Molecular interactions and electronic spectra. Dekker, New York
10. Suppan PJ (1990) J Photochem Photobiol 50: 293
11. Lombardi JR (1998) J Phys Chem A 102: 2817
12. Luzkhov V, Warshel A (1992) J Comput Chem 13: 199
13. Böttcher CJF (1938) Physica 5: 635
14. Chipman DM (1997) J Chem Phys 106: 10194
15. Lindner B (1960) J Chem Phys 33: 668
16. Tomasi J, Persico M (1994) Chem Rev 94: 2027

Theor Chem Acc (2000) 103:200–204
DOI 10.1007/s002149900046

Theoretical
Chemistry Accounts
© Springer-Verlag 2000

*Perspective*

# Perspective on "The transition state method"

## Wigner E (1938) Trans Faraday Soc 34: 29–41

**Bruce C. Garrett**

Environmental Molecular Sciences Laboratory, Pacific Northwest National Laboratory, Richland, WA 93352, USA

Received: 26 February 1999 / Accepted: 6 April 1999 / Published online: 9 September 1999

**Abstract.** A perspective is provided on Wigner's classic paper on transition-state theory (TST). After providing a brief review of the historical context of this work, we review its key contributions including Wigner's dynamical perspective on TST, the fundamental assumption of TST, and the upper-bound property of classical TST. A discussion is also presented of subsequent progress in the field, which was stimulated by this work. This progress includes the following:

1. Demonstrations of the validity of the fundamental assumption for classical systems.
2. Further investigations into the classical foundations of TST that helped elucidate relationships between classical trajectories and TST.
3. The development of a variational form of the theory.
4. The development of variational TST into a quantitative tool for predicting rate constants.
5. The search for an "exact" quantum mechanical version of TST.
6. The development of TST-like expressions for the exact quantum mechanical rate constant.
7. The extension of TST to reactions in condensed phases.

**Key words:** Transition-state theory (TST) – Variational TST (VTST) – Fundamental assumption of TST – Quantum mechanical TST

## 1 Historical context and perspective

The early to mid 1930s was a time of intensive activity in the formulation of transition-state theory (TST). Laidler and King [1] have provided an excellent review of the early history of TST, tracing the development of rate theories using treatments based upon thermodynamics, kinetic theory, and statistical mechanics, and focusing on Eyring's 1935 contribution to the formulation of TST [2]. A snapshot of the state of the development of TST and some of the controversy surrounding it in 1937 is captured in volume 34 of the *Transactions of the Faraday Society*, which was published in 1938. This volume is a compilation of papers and discussion comments from the 1937 General Discussion of the Faraday Society on "Reaction kinetics" and it includes contributions from many of the founders of TST (e.g., E. Wigner, H. Eyring, M. Polanyi, and M.C. Evans). Wigner's contribution [3] presented one of two prevailing perspectives on TST at that time – a dynamical perspective that is firmly based on a statistical mechanical approach to formulating the theory. The subsequent paper in this volume was a presentation of Eyring's alternative perspective on "The theory of absolute reaction rates" [4], which is based upon quasiequilibrium thermodynamics. A historical discussion of the 1937 General Discussion of the Faraday Society has recently been presented by Miller [5], including a detailed analysis of Wigner's seminal contribution. The current paper parallels Miller's review and presents a complementary view that should help provide an even broader perspective on Wigner's important contribution.

In the early days of TST, it was apparent that there was great hope that the method would allow quantitative predictions of reaction rate constants. This hope was reflected in the title of Eyring's contribution to the 1937 General Discussion of the Faraday Society which refers to the ability of the theory to provide the absolute magnitude of the pre-exponential factor and thereby the absolute value of the rate constant. In addition, there was much activity comparing TST with recent experimental studies. As stated by Wigner in the discussion comments to this volume "there seems to be, however, a discrepancy between theory and experiment in practically all cases in which a numerical comparison is possible" [6]. In the same discussion comment, Wigner made the prescient observation that energy surfaces may be more complicated than assumed. Largely because of the inability to accurately predict reaction energetics (particularly barrier heights), the realization of the hope of using TST to quantitatively predict rate constants waited nearly 40 years.

The popularity of TST during the intervening years (1940s–1960s) was largely due to Eyring's thermodynamic formulation that provided the basis for correlating and interpreting kinetic data (including isotope effects). A testament to the popularity of the thermodynamic formulation of TST was the development of the field of thermochemical kinetics [7]. In the 1970s, as it became possible to accurately predict potential-energy surfaces using ab initio electronic structure methods and to perform accurate quantum dynamical calculations, interest was revived in developing methods to calculate absolute rate constants from first principles. During this time the approximations in TST were examined more carefully to understand when TST could be expected to give quantitative predictions of rate constants. Wigner's 1938 work played a crucial role in the revival of the hope to develop TST into an accurate, predictive tool and in advancements in this area over the last 25 years.

## 2 Key contributions

The major contribution of the title work was an exceptionally clear exposition of the approximations inherent in TST. Wigner first stated that elementary chemical reactions were considered in which the (equilibrium) Maxwell–Boltzmann velocity and energy distribution was maintained, and for which the potential-energy surface was known. He then went on to state "that the transition state method is based, in addition to well-established principles of statistical mechanics, on only three assumptions, two of which are generally accepted." The first two assumptions were those that Wigner categorized as "generally accepted": the electronic adiabaticity of the reaction and the adequacy of classical mechanics to treat the motion of the nuclei. The third assumption has become known as the fundamental assumption, the fundamental dynamical assumption, or the no-recrossing assumption of TST.

### 2.1 The fundamental assumption of TST

Wigner first defined a dividing surface, through the saddle point of the potential-energy surface and perpendicular to the direction of steepest descent, through which all reactive trajectories must pass. Wigner then identified the fundamental dynamical assumption as follows: a reactive trajectory originating in reactants must cross the dividing surface only once and proceed to products. The TST expression for the rate constant could then be expressed using equilibrium statistical mechanics without the need to calculate classical trajectories.

### 2.2 The upper-bound principle

In analyzing the validity of the fundamental dynamical assumption, Wigner pointed out the effect of trajectories that recross the dividing surface: the TST rate constant "will lead, in general, to too high values of the reaction rate and should be corrected by a factor $\gamma$, smaller than 1, ...". He also provided a compelling argument for why $\gamma$ will go to 1 with decreasing temperature for activated reactions. Thus, he showed that classical TST is accurate (for reactions with barriers) at sufficiently low temperatures and its overestimation of the exact classical rate constant generally increases with increasing temperature.

### 2.3 Variational TST

In the final section of the paper, Wigner discussed the application of TST to three-atom recombination reactions, $A + B + C \rightarrow AB + C$. For these barrierless reactions, Wigner suggested using dividing surfaces that are more complicated functions of both coordinates and momentum (e.g., dividing surfaces depending on an energy), instead of just spatial coordinates. This idea was developed in more detail in a slightly earlier paper by Wigner [8] (the "received" dates on the two publications differ by only 11 days). In the 1937 paper, Wigner presented the obvious corollary to the upper-bound principle, that the best estimate of the rate constant can be obtained by optimizing the dividing surface to minimize the rate constant. In the 1937 paper, Wigner went further to state that the accurate (classical) rate constant is the minimum value of the TST rate constant for a properly chosen dividing surface.

## 3 Subsequent progress in the field

Advancements in TST have been well documented in the literature over the past 23 years [9–16]. Much of the work on TST has focused on understanding the dynamical foundations of the theory and the extension of the theory to allow for quantitatively accurate estimates of rate constants. Advancements in these areas can be attributed to the fact that the TST expression for the classical equilibrium rate constant can be formulated by making a single approximation, Wigner's fundamental assumption.

### 3.1 Accuracy of the fundamental assumption of TST

As computational capabilities increased in the 1960s and 1970s, it became possible to test the accuracy of Wigner's fundamental assumption by comparing classical TST with the results of accurate classical trajectory studies. The first such comparisons can be attributed to Bunker's work on unimolecular reactions [17, 18]. Karplus and coworkers [19–21] performed the first comparisons for bimolecular reactions. Many tests of conventional TST were performed on atom–diatom (A + BC) reactions [22–30], and these tests confirmed Wigner's argument that classical TST is accurate at low temperatures (or equivalently at energies close to and above the barrier). An interesting observation was that the validity of TST extended over a broader energy range for the three-dimensional $H + H_2$ reaction compared to the collinear reaction [23]. The implication is that classical trajectories in the higher-dimensional phase space are less likely to make it back to the dividing surface to recross.

Wigner's dynamical perspective on TST also led naturally to the development of efficient methods to calculate the transmission coefficient (or Wigner's recrossing factor $\gamma$). This approach is outlined by Keck [31, 32] for recombination reactions and by Anderson and coworkers [33, 34] for bimolecular reactions. This work helped elucidate the connection between classical TST and classical trajectory calculations.

The fact that classical TST can be derived from a single dynamical assumption was the impetus for researchers in the 1970s to begin examining in detail the classical foundations of TST. This area of research was initiated by the paper by Pechukas and McLafferty [22]. In this work, they described conditions (e.g., features of the potential and energy range above threshold) such that classical trajectories will not recross the dividing surface and classical TST will reproduce the exact classical rate constant. Subsequent research by Pechukas and Pollak [10, 35–37] further elucidated the relationship between classical trajectories and TST. This work showed that for collinear A + BC reactions, the best possible TST dividing surfaces for a microcanonical ensemble are so-called periodic-orbit dividing surfaces, which are generated by classical trajectories that vibrate back and forth between two equipotentials in the interaction region. Furthermore, Pechukas and Pollak [37] demonstrated that for collinear A + BC reactions TST is exact for energies at which the potential surface admits only one periodic-orbit dividing surface.

## 3.2 Variational TST

Although a variational formulation of TST was not presented in Wigner's 1938 paper, it is clear that the development of a variational approach to TST follows naturally from the upper-bound principle of TST. This was also recognized in the 1930s by Horiuti [38], who adopted the dynamical perspective and developed a variational approach to TST. These ideas were extensively developed by Keck [39, 40] to formulate a more formal variational procedure for finding the best dividing surface. As already mentioned for collinear A + BC reactions periodic-orbit dividing surfaces are the optimum configuration-space dividing surfaces for a microcanonical ensemble. For reactions with larger numbers of degrees of freedom, practical techniques have been developed based upon dividing surfaces that are required to be orthogonal to the minimum energy path [27] (see Ref. [15] for further discussion and review). This approach has also proved useful for including approximate quantum mechanical effects (see later). In numerical tests, the variational TST (VTST) procedures were seen to give practical improvement over conventional TST, in which the dividing surface is constrained to pass through the saddle point (see Ref. [15] for a review of tests of classical VTST).

## 3.3 Practical methods for predicting absolute rate constants

As computational capabilities continued to improve in the 1970s and 1980s, it became possible to calculate potential-energy surfaces sufficiently accurately to allow accurate predictions of rate constants. This advancement, as well as progress in the development of TST itself, rekindled hope of developing TST into a tool that would provide quantitative accuracy for computed rate constants. For many reactions of practical interest, particularly those involving hydrogen-atom transfer, quantitative accuracy in computed rate constants requires that quantum mechanical effects be included in the theory; however, as Wigner already realized in 1938, the rigorous implementation of the fundamental dynamical assumption in a quantum mechanical theory is difficult because "one cannot speak about the mean velocity at the activation point. (Heisenberg's indetermination principle)" [3]. The earliest attempt to develop an approximate treatment of quantum effects in chemical reaction rate constants is that due to Wigner [41]. By considering the lowest-order terms in an expansion in $\hbar$ of the phase-space probability distribution function around the saddle point, Wigner developed a separable approximation, in which partition functions for bound modes are quantized and a correction is included for quantum motion along the reaction coordinate. Eyring [2] systematized the procedure of quantizing the partition functions for the bound modes of the reactants and the transition state, and this became the standard approach to including quantum mechanical effects. In the 1960s and 1970s, tests of TST against accurate quantum mechanical results indicated the deficiency of this separable approach. The failure of the standard approach was attributed largely to nonseparable effects, particular on quantum mechanical tunneling [42, 43].

Encouraged by the validity of the fundamental assumption of TST for classical TST and VTST, efforts were made to develop improved methods for including quantum mechanical effects into TST. One approximate approach is to quantize the partition functions for bound modes and focus on developing improved methods for tunneling that include some aspects of the multidimensional nature of the tunneling process. The development of tunneling methods that are consistent with VTST was greatly facilitated by the realization that the adiabatic theory of reactions is equivalent to one form of variational TST (microcanonical VTST) when the reaction coordinate is treated classically [27, 44]. In addition, an important advancement was the development of methods that consistently treat threshold contributions in the quantized partition functions and tunneling correction factors [45]. The first successful nonseparable tunneling correction was that developed by Marcus and Coltrin [46] for the collinear H + $H_2$ reaction. Marcus and Coltrin sought the tunneling path with the smallest tunneling action integral and thereby least exponential damping. They found the optimum tunneling path for the H + $H_2$ reaction was the path of turning points of the quantized adiabatic vibrations, where the turning points are chosen so that the path "cuts the corner" and reduces the length of the tunneling path. This method was subsequently extended to other systems with small-to-moderate reaction-path curvature [44, 47, 48]. The idea of finding the optimum tunneling path was further extended to systems with large reaction-path curvature in the least-action tunneling method [49]. A more detailed review of tunneling methods implemented within this framework is presented elsewhere [48, 50]. With the implementation of these new tunneling methods and demonstration of their accuracy [51, 52], quantum mechanical VTST has turned into a standard tool for studies of the kinetics of gas-phase reactions.

The success of the fundamental assumption of TST also provided impetus to develop a rigorous quantum mechanical TST, i.e., a quantum mechanical theory that employs the fundamental assumption as its only approximation and treats the problem as nonseparable [10, 53–55]. Although semiclassical approximations to the quantum mechanical TST expression of Miller were developed and successfully applied to the H + $H_2$ reaction [56, 57], by 1993 it was concluded that no rigorous quantum mechanical version of TST exists which does not require a solution of the full multidimensional reaction dynamics [58]. The effort to develop a rigorous quantum mechanical TST was productive as it provided the foundation for the development of accurate quantum mechanical methods to directly calculate thermal rate constants. (See Miller [58, 59] and references therein.) These methods are not a form of TST (since they do not invoke the fundamental assumption), but exploit the use of short-time quantum dynamics in the interaction region to calculate the reactive flux through the transition-state dividing surface. In this sense, this approach is the quantum analog of the classical approach of Keck and Anderson [31–34]. Further advances in reaction rate theory, which have benefited from research and advances in TST, are documented in Miller's recent paper [5].

### 3.4 Reactions in liquids

TST has also been widely used to treat reactions in condensed phases. Wigner's dynamical perspective has particularly had an impact on the extension of TST to reactions in liquids. Most applications to liquid-phase reactions have used the thermodynamic formulation of TST [60], which includes the effects of the condensed phase on reaction free energies in an approximate manner. Chandler [61] provided a more rigorous formulation of classical TST for liquids. The new element introduced by the liquid phase is collisions of solvent molecules with the reacting species that can lead to recrossings of the dividing surface and a breakdown of the fundamental assumption. A recent review [16] documents many more advances in the extension of TST to the kinetics of condensed-phase processes.

### Summary

In 1981 Pechukas wrote [12] "Transition state theory (TST) is 50 years old, and it is a tribute to the power and subtlety of the theory that work on the foundations of it is still a respectable and popular activity". To a large extent, this is still true today. The development and advancement of TST can be credited to many workers over the years. H. Eyring, M. Polanyi, and E. Wigner were all involved with early development of the theory and Eyring's work in particular played a major role in popularizing TST in the early years. Many of the advancements in TST over the last 20-plus years were aimed at understanding the foundations of the theory and developing TST into a more rigorous framework for accurately predicting rate constants. These advancements have drawn from many contributions in the field, some of which are described in this special issue; however, the foundation for these advancements can be traced back to Wigner's classic work that clearly described the dynamical nature of the theory and the one, fundamental approximation needed to derive the classical TST rate constant.

*Acknowledgements.* The author is supported by the Division of Chemical Sciences, U.S. Department of Energy, under contract DE-AC06–76RLO 1830 with Battelle, which operates the Pacific Northwest National Laboratory. The author wishes to thank Bill Miller and Don Truhlar for thoughtful reviews of the paper and for helpful comments.

### References

1. Laidler KJ, King MC (1983) J Phys Chem 87: 2657
2. Eyring H (1935) J Chem Phys 3: 107
3. Wigner E (1938) Trans Faraday Soc 34: 29
4. Eyring H (1938) Trans Faraday Soc 34: 41
5. Miller WH (1998) Faraday Discuss 110: 1
6. Wigner E (1938) Trans Faraday Soc 34: 70
7. Benson SW (1968) Thermochemical kinetics: methods for the estimation of thermochemical data and rate parameters, 1st edn. Wiley, New York
8. Wigner E (1937) J Chem Phys 5: 720
9. Truhlar DG, Wyatt RE (1976) Annu Rev Phys Chem 27: 1
10. Pechukas P (1976) In: Miller WH (ed) Dynamics of molecular collisions, part B, Plenum, New York, p 269
11. Truhlar DG, Garrett BC (1980) Acc Chem Res 13: 440
12. Pechukas P (1981) Annu Rev Phys Chem 32: 159
13. Truhlar DG, Hase WL, Hynes JT (1983) J Phys Chem 87: 2664
14. Hase WL (1983) Acc Chem Res 16: 258
15. Truhlar DG, Garrett BC (1984) Annu Rev Phys Chem 35: 159
16. Truhlar DG, Garrett BC, Klippenstein SJ (1996) J Phys Chem 100: 12771
17. Bunker DL (1962) J Chem Phys 37: 393
18. Bunker DL (1964) J Chem Phys 40: 1946
19. Karplus M, Porter RN, Sharma RD (1965) J Chem Phys 43: 3259
20. Morokuma K, Karplus M (1971) J Chem Phys 55: 63
21. Koeppl GW, Karplus M (1971) J Chem Phys 55: 4667
22. Pechukas P, McLafferty FJ (1973) J Chem Phys 58: 1622
23. Chapman S, Hornstein SM, Miller WH (1975) J Am Chem Soc 97: 892
24. Grimmelmann EK, Lohr LL (1977) Chem Phys Lett 48: 487
25. Chesnavich WJ (1978) Chem Phys Lett 53: 300
26. Sverdlik DI, Koeppl GW (1978) Chem Phys Lett 59: 449
27. Garrett BC, Truhlar DG (1979) J Phys Chem 83: 1052
28. Garrett BC, Truhlar DG (1980) J Phys Chem 84: 805
29. Garrett BC, Truhlar DG, Grev RS (1981) J Phys Chem 85: 1569
30. Garrett BC, Truhlar DG (1982) J Chem Phys 76: 1853
31. Keck JC (1962) Discuss Faraday Soc 33: 173
32. Keck JC (1972) Adv At Mol Phys 8: 39
33. Anderson JB (1973) J Chem Phys 58: 4684
34. Jaffe RL, Henry JM, Anderson JB (1973) J Chem Phys 59: 1128
35. Pollak E, Pechukas P (1977) J Chem Phys 67: 5976
36. Pollak E, Pechukas P (1978) J Chem Phys 69: 1218
37. Pechukas P, Pollak E (1979) J Chem Phys 71: 2062

38. Horiuti J (1938) Bull Chem Soc Jpn 13: 210
39. Keck JC (1960) J Chem Phys 32: 1035
40. Keck JC (1967) Adv Chem Phys 13: 85
41. Wigner E (1932) Z Phys Chem B 19: 203
42. Truhlar DG, Kuppermann A (1971) Chem Phys Lett 9: 269
43. Miller WH (1976) Acc Chem Res 9: 306
44. Garrett BC, Truhlar DG (1979) J Phys Chem 83: 1079
45. Garrett BC, Truhlar DG, Grev RS, Magnuson AW (1980) J Phys Chem 84: 1730
46. Marcus RA, Coltrin ME (1977) J Chem Phys 67: 2609
47. Skodje RT, Truhlar DG, Garrett BC (1981) J Phys Chem 85: 3019
48. Lu DH, Truong TN, Melissas VS, Lynch GC, Liu YP, Garrett BC, Steckler R, Isaacson AD, Rai SN, Hancock GC, Lauderdale JG, Joseph T, Truhlar DG (1992) Comput Phys Commun 71: 235
49. Garrett BC, Truhlar DG (1983) J Chem Phys 79: 4931
50. Truhlar DG, Isaacson AD, Garrett BC (1985) In: Baer M (ed) Theory of chemical reaction dynamics, vol IV. CRC Press, Boca Raton, p 65
51. Garrett BC, Truhlar DG (1984) J Chem Phys 81: 309
52. Allison TC, Truhlar DG (1998) In: Thompson DL (ed) Modern methods for multidimensional dynamics computations in chemistry. World Scientific, Singapore, p 618
53. McLafferty FJ, Pechukas P (1974) Chem Phys Lett 27: 511
54. Miller WH (1974) J Chem Phys 61: 1823
55. Miller WH (1976) Acc Chem Res 9: 306
56. Miller WH (1975) J Chem Phys 62: 1899
57. Chapman S, Garrett BC, Miller WH (1975) J Chem Phys 63: 2710
58. Miller WH (1993) Acc Chem Res 26: 174
59. Miller WH (1998) J Phys Chem A 102: 793
60. Evans MG, Polanyi M (1935) Trans Faraday Soc 31: 875
61. Chandler D (1978) J Chem Phys 68: 2959

Theor Chem Acc (2000) 103:205–208
DOI 10.1007/s002149900035

Theoretical
Chemistry Accounts
© Springer-Verlag 2000

*Perspective*

# Perspective on "Conduction in polar crystals. I. Electrolyte conduction in solid salts"

## Mott NF, Littleton MJ (1938) Trans Faraday Soc 34: 485

**C.R.A. Catlow**

Davy Faraday Research Laboratory, The Royal Institution of Great Britain, 21, Albemarle Street, London W1X 4BS, UK

Received: 8 March 1999 / Accepted: 29 March 1999 / Published online: 28 June 1999

**Abstract.** We survey the consequences of the landmark paper in the development of the contemporary theories of defects in solids – a core area of modern solid-state science. We summarise the basic concepts behind the "Mott–Littleton" approach and the developments to which it has led.

**Key words:** Defects in solids – Iouic transport – Nonstoichiometry

## 1 Introduction

The structures, energies and mobilities of defects in crystalline solids play a central role in controlling the thermodynamic, transport, electrical and chemical properties of materials. The science of defects in solids began to develop in the 1920s and 1930s with the pioneering work of Frenkel, Schottky, Jost and Wagner which established the basic nature of point defects and defect reactions, and of their generation by thermal or chemical means.

An understanding of the behaviour and consequences of defects in solids requires knowledge of their formation and migration energies. The derivation of these crucial parameters from experimental data is frequently difficult, and in many cases impossible. Independent estimates from theory were therefore necessary for the development of the field. The pioneering work of Mott and Littleton [1] laid the foundation for the modern quantitative studies of defects in polar solids, which have had a major impact on contemporary solid-state chemistry and physics.

## 2 The Mott–Littleton method

The basic concept behind the technique advanced in this seminal paper is simple. In the region close to a defect, the forces exerted on the surrounding lattice are strong;

this region is treated explicitly, i.e. atomic positions are adjusted to the equilibrium positions using knowledge of the interatomic forces. For more distant regions, the forces are weaker and in polar materials will essentially correspond to the polarisation of the crystal by the effective charge of the defect and can be estimated on the basis of the macroscopic dielectric constant. This approach, in its modern formulation, is summarised in Fi.g. 1, in which an interface region is used between the two regions. The defect formation energy depends on the position of the atoms ($x$) in the inner region and the displacements ($y$) in the outer region, and can be written as

$$E_D(x, y) = E_1(x) + E_{1,2}(x, y) + E_2(y) \;, \qquad (1)$$

where the first and last terms on the right-hand side of the equation arise from the inner and outer regions, respectively, and the second term represents the interaction between the inner region and the displacements in the outer region. $E_2$ is generally considered to be quadratic in $y$ and invocation of the equilibrium condition allows $y$ to be written in terms of the

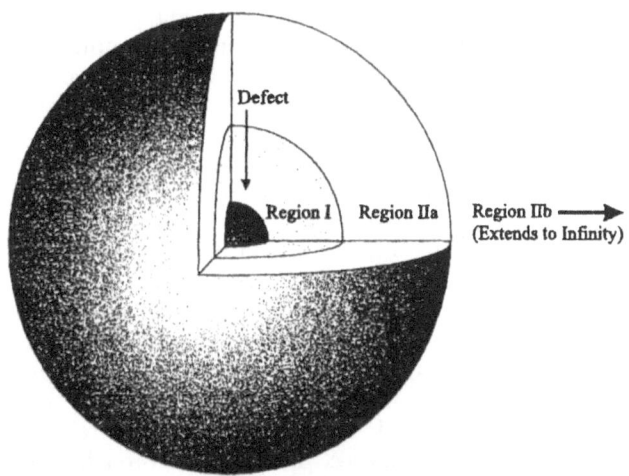

**Fig. 1.** Two-region strategy for defect calculations

derivative of $E_{1,2}$ and hence the explicit dependence of the total energy on $E_2$ is removed.

Details of the theory are given in Refs. [1–3]. Ref. [3] provides a good survey of the field up to the 1980s.

## 3 The development of the method

Since the application of the method to all but the simplest and most-restricted models requires intensive numerical calculations, its widespread use had to wait until the advent of modern computation. Useful progress was made by, for example, Barr and Lidiard [4] in the 1960s in applying the method to the classic problem of vacancy defects in ionic crystals. The major breakthrough was, however, the development by Norgett [5] of the first general-purpose computer code, HADES, for modelling defects in ionic crystals employing a methodology based on the Mott–Littleton approach. A crucially important feature of the implementation of the method in HADES was the use of the shell-model treatment of ionic polarisation [6], which is based on a simple mechanical model of polarisability in which a core is coupled to a shell (representing the polarisable valence shell electron) by a harmonic spring, which has proved very effective in modelling polarisation properties of these materials.

The field developed rapidly in the 1970s, with applications of the technique to both halide [7, 8] and oxide [9–11] crystals which demonstrated the qualitative reliability of the method for calculating both formation and migration energies of defects, given reliable interatomic potentials.

During the same period, it also became apparent that the method could be used to provide qualitative guidance as to the nature of complex defect structures in heavily doped and nonstoichiometric materials. One of the best examples was the study [12] of defect aggregation in nonstoichiometric wustite ($Fe_{1-x}O$) – a problem that still remains controversial. This study, however, established the stability of the defect aggregate shown in Fig. 2, in which four vacancies surround a central $Fe^{3+}$ interstitial site moreover it suggested ways in which these "4:1" clusters could aggregate.

During the 1980s, there were four major developments. First, the method was extended to noncubic crystals in the HADES3 [13] and CASCADE [14] codes. Secondly, Gillan and Jacobs [15] and Harding [16] introduced methods for calculating defect entropies based on the perturbation of the vibrational density of states by the defect. Thirdly, the method was adapted by Tasker [17] and Mackrodt [18] to treat surface defects. Finally, work started on interfacing the Mott–Littleton approach with quantum cluster calculations. In such studies, the core of region I, comprising the defect and one or two surrounding shells of atoms, is treated quantum mechanically; care must be taken to ensure consistency between the relaxation of the quantum mechanically described component and the relaxed classically modelled lattice. Such an approach was implemented in the ICECAP code [19], and developments of the approach are still being actively pursued.

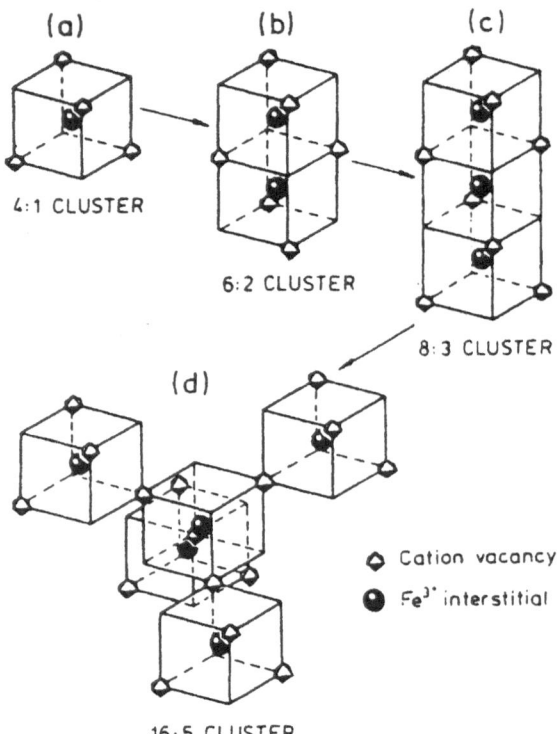

**Fig. 2a–d.** Vacancy–interstitial clusters in $Fe_{1-x}O$. **b, c** and **d** are different forms of linking the basic "4:1" cluster shown in **a**

During the same period, the range and sophistication of the applications developed. Representative new areas included the extensive range of simulation studies of defects in superionic conductors [20] in high-temperature superconductors (see, for example, Refs. [21, 22]) and the detailed studies of the complex defect chemistry arising from the creation of fission products in $CUO_2$ nuclear fuels [23].

## 4 Present status and future development

The Mott–Littleton method is now a routine tool in computational solid-state chemistry and physics, and is implemented, together with other static lattice modelling tools, in the GULP code [24], written by Gale. Two recent applications serve to illustrate the range and diversity of current applications.

The first relates to the crucial and controversial question of the nature of charge carriers in high-temperature superconductors. Early work [25] had established the usefulness of the Mott–Littleton approach in modelling localised electron states in ionic crystals. Recent work of Catlow, Islam and Zhang [26] (which built on earlier studies of the same authors) examined the stability of hole pairs or "bipolarons" in the $La_2CuO_4$ superconductor. It is generally accepted that electron holes in these materials are $O(2p)$ in character, i.e. to a very simple approximation they correspond to localised $O^-$ ions. Hole pairs would therefore correspond to $O_2^{2-}$ i.e. peroxyanions, located at a pair of lattice sites. Mott–Littleton calculations indeed suggest that such species are weakly bound with respect to isolated hole species, in both configurations shown in Fig. 3. These results do not, of course, necessarily show that such species are the charge carriers in high-temperature superconductors. They do, however, suggest that peroxy bipolarons may play a significant role in the electronic behaviour of the material, either above or below the critical temperature.

The second example concerns the mobility of oxygen vacancies in doped perovskite-structured materials – a problem of substantial technological importance in view of the widespread use of these materials as oxygen-ion-conducting solid electrolytes. The materials are normally "acceptor doped", i.e. low-valence ions substitute for either the A or B metal ion of the $ABO_3$ perovskite–structured material. The work of Cherry et al. [27] established that the vacancies which are created as charge compensators for the dopant have a relatively low activation energy and migrate by the mechanism shown in Fig. 4a. More importantly, Cherry et al. [27] investigated the variation of the oxygen vacancy activation energy with the A/B cation radius ratio (expressed as the tolerance factor $t = (r_A + r_O)/\sqrt{2}(r_B + r_O)$, where $r_A$, $r_B$ and $r_O$ are the A, B and oxygen ion radii, respectively). The results, illustrated diagrammatically in Fig. 4b, show the strong effect of $t$ on the activation energy and indicate that a value for $t$ of about 0.8 will give the lowest activation energy and the maximum conductivity. The result illustrates the increasingly predictive nature of these calculations and also their growing potential in materials design.

The future of the field will unquestionably be in this kind of predictive and design application. We can also anticipate rapid growth in the use of embedded-cluster techniques in which, as described earlier, the core region surrounding the defect is treated by a high-level quantum mechanical method. With these and other developments, methods building on the approach established in Mott and Littleton's remarkable paper are likely to continue to play a productive role in simulating the complex solid-state chemistry of defective compounds.

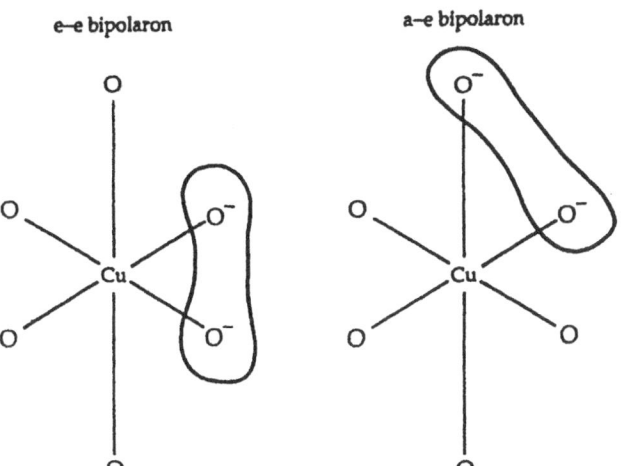

**Fig. 3.** Peroxy bipolaron in the axially distorted $CuO_6$ octahedron of $La_2CuO_4$. In the "e–e" configuration, both component holes are in equatorial sites; in the "a–e" configuration they are distributed between axial and equatorial sites

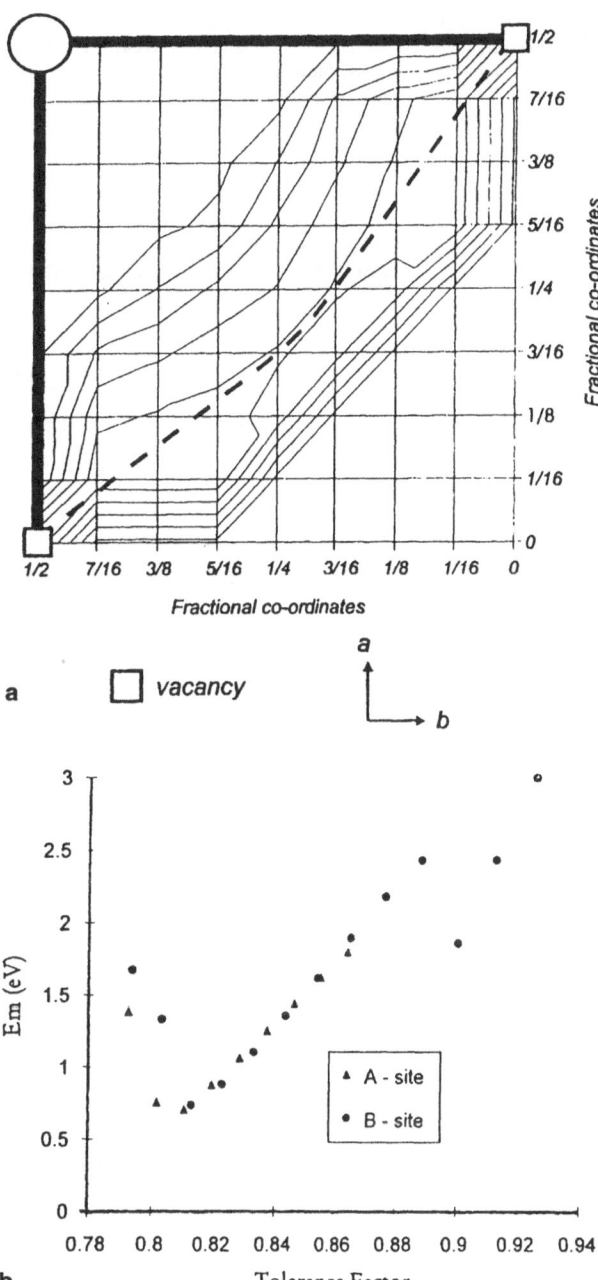

**Fig. 4.** Contour plot of the potential-energy surface for oxygen-vacancy migration, showing the curved path between adjacent anion sites of a $BO_6$ octahedron (in the *ab* plane). **b** Calculated migration energy as a function of the tolerance factor (from both the A and B site simulations)

*Acknowledgements.* I am grateful to many colleagues for collaboration and discussion relating to the Mott–Littleton method, but perhaps most notably to A.B. Lidiard, A.M. Stoneham, M.J. Norgett, J.H. Harding, R.W. Grimes, M.S. Islam, R.A. Jackson, P.W.M. Jacobs, J. Corish and A.V. Chadwick.

## References

1. Mott NF, Littleton MJ (1938) Trans Farady Soc 34: 485
2. Catlow CRA, Mackrodt WC (1982) Lect Notes Phys 166: 3
3. (1989) J Chem Soc Faraday Trans 2: 85

208

4. Barr LW, Lidiard AE (1970) In: Eyring H, Henderson D, Joist W (eds) Physical chemistry – an advanced treatise, vol 10, chapter 3, Academic Press, New York
5. Norgett MJ (1974) UKAEA Report AERE–R7650
6. Dick BG, Overhauser AW (1958) Phys Rev 112: 90
7. Catlow CRA, Norgett MJ, Ross TA (1977) J phys C 10: 1627
8. Catlow CRA, Corsih I, Diller KM, Jacobs PWM, Norgett MJ (1979) J Phys C 12: 451
9. (a) Catlow CRA (1975) Proc R Soc Lond Ser A 353; (b) Catlow CRA (1977) Proc R Soc Lond Ser A 533
10. Catlow CRA, Mackrodt WC, Norgett MJ, Stoneham AM (1977) Philos Mag 35: 177
11. Mackrodt WC, Stewart RE (1979) J Phys C 12: L131
12. Catlow CRA, Fender BEF (1975) J Phys C 8: 3267
13. Catlow CRA, James R, Mackrodt WC, Stewart RE (1982) Phys Rev B Condens Matter 25: 1006
14. Leslle MJ (1984) Daresbury Laboratory Report DL/SCI/TM36T
15. Gillan MJ, Jacobs PWM (1983) Phys Rev B 28: 159
16. Harding JH (1989) J Chem Soc Faraday Trans 2 85: 351
17. Tasker PW (1979) Philos Mag 39: 119
18. Mackrodt WC (1989) J Chem Soc Faraday Trans 2 85: 541
19. Harding JH, Harker AH, Keegstra PB, Pandey P, Vail JM, Woodward C (1985) Physica B 131: 151
20. Catlow CRA (1983) Solid State Ionics 8: 89
21. Allan NL, Mackrodt WC (1988) Philos Mag A 58: 555
22. Allan NL, Mackrodt WC (1989) J Chem Soc Faraday Trans 2 85: 385
23. Grimes RW, Catalow CRA (1991) Philos Trans R Soc Lond, Ser A 335: 609
24. Gale JD (1997) J Chem Soc Faraday Trans 93: 629
25. Tasker PW, Stoneham AM (1977) J Phys Chem Solids 38: 1185
26. Catlow CRA, Islam MS, Zhang X (1998) J Phys Condensed Matter, 3: L49
27. Cherry M, Islam MS, Catlow CRA (1995) J Solid State Chem 118: 125

Theor Chem Acc (2000) 103:209–211
DOI 10.1007/s002149900074

Theoretical
Chemistry Accounts
© Springer-Verlag 2000

## Perspective

# Reaction rates in condensed phases.
# Perspective on "Brownian motion in a field of force
# and the diffusion model of chemical reactions"

## Kramers HA (1940) Physica 7:284

### Susan C. Tucker

Department of Chemistry, University of California at Davis, CA 95616, USA

Received: 21 May 1999 / Accepted: 10 June 1999 / Published online: 4 October 1999

**Abstract.** The key concepts from H.A. Kramers' influential work on noise-assisted escape of a particle bound in a potential well are summarized, as is the extensive impact that these ideas have had on the development of condensed-phase reaction-rate theories in the twentieth century.

**Key words:** Kramers theory – Reaction dynamics – Solvent effects – Stochastic processes – Escape over a potential barrier

In April 1940, H.A. Kramers published his work entitled "Brownian motion in a field of force and the diffusion model of chemical reactions" [1]. This work, which appears to have generated little activity at the time [2], was, many years later, to become one of the most influential contributions to reaction-rate theory in the twentieth century. The prominence of Kramers' work is reflected in the titles of two recent reviews on the current status of condensed-phase reaction-rate theory: "Reaction-rate theory: fifty years after Kramers" and "New trends in Kramers' reaction-rate theory".

In his seminal work (as it is frequently called), Kramers treated the escape over a potential barrier by a particle undergoing Brownian motion, i.e. thermal noise-assisted escape [1]. Hence, his focus was on the effect of the medium – solvent or bath gas – on the solute reaction rate. While much of the physical chemistry community was at the time focused on the rate of reaction of an isolated molecule – and would remain so occupied for many years to come, Kramers' work was not completed in a vacuum. Indeed, Lindemann, Rice and Ramsperger, Kassel, Slater, Christiansen, and others had already published their collision-rate-based theories of the role of the bath gas in promoting chemical reactions in low-density gases [3, 4]. Thus, one must ask,

what is it that distinguishes Kramers' article? What new ideas and new results within his article have given his work such a prominent place in modern rate theory?

First, Kramers recognized that the motion of the solvent molecules could be modeled according to the theory of Brownian motion, i.e. according to the Langevin equation, within which the solvent is described by its viscosity and a random force term. Such a stochastic equation for the solute variables $(p, q)$ can be reformulated as either a diffusion or a master equation for the probability distribution of the solute variables $[\mathscr{P}(p, q)]$, leaving open many routes to the solution of the reaction problem [5]. Indeed, Kramers himself chose the diffusion-equation approach, devising a specific Fokker–Planck equation, now known as the Kramers–Klein equation, for the purpose. Yet, Kramers' choice of a Brownian description for the reacting solute had much more far-reaching consequences in that it enabled Kramers to describe the effect of the solvent over a broad range of conditions, from the very low viscosities expected in low-density gases to the high viscosities expected in liquids. This description stands in contrast to the earlier collision theories, which are applicable only under the single-collision conditions of low-density gases. Kramers' work thus introduced into the reaction dynamics community the novel idea that solvent motion, i.e. thermal noise, could affect solute reaction rates in high-density solvents, as well as in low-density gases.

Although Kramers' description applies to all viscosity regimes, Kramers was only able to solve for the reaction rate in the limits of very low and very high viscosities (which, in itself, was no small feat). Kramers found that at low viscosities the reaction rate is controlled by the rate of energy diffusion (consistent with collision theories) and rises as $k \propto \eta$, where $k$ is the rate constant and $\eta$ is the viscosity. At high viscosities, Kramers indeed uncovered a thermal-noise effect, finding that the rate is controlled by spatial diffusion of the particle (i.e. of the reaction coordinate) and falls off as $k \propto \eta^{-1}$ with increasing viscosity. These two findings comprise the main, and far-reaching, result of Kramers' paper. They suggest that with increasing solvent viscosity the reaction rate should first increase (energy-diffusion-limited regime), pass through a turnover, and then subsequently decrease with further increases in viscosity (spatial-diffusion-limited regime), a behavior which has been verified via experiment and simulation, and is now referred to as the "Kramers turnover" [3, 6, 7].

Kramers final coup d'état in this work was a recasting of his rate expressions in terms of the then newly developed transition-state theory [8, 9], which has since become the most prominent rate theory in chemistry. In both limits Kramers was able to cast his result in terms of a multiplicative prefactor to the transition-state theory result. I note that the "transition-state method" to which Kramers compared takes only the solute degrees of freedom into consideration. Only some 40 years later was it recognized that multidimensional variational-transition-state theory [10], inclusive of all the solvent degrees of freedom, can reproduce the Kramers result in the high-viscosity, spatial-diffusion-limited regime [11–13].

In spite of, or perhaps because of, the novel, far-reaching conclusions of Kramers' work, it was largely neglected for 30–40 years. Indeed, authorative books on chemical reaction dynamics, such as that by Moore and Pearson [4], make no mention of Kramers' work or the ideas contained therein. An important exception to this neglect was Chandrasekhar's application of Kramers' work to the astrophysical problem of star clusters in 1943 [14], as much of Chandrasekhar's work was, in turn, to later influence the chemical dynamics community. The works of Bak in the 1960s, pursuing Kramers' idea of applying diffusion equations to activated chemical reactions, also appear to have been influential [15, 16]. Yet, during the period 1940–1970, most chemical studies of reactions in low-density gases concentrated on the aforementioned strong-collision models and the development of important extensions thereof, most importantly RRKM theory [4, 17]. It was only much later that Kramers' "weak collision" energy-diffusion model was considered and the relationship between the strong- and weak-collision models was elucidated [18]. Similarly, in the high-viscosity, solution-phase limit, the chemical community (with the exception of Brinkman [19] and Takeyama [20]) primarily concerned itself with the effect of solvation free energies on the reaction activation barrier, an effect which, the astute reader will notice, is completely absent from the treatment of Kramers. Within Kramers' formalism, such solvation-induced changes in the barrier height would be hidden within the simple transition-state-theory rate to which Kramers' results provide a prefactor correction. However, it was many years before it was recognized that solvation effects on solute reaction rates can be formally separated into activation free-energy (potential of mean force) and dynamical (thermal-noise induced) effects [21], both of which will typically vary with changing solvent viscosity.

Extensive interest in Kramers' work by the chemical community had to wait until the late 1970s, early 1980s, when technological advances – fast pulsed lasers and computers – provided the ability to probe the microscopic details of solution-phase reaction dynamics, thus enabling the study of noise-induced "dynamical" effects in liquids. A key paper in stimulating this renewed interest was that of Grote and Hynes [16], in which the Kramers result in the high-viscosity limit was rederived (along with an extension to time-dependent friction).

Once Kramers' work returned to the public eye, the approximations and restrictions inherent in Kramers theory came under close scrutiny. These are

1. An assumption that the microscopic friction $\gamma(t)$, exerted by the solvent on the solute reaction coordinate, can be approximated by the viscosity of the bulk fluid, $\eta$.
2. An assumption that the solvent moves rapidly with respect to the solute, such that the solvent friction acts effectively instantaneously, i.e. $\gamma(t) = \eta\delta(t)$, with $\delta(t)$ the Dirac delta function.
3. The neglect of anharmonic terms in the solute potential of mean force (also known as nonlinearity and finite-barrier effects).
4. The restriction to a 1-dimensional (1D) solute.
5. The lack of a solution to the rate expression in the intermediate-viscosity "turnover" region.
6. The neglect of quantum effects.

By the late 1980s, development of theoretical extensions of Kramers' result, along with numerical (via computer simulation) and experimental tests, had become a cottage industry, as is evidenced by the 700-plus references cited in the review in Ref. [3]. These extensions came from both the chemistry and the physics communities and were developed from an extremely broad arsenal of formalisms [2, 3, 7].

Here I list just a small sample from the number of things about condensed-phase rate theory that have been learned from the past two decades of intense activity on this problem (I give only representative references, as there are far too many to list here). From studies of 1D model solutes coupled to a solvent bath, for example, we have learned that at intermediate viscosities, the spatial-diffusion effect typically becomes important before the energy-diffusion effect becomes negligible, a behavior which is greatly magnified in slowly relaxing solvents [22]. Consequently, the simple 1D transition-state-theory estimate of the rate is rarely, if ever, accurate for these model systems, even at intermediate viscosities [22]. Additionally, new methods have been developed for predicting rates in the difficult intermediate-viscosity regime [23, 24], where simple interpolation schemes have been shown to fail [25]. We now know that the viscosity of the bulk fluid is often a poor approximation to the microscopic friction [26], although how to determine the microscopic friction without resorting to computer simulation [27] remains an open question. We have also learned that Kramers' delta-function friction, which assumes rapid solvent motion, is a poor approximation for many chemical systems [28, 29], and that a slow solvent relaxation can greatly reduce the effect of the solvent, either in promoting (low viscosity) [22, 33] or hindering (high viscosity) [16, 30] the reaction. Numerous methods have been developed for the incorporation of nonlinear, finite-barrier effects [3, 31], as have been methods to enable the microscopic friction to depend upon the reaction coordinate [32]. The effect of strongly coupled solute degrees of freedom has been examined, demonstrating that such strongly coupled modes enhance the rapidity with which the energy-diffusion-limited rate increases with increasing solvent friction [18] and can also lead to interesting spatial diffusion behaviors, such as saddle-point avoidance [33]. The transition from weakly to strongly coupled solute modes is an area of current investigation [34, 35], as is the incorporation of quantum effects, even though this latter topic has already seen extensive attention for well over a decade [36]. Finally, on a more practical note, it has become clear that only the spatial-diffusion regime is accessed in liquids, whereas only energy-diffusion behavior is typically found in gases; it is only in intermediate-density supercritical fluids that the turnover behavior is observed [26, 37]. Another hard-earned lesson has been that uncharacterized changes in the activation barrier, which generally occur simultaneously with changes in the viscosity (be they due to density changes or to changed chemical interactions), almost always inexorably cloud the extraction of dynamic solvent effects from the data. Indeed, spatial-diffusion effects in liquids typically alter the rate by a factor less than 10, whereas (at 298K) activation-energy changes alter the rate by an order of magnitude per kilocalorie/mole. Yet, if a factor of 2 accuracy is to be achieved in the prediction of reaction rates in condensed phases [7] such Kramers effects will require proper evaluation.

Given the successes of the past two decades, what does the future hold for Kramers' reaction-rate theories? In addition to continued work on the difficult problem of quantum effects in condensed media, I believe that the need to understand reactivity in a broad range of complex systems will drive the development of new extensions to the Kramers' method. In fact, such application-driven extensions have already begun to appear, such as the extension to nonstationary friction kernels, $\gamma[(t' + \tau) - (t + \tau)] \neq \gamma(t' - t)$, for thermosetting polymers [38], to multiple correlated barrier crossings for surface reactions in the energy-transfer-limited regime [39, 40], and to fluctuating barriers for modeling the kinetics of ion channels [41]. Thus, the ideas of Kramers are likely to be moved forward into the next century like a coat of many colors, being continually directed into new dimensions which have not yet entered our thoughts, but will arise naturally from new and varied complex applications.

# References

1. Kramers HA (1940) Physica 7: 284
2. Talkner P, Hänggi P (1995) New trends in Kramers' reaction-rate theory. Understanding chemical reactivity. Kluwer, Dordrecht
3. Hänggi P, Talkner P, Borkovec M (1990) Rev Mod Phys 62: 250
4. Moore JW, Pearson RG (1981) Kinetics and mechanism. Wiley, New York
5. Risken H (1989) The Fokker–Planck equation, 2nd edn. Springer, Berlin Heidelberg New York
6. Hänggi P, Fleming G (1992) Activated barrier crossing. World Scientific, Singapore
7. Voth GA, Hochstrasser RM (1996) J Phys Chem 100: 13034
8. Eyring H (1935) J Chem Phys 3: 107
9. Wigner EP (1938) Trans Faraday Soc 34: 29
10. Tucker SC (1995) In: Talkner P, Hänggi P (eds) New trends in Kramers' reaction-rate theory. Understanding chemical reactivity. Kluwer, Dordrecht, p 5
11. van der Zwan G, Hynes JT (1983) J Chem Phys 78: 4174
12. Pollak E (1986) J Chem Phys 85: 865
13. Dakhnovskii YI, Ovchinnikov AA (1985) Phys Lett A 113: 147
14. Chandrasekhar S (1943) Astrophys J 98: 54
15. Bak TA (1967) Ind Eng Chem 59: 51
16. Grote RF, Hynes JT (1980) J Chem Phys 73: 2715
17. Marcus RA (1952) J Chem Phys 20: 359
18. Borkovec M, Berne BJ (1985) J Chem Phys 82: 794
19. Brinkman HC (1956) Physica 22: 149
20. Takeyama N (1971) Experentia 17: 425
21. Chandler D (1978) J Chem Phys 68: 2959
22. Tucker SC (1993) J Phys Chem 97: 1596
23. Mel'nikov VI, Meshkov SV (1986) J Chem Phys 85: 1018
24. Pollak E, Grabert H, Hänggi P (1989) J Chem Phys 91: 4073
25. (a) Straub JE, Borkovec M, Berne BJ (1986) J Chem Phys 84: 1788; (b) Straub JE, Borkovec M, Berne BJ (1986) J Chem Phys 86: 1079 (addendum and erratum)
26. Troe J (1991) Ber Bunsenges Phys Chem 95: 228
27. Berne BJ, Tuckerman ME, Straub JE, Bug ALR (1990) J Chem Phys 93: 5084
28. Straub JE, Borkovec M, Berne BJ (1988) J Chem Phys 89: 4833
29. Gertner BJ, Wilson KR, Hynes JT (1989) J Chem Phys 90: 3537
30. Grote RF, van der Zwan G, Hynes JT (1984) J Phys Chem 88: 4676
31. Talkner P (1995) In: Talkner P, Hänggi P (eds) New trends in Kramers' reaction-rate theory. Understanding chemical reactivity. Kluwer, Dordrecht, p 47
32. Haynes GR, Voth GA (1995) J Chem Phys 103: 10176
33. Berezhkovskii AM, Zitserman VY (1992) J Phys A 25: 2077
34. Hershkowitz E, Pollak E (1997) J Chem Phys 106: 7678
35. Reese SK, Tucker SC (1998) Chem Phys 235: 171
36. Voth GA (1995) In: Talkner P, Hänggi P (eds) New trends in Kramers' reaction-rate theory. Understanding chemical reactivity. Kluwer, Dordrecht, p 197
37. Hara K, Kiyotani H, Kajimoto O (1995) J Chem Phys 103: 5548
38. Hernandez R, Sommer FL (1999) J Phys Chem 103: 1070
39. Haug K, Metiu H (1991) J Chem Phys 94: 3251
40. Pollak E, Bader J, Berne BJ, Talkner P (1993) Phys Rev Lett 70: 3299
41. Reimann P, Bartussek R, Hänggi P (1998) Chem Phys 235: 11

Theor Chem Acc (2000) 103:212–213
DOI 10.1007/s002149900018

Theoretical
Chemistry Accounts
© Springer-Verlag 2000

*Perspective*

# Perspective on "The effect of shape on the interaction of colloidal particles"

## Onsager L (1949) Ann NY Acad Sci 51: 627

**Daan Frenkel**

FOM Institute for Atomic and Molecular Physics, Kruislaan 407, 1098 SJ Amsterdam, The Netherlands

Received: 24 February 1999 / Accepted: 3 March 1999 / Published online: 21 June 1999

**Abstract.** Onsager's paper on the effect of shape on the interaction of colloidal particles is seminal in many ways. I shall focus on two aspects: it is (to my knowledge) the earliest classical density functional theory, and it demonstrates the possibility of ordering transitions driven by entropy

**Key words:** Entropic phase transition – Density functional theory – Nematic liquid crystal – Freezing

If ever the description "multifaceted" should apply to a paper, it is to Onsager's analysis of the effect of shape on the interaction of colloidal particles. The primary aim of this paper was to explain the physics behind the formation of an anisotropic (nematic) phase in a relatively dilute suspension of rigid, linear colloids. Here I shall not discuss those aspects of the paper that focus mainly on the analysis of the forces acting between charged colloidal rods (actually not just rods) in solution. Although this analysis makes the present paper one of the milestones in theoretical colloid physics, I would like to start where this part of the analysis ends – namely with the conclusion that, to a good approximation, the interaction between charged rods in a not too dilute salt solution is short-ranged repulsion. This implies that, as a first approximation, it is permissible to treat such rodlike particles as slender, hard rods. Onsager then goes on to consider a limiting case: rods for which the ratio of length ($L$) to the diameter ($D$) tends to infinity. Using this description of rodlike colloids, Onsager then proceeds to show that in such a model, a transition from the isotropic to the nematic phase must occur above a certain density (of order $(L^2 D)^{-1}$).

There are several features of Onsager's analysis that make this paper crucial for subsequent developments. The first is the way in which Onsager derives an expression for the free energy of a system of thin hard rods. What Onsager does is nothing less than write down the free energy of the system as a functional of the single-particle distribution function – deriving such an expression is the central aim of all subsequent density-functional theories, be they classical or quantum-mechanical (see e.g. Ref. [1]). Unfortunately, we usually do not know the true density functional, and we are forced to make approximations. Not so in Onsager's case: the density functional that Onsager writes down for infinitely thin hard rods is almost certainly exact. Personally, I would say that it is exact – it relies on the fact that all virial coefficients higher than the second are negligible: Onsager showed that this is plausible, and subsequent numerical work [2] supports this claim. This makes the Onsager density functional one of the very few that is both nontrivial and exact. Although the density functional that Onsager derived is exact, the free-energy minimization cannot be carried out analytically. Onsager found an approximate solution, using an ingenious ansatz for the single-particle distribution function; however, numerically, the minimization can be carried out to any desired accuracy. In this sense, we now know the "exact" density dependence of the free energy of thin hard rods. The Onsager approach has proven extremely fruitful. It has been applied to a variety of model systems, including semiflexible particles [3, 4]. However, here I will not review the subsequent work, but rather discuss another aspect of Onsager's paper.

The transition from the isotropic to the nematic phase involves a partial ordering of the molecular orientations. It is common practice to consider entropy as a measure for the "disorder" in a system. A naive observer would therefore conclude that the isotropic phase must always have a higher entropy than the nematic phase at the same density and energy. In this picture, ordering can only occur due to a lowering of the energy, such that the free energy of the system decreases upon going from the metastable isotropic to the stable nematic phase, as required by thermodynamics. However, the Onsager model is a hard-core model: this means that, at a given temperature, the energy does not depend on density.

Hence, all spontaneous phase transformations must involve an increase in entropy. In other words at a given density, the "ordered" nematic phase has a higher entropy than the "disordered" isotropic phase. The ordering of hard rods is an example of an entropic phase transition. Although counterintuitive at first, no laws of physics are violated: sure enough, the orientational entropy of the nematic phase is lower than that of the isotropic phase; however, this loss in orientational entropy is more than offset by the increase in translational entropy. As the particles align, their excluded volume decreases and hence they are able to explore a larger fraction of the volume without overlapping with other particles – this causes the translational entropy to increase.

A second important example of an entropic phase transition is the freezing transition of hard spheres. The first direct evidence for this transition came from the computer simulations of Wood and Jacobson, and Alder and Wainwright [5]. Initially, these results were received with a lot of scepticism (see, e.g. Ref. [6]); however, in recent years, it has become apparent that entropic phase transitions are quite common, not only in computer models, but also in real systems (mainly colloidal dispersions: see e.g. Ref. [7]). However, the concepts, the basic physics, were all contained in Onsager's paper.

## References

1. Hansen JP, McDonald IR (1986) Theory of simple liquids, Academic Press, London
2. (a) Frenkel D (1987) J Phys Chem 91: 4912, (b) Frenkel D (1988) J Phys Chem 92: 5314
3. Khokhlov AR, Semenov AN (1981) Physica A 108: 546
4. Odijk T (1983) Macromolecules 16: 1340
5. (a) Wood WW, Jacobson JD (1957) J Chem Phys 27: 1207; (b) Alder BJ, Wainwright TE (1957) J Chem Phys 27: 1208
6. Percus JK (ed) (1963) The many-body problem. Interscience, New York
7. Pusey P (1991) In: Hansen JP, Levesque D, Zinn-Justin J (eds) Proceedings of the Les Houches summerschool session LI on: liquids, freezing and glass North-Holland, Amsterdam, pp 763

Theor Chem Acc (2000) 103:214–216
DOI 10.1007/s002149900011

Theoretical
Chemistry Accounts
© Springer-Verlag 2000

*Perspective*

# Perspective on Norman Ramsey's theories of NMR chemical shifts and nuclear spin–spin coupling

## Phys Rev 77: 567 (1950) to Phys Rev 91: 303 (1953)

**Pekka Pyykkö**

Department of Chemistry, University of Helsinki, P.O. Box 55, FIN-00014 Helsinki, Finland
e-mail: pekka.pyykko@helsinki.fi

Received: 9 February 1999 / Accepted: 22 February 1999 / Published online: 7 June 1999

**Abstract.** The theories connecting the observed NMR chemical shifts and nuclear spin–spin coupling constants to electronic wave functions were published by Norman Ramsey in eight connected *Physical Review* papers from 1950 to 1953. At the nonrelativistic limit these expressions still stand as the final answer.

**Key words:** Ramsey's theories – NMR spin – spin coupling – NMR chemical shift – Relativistic effects

## 1 Introduction

NMR is currently one of the most important forms of chemical spectroscopy. The two main spectroscopic parameters are the chemical-shift tensor, $\sigma$, and the nuclear spin–spin coupling tensor, $\mathbf{J}$. The connection between these two quantities and the electronic wave functions of the molecule was described in a series of five papers [1–5] published between 1950 and 1953. These five papers should be regarded as a whole. Together they are among the most influential ones in the quantum chemistry of the twentieth century. A sixth paper [6] pointed out a second-order magnetic self-coupling term, looking like the electric quadrupole interaction. The spin–spin coupling paper [5] had a precursor by Ramsey and Purcell [7]. In rotating molecules further chemical-shift terms occur from zero-point vibrations and "Thomas precession" [8].

## 2 The original papers

### 2.1 The NMR Hamiltonian

The NMR spectrum of a molecule containing the magnetic nuclei $\mathbf{I}_i$ is determined in the absence of electric quadrupole effects by the magnetic spin Hamiltonian

$$H = -\mathbf{B} \cdot \sum_i \hbar\gamma_i(1 - \sigma_i) \cdot \mathbf{I}_i + \sum_{i>j} h\mathbf{I}_i \cdot \mathbf{J}_{ij} \cdot \mathbf{I}_j \ . \quad (1)$$

Usually the components of $\sigma$ are expressed in parts per million (ppm) and those of $\mathbf{J}$ in hertz. The nuclear magnetic moment is $\mu = \hbar\gamma\mathbf{I} = g_N\beta_n\mathbf{I}$; $\beta_n = e\hbar/2m_\mathrm{p}c$ is the nuclear magneton.

We consider closed-shell molecules and also neglect here the direct, through-space nuclear spin–spin interactions. For a rotating molecule in vacuum one has to add the spin-rotation term

$$H_\mathrm{sr} = -h \sum_N \mathbf{I}_N \cdot C_N \cdot \mathbf{K} \ , \quad (2)$$

where $C_N$ (in hertz) is the spin-rotation coupling tensor of nucleus $N$ and $\mathbf{K}$ is the rotational angular momentum operator of the molecule.

### 2.2 Chemical shifts

In his first paper Ramsey [1] showed that the isotropic chemical-shift formula of Lamb [9],

$$\sigma_\mathrm{d} = \left(e^2/3mc^2\right)\langle 0|\tfrac{1}{r}|0\rangle \ , \quad (3)$$

where $|0\rangle$ is the total electronic wave function, had to be completed by a term related to the spin-rotation constant:

$$\sigma_\mathrm{p} = \frac{1}{6} \frac{\alpha^2(a_0a^2)}{\beta_n m/M} \left[ \frac{2Z(\beta_n m/M)}{a^3} - \frac{m_\mathrm{r}}{M}B_N \right] \ . \quad (4)$$

Here $m$ and $M$ are the electronic and nuclear masses, respectively, while $m_\mathrm{r}$ is the reduced mass, $\alpha = e^2/hc$ is the fine-structure constant and $a$ is the bond length of the molecule. $B_N$ is the magnetic flux density at nucleus

$N$ arising from molecular rotation. For the proton in $H_2$ $\sigma_d$ and $\sigma_p$ were estimated to be 32.4 and $-5.3$ ppm respectively [1].

In the derivation of the latter term the infinite summation over excited electronic states could be avoided by expressing it in terms of the experimentally known constant, $B_N$.

The full paper on chemical shifts [2] builds on an analogy of Van Vleck's theory of magnetism. The theory starts from the substitution

$$\mathbf{p} \rightarrow \pi = \mathbf{p} + e\mathbf{A}/c \ , \tag{5}$$

with $e < 0$ and the magnetic vector potential

$$\mathbf{A} = \frac{1}{2}\mathbf{B} \times (\mathbf{r}_{kn} - \mathbf{R}_{kl}) + \mu \times \mathbf{r}_{kN}/r_{kN}^3 \ . \tag{6}$$

Here $\mathbf{r}_{kN}$ is the distance to electron $k$ from the nucleus $N$ and $\mathbf{R}_{kl}$ is the distance from a gauge origin. The final expression is

$$\sigma_{zz} = (e^2/2mc^2)\langle 0 | (x^2 + y^2)/r^3 | 0\rangle$$
$$+ 2\sum_n{}'[\langle 0|m_z|n\rangle\langle n|m_z/r_k^3|0\rangle + \text{c.c.}]/(E_0 - E_n) \ , \tag{7}$$

with $m_z = \beta l_z$. This equation still stands as the correct one at the nonrelativistic limit. The connection with the spin-rotation constant is now discussed in more detail and the result for a linear molecule becomes

$$\sigma = (e^2/3mc^2)\langle 0|\frac{1}{r}|0\rangle - \frac{\alpha^2 a_0 I}{6\beta_n}\left(\sum_N \frac{2Z_N\beta_n}{Ia_N} - \frac{B_N}{KM}\right) \ . \tag{8}$$

Here $I$ is the moment of inertia and $M$ is the nuclear mass. The dependence of the shielding on the $M_K$ quantum number of the rotational state $|KM_K\rangle$ for a linear molecule is discussed in Ref. [3].

In Ref. [4] Ramsey discussed the chemical shifts of a general polyatomic molecule or a solid. It was further shown that the paramagnetic term arising from tightly bound electrons on a distant atom, $L$, can be made to vanish by setting the gauge origin $\mathbf{R}_{kL} = 0$. The possible contributions from thermally accessible excited states were also discussed.

## 2.3 Spin–spin coupling

The fifth paper [5] formulates the theory of spin–spin coupling in molecules starting from the hyperfine Hamiltonians

$$h_1 = \sum_k \frac{1}{2m}\left[\left(\mathbf{p}_k + \frac{e}{c}\sum_N \mathbf{A}_N\right)^2 - \mathbf{p}_k^2\right] \ , \tag{9}$$

$$h_2 = \sum_{kN} gg_N\beta\beta_n[3(\mathbf{S}_k \cdot \mathbf{r})(\mathbf{I}_N \cdot \mathbf{r})r^{-5} - \mathbf{S}_k \cdot \mathbf{I}_N r^{-3}] \ , \tag{10}$$

$$h_3 = \sum_{kN} -\frac{8\pi}{3}gg_N\beta\beta_n\mathbf{S}_k \cdot \mathbf{I}_N \delta(\mathbf{r}) \ . \tag{11}$$

Here $\mathbf{r}$ is the vector from the nucleus $N$ to the electron $k$. Ramsey derived the expressions for the full spin–spin

coupling tensor $\mathbf{J}$ and its scalar part $J$ between the nuclei $N$ and $N'$.

$$h\mathbf{J}^{(1a)} = \frac{e^2\beta_n^2 g_N g_{N'}}{mc^2}\left\langle 0\left|\frac{\mathbf{1}\mathbf{r}_{kN} \cdot \mathbf{r}_{kN'} - \mathbf{r}_{kN}\mathbf{r}_{kN'}}{r_{kN}^3 r_{kN'}^3}\right|0\right\rangle \ , \tag{12}$$

$$hJ^{(1a)} = \frac{2}{3}\frac{e^2\beta_n^2 g_N g_{N'}}{mc^2}\left\langle 0\left|\frac{\mathbf{r}_{kN} \cdot \mathbf{r}_{kN'}}{r_{kN}^3 r_{kN'}^3}\right|0\right\rangle \ , \tag{13}$$

$$hJ^{(1b)} = -\frac{8}{3}\beta^2\beta_n^2 g_N g_{N'}\sum_n{}'\left\langle 0\left|\frac{\mathbf{r}_{kN} \times \nabla}{r_{kN}^3}\right|n\right\rangle$$
$$\cdot \left\langle n\left|\frac{\mathbf{r}_{kN'} \times \nabla}{r_{kN'}^3}\right|0\right\rangle/(E_0 - E_n) \ , \tag{14}$$

$$h\mathbf{J}^{(2)} = 2g^2\beta^2\beta_n^2 g_N g_{N'}\sum_n{}'\left\langle 0\left|\frac{\mathbf{S}_k}{r_{kN}^3} - \frac{3(\mathbf{S}_k \cdot \mathbf{r}_{kN})\mathbf{r}_{kN}}{r_{kN}^5}\right|n\right\rangle$$
$$\cdot \left\langle n\left|\frac{\mathbf{S}_k}{r_{kN'}^3} - \frac{3(\mathbf{S}_k \cdot \mathbf{r}_{kN'})\mathbf{r}_{kN'}}{r_{kN'}^5}\right|0\right\rangle/(E_0 - E_n) \ , \tag{15}$$

$$h\mathbf{J}^{(3)} = 2\left(\frac{8\pi}{3}\right)^2 g^2\beta^2\beta_n^2 g_N g_{N'}\sum_n{}'\langle 0|\mathbf{S}\delta(\mathbf{r} - \mathbf{r}_N)|n\rangle$$
$$\times \langle n|\mathbf{S}\delta(\mathbf{r} - \mathbf{r}_{N'})|0\rangle/(E_0 - E_n) \ . \tag{16}$$

These four contributions 1a, 1b, 2, and 3 are called the diamagnetic, orbital, dipole, and Fermi contact terms, respectively. For the anisotropic part of the spin–spin coupling tensor, a dipolar-contact cross term $\mathbf{J}^{(2,3)}$ was obtained. The gauge origin for $\mathbf{A}_N$ was here placed on the nucleus $N$. Numerical values were estimated for $J$ in HD using the closure approximation and an effective energy denominator, $\Delta E$, for $J^{(2)}$ and $J^{(3)}$. Again, Eqs. (12)–(16) remain exact at the non-relativistic limit.

## 2.4 Later developments

The quoted papers still form the basis of our understanding of $\sigma$ and $\mathbf{J}$ in molecules studied by condensed-phase NMR. No new terms were found later. Ramsey [2] had cautiously thought that his terms "at least partially and perhaps completely" explained the chemical shifts. For freely rotating molecules in the gas phase, the small "Thomas precession" chemical-shift term of Ramsey [8] was later improved by Reid and Chu [10] and by Rebane and Volodicheva [11]. For the latest references to it, see Ref. [12].

An important aspect for practical numerical calculations by the practising quantum chemist turned out to be the effect of the chosen gauge origin, $\mathbf{R}_{kl}$, on the chemical shift $\sigma$. A common way to secure this is to use the so-called London orbitals [13] or "gauge-including atomic orbitals (GIAO)". Initially the acronym stood for "Gauge-invariant" ones. Other, potentially more economical alternatives exist, such as the "individual gauge origins for local orbitals" by Kutzelnigg [14] or the "localized orbital/local origin" one [15]. These methods are now a standard part of quantum chemical NMR chemical-shift packages. Without them correct results can be obtained but the basis-set convergence would be much slower.

## 3 Relativistic effects

Ramsey's theories, like their Van Vleck analogues for magnetism, were entirely nonrelativistic. They were originally only applied to compounds of light elements. In treatises on NMR, such as that of Abragam [16], only this nonrelativistic theory was discussed.

It had been shown early on by Breit [17] and Racah [18] that the relativistic corrections to magnetic dipole hyperfine interactions can be substantial. For later references, see, for instance, Refs. [19–21]. For the $6s$ valence orbital of an element such as mercury this correction is roughly a factor of 3. Hence the $J$(HgHg) coupling constant is increased by an order of magnitude due to relativistic effects.

More subtle are the spin–orbit-induced "heavy-atom chemical shifts" at the atom nearest to the heavy one, or at more remote nuclei. If one uses the same Hamiltonians as Ramsey, one must go to third-order perturbation theory, with one Zeeman, one hyperfine, and one spin–orbit matrix element [22]. For a recent discussion on the nature of this shift, see Ref. [23]. It was also noted that an analogous effect, a "heavy-atom shift on the heavy atom" can occur, for instance on the Pb(II) nucleus in $PbR_3^-$ compounds. The early semiempirical calculations suggested that the Zeeman–SO–Fermi contact cross term, zero in Ramsey's theory, could then become the dominant contribution to the Pb chemical shift [24].

If relativistic wave functions are used, second-order perturbation theory is enough. The ultimate goal, of course, remains a fully relativistic theory of both **J** and $\sigma$. The analogues to Ramsey's theories for them using the Dirac equation were formulated in Ref. [25] and in Refs. [26–28], respectively. The first numerical applications are now starting to appear, see, for example, the two conference proceedings [29, 30].

The quantum electrodynamical corrections to these Dirac-level results are small. The leading one is the correction factor of 1.001 159 652 193(4) to the free-electron $g$-factor of $-2$.

## 4 Conclusion

Ramsey's theories turned out to be virtually complete, when applied to compounds of light elements. A huge number of applications have been built on them. Among the excellent reviews, we mention the latest ones [31, 32]. Ramsey's theories undoubtedly belong to the classics of twentieth-century science. Corrections to them are required for compounds containing heavier elements.

*Acknowledgements.* The author apologizes for the rather personal tone in Sect. 3 but, having started as an experimental NMR spectroscopist, he was propelled to relativistic quantum chemistry by the lack of it in earlier discussions of NMR parameters, *par ricochet*, as it were. This work was supported by The Academy of Finland.

## Appendix

In order to maintain historical continuity we have used the Gauss-cgs systems of units, used in the original papers of Ramsey. In this system $\epsilon_0 = \mu_0 = 1$ with a Coulomb potential $V = Q/r$. The transition to atomic units is made with $e = m_e = \hbar = 1$. Apart from units of cm for $l$, g for $m$, and s for $t$, yielding the energy unit erg $= cm^2\, gs^{-2} = 10^{-7}$ J some electromagnetic units are as follows:

charge Q: $cm^{3/2}\, g^{1/2}\, s^{-1} = 1$ esu; magnetic moment $\mu$: $cm^{5/2}\, g^{1/2}\, s^{-1}$; magnetic flux density **B**: $g^{1/2}\, cm^{-1/2}\, s^{-1} = 1$ gauss; magnetic vector potential **A**: $g^{1/2}\, cm^{1/2}\, s^{-1}$.

## References

1. Ramsey NF (1950) Phys Rev 77: 567
2. Ramsey NF (1950) Phys Rev 78: 699–703
3. Ramsey NF (1951) Phys Rev 83: 540–541
4. Ramsey NF (1952) Phys Rev 86: 243–246
5. Ramsey NF (1953) Phys Rev 91: 303–307
6. Ramsey NF (1953) Phys Rev 89: 527
7. Ramsey NF, Purcell EM (1952) Phys Rev 85: 143–144
8. Ramsey NF (1953) Phys Rev 90: 232
9. Lamb W (1941) Phys Rev 60: 817
10. Reid RV Jr, Chu AH-M (1974) Phys Rev A 9: 609
11. Rebane TK, Volodicheva MI (1974) Vestn Leningr Univ 22: 55
12. Sundholm D, Gauss J, Schäfer A (1996) J Chem Phys 105: 11051
13. London F (1937) J Phys Radium 8: 397
14. Kutzelnigg W (1980) Isr J Chem 19: 193
15. Hansen AaE, Bouman TD (1985) J Chem Phys 82: 5035
16. Abragam A (1961) Principles of nuclear magnetism, Clarendon, Oxford
17. Breit G (1930) Phys Rev 35: 1447
18. Racah G (1931) Z Phys 71: 431
19. Pyykkö P, Pajanne E, Inokuti M (1973) Int J Quantum Chem 7: 785
20. (a) Lindgren I, Rosén A (1974) Case Stud At Phys 4: 93; (b) Lindgren I, Rosén (1974) Case Stud At Phys 4: 197
21. Pyykkö P, Wiesenfeld L (1981) Mol Phys 43: 557
22. Nomura Y, Takeuchi Y, Nakagawa N (1969) Tetrahedron Lett 8: 639
23. Kaupp M, Malkina OL, Malkin VG, Pyykkö P (1998) Chem Eur J 4: 118
24. Edlund U, Lejon T, Pyykkö P, Venkatachalam TK, Buncel E (1987) J Am Chem Soc 109: 5982
25. Pyykkö P (1977) Chem Phys 22: 289
26. Pyykkö P (1983) Chem Phys 74: 1
27. (a) Pyper NC (1983) Chem Phys Lett 96: 204 (b) Pyper NC (1983) Chem Phys Lett 96: 211
28. Zhang ZC, Webb GA (1983) J Mol Struct 104: 439
29. ACS Conf Proc (1999)
30. J Comput Chem (1999)
31. Jameson CJ, de Dios AC (1999) In: Webb GA (ed) Specialist periodical reports. Nuclear magnetic resonance, vol 28. The Royal Society of Chemistry, Cambridge (in press)
32. Helgaker T, Jaszuński M, Ruud K (1999) Chem Rev 99: 293

Theor Chem Acc (2000) 103:217–218
DOI 10.1007/s002149900010

Theoretical
Chemistry Accounts
© Springer-Verlag 2000

*Perspective*

# Perspective on "New developments in molecular orbital theory"

## Roothaan CCJ (1951) Rev Mod Phys 23: 69–89

**Michael C. Zerner**

Quantum Theory Project, Department of Chemistry, University of Florida, Gainesville, FL 32611, USA

Received: 12 February 1999 / Accepted: 26 February 1999 / Published online: 7 June 1999

**Abstract.** This paper reviews the title article by Clemens Roothaan and the huge impact that his paper has had in modern chemistry. In his paper Roothaan converts the molecular Schödinger equation into a matrix equation by systematically introducing the linear combination of atomic orbitals–molecular orbital approximation and by invoking the variational principle.

**Key words:** Roothaan equation – Roothaan–Hall – Hartree–Fock – Roothaan self-consistent field

## 1 Introduction

Quantum chemistry has certainly come of age, and quantum chemical concepts appear in nearly all papers published in the chemical literature today. Actual quantum chemical calculations are now reported in many experimental papers, and computer codes that perform these calculations are now often considered as another piece of chemical apparatus. Various experimental groups now train experts in computational chemistry, along with experts in NMR spectroscopy, mass spectroscopy, and so on. Nearly all molecular electronic structure calculations today start with molecular orbital (MO) calculations, but the history of the development of this methodology is often forgotten. Today's heroes have become the writers of useful computer code, but the basic underpinnings of these codes, the ideas that let these codes develop and become useful and those who developed these ideas, are often forgotten. Who is Roothaan? What did he do that so influenced MO theory? I can make my distinction of theoretical chemist versus computational chemist, should such a distinction be appropriate, on the basis of this answer. This short manuscript reviews the 1951 paper by C.C.J. Roothaan entitled "New developments in molecular orbital theory" [1], hopefully putting this

paper in some perspective. When I was a graduate student interested in theoretical chemistry this paper was required reading, and it is interesting to note that this paper is the first reprint in the book by Parr [2] "Quantum theory of molecular electronic structure" published in 1963, which itself had a large impact on the development of quantum chemistry as a useful tool. Slater in his 1963 book "Quantum theory of molecules and solids" [3], which had a similarly strong influence in the physics community, also assigns much of the credit of the "modern" Hartree–Fock MO method to Roothaan's paper. I had a good deal of trouble with this paper, as did my contemporaries, for it is full of the statement infamous to all students "The proof of Eq. X is elementary, and will be omitted here." (Looking back at this paper, the proofs are obvious, but I did not think so at the time!) Regardless, understanding this paper is fundamental to really understanding MO theory.

## 2 New developments in MO theory, 1951

To put the reader in the proper perspective of the time, I quote below part of the first paragraph of Roothaan's paper:

"For dealing with the problems of molecular quantum mechanics, two methods of approximation have been developed which are capable of handling many-electron systems. The Heitler–London–Pauling–Slater or valence bond (VB) method [1–3] originated from a chemical point of view. The atoms are considered as the material from which the molecule is built; accordingly, the molecular wave function is constructed from the wave functions of individual atoms. The Hund–Mulliken or molecular orbital (MO), method [4] is an extension of the Bohr theory of electronic configurations from atoms to molecules. Each electron is assigned to a one-electron wave function or molecular orbital, which is the quantum mechanical analog of an electron orbit. Each of the two fundamentally so different approaches has its merits: ..." [4].

This last statement might have been so prior to Roothaan's 1951 paper, but this paper clearly gave the advantage to MO theory. Prior to this work, the extension of VB and MO theory to molecules seemed

218

rather ad hoc. This paper systematized MO theory by converting the many-dimensional molecular Schödinger equation (non-relativistic, time-independent and fixed nuclei)

$$\mathbf{H}\Psi_I = E_I\Psi_I \qquad (1)$$

$$\mathbf{H} = -\hbar^2/2m \sum_i \nabla_i^2 - \sum_i \sum_A Z_A e^2/R_{A,i} + \sum_i \sum_{j<i} e^2/r_{i,j} , \qquad (2)$$

in which the symbols take on their usual meanings, a seemingly impossible differential equation, into the now familiar matrix equation, often called the Roothaan equation (or Roothaan–Hall [5] equation)

$$\mathbf{FC} = \mathbf{SC}\varepsilon \qquad (3)$$

first by assuming a determinental antisymmetrized product (AP) wave function for the closed shell system [6]

$$\Psi_I = |\Phi_1(1)\bar{\Phi}_1(2)\Phi_2(3)\bar{\Phi}_2(4)\cdots\Phi_n(N-1)\bar{\Phi}_n(N)| \qquad (4)$$

and then, the crucial step, by systematically introducing the linear combination of atomic orbitals $\{\chi_\mu\}$ (LCAO-MO) approximation

$$\Phi_i = \Sigma_\mu \chi_\mu C_{\mu i} \qquad (5)$$

The bars over orbitals in equation (4) designate $\beta$-spin. By carefully invoking the variational principle for the MO coefficients $C_{\mu i}$ an equation is obtained for the molecular orbitals in terms of the "Fock" operator $\mathbf{f}$

$$\mathbf{f}\Phi_i = \varepsilon_i\Phi_i , \qquad (6)$$

with $\mathbf{f}$ given in terms of a Coulomb operator $\mathbf{J}_i$ and exchange operator $\mathbf{K}_i$

$$\mathbf{f} = \mathbf{h} + \Sigma_i(2\mathbf{J}_i - \mathbf{K}_i) \qquad (7)$$

and few of us have not gone through this same derivation, first as students and then, perhaps, as teachers. In addition, few of us have modified this eloquent derivation. The matrix $\mathbf{F}$ of Eq. (3) was developed in terms of integrals over the one-electron operators and the two-electron Coulomb and exchange operators. The molecular differential equation was successfully transformed into a matrix equation involving molecular integrals. Although solving matrix equations was, and still is, certainly much more familiar than solving the original differential equations by numerical techniques or other various somewhat awkward procedures, routine use of MO theory by the community of chemists and physicists at large awaited the development of fast methods for diagonalization, the evaluation of integrals from those over exponential atomic functions to those over Gaussian atomic functions, and faster computers; however, the basic framework used today in all MO calculations is that described by Roothaan. Open-shell methods of MO theory were also systematized first by Roothaan in 1960 with his introduction of the vector coupling coefficient [7].

The Roothaan paper contains more than the development of the Roothaan equations. There is the description of the use of molecular symmetry. Three points are developed for the closed-shell ground state:

"(1) The LCAO AP which minimizes the energy is necessarily a singlet and is totally symmetric with respect to the symmetry point group of the molecule.
(2) The best LCAO MO's can be chosen so that they belong in sets to irreducible representations of the symmetry group of the molecule.
(3) The best LCAO MO's can all be chosen real."

The assumption here is that if the total many-electron Hamiltonian commutes with various symmetry operations of the group $\mathscr{C}$ then the Fock operator that leads to $\mathbf{F}$ can also be chosen in such a fashion. In such a case each MO will transform as one of the irreducible representations of the group. The ideas presented in the Roothaan paper are very important, and allow factorization of the Roothaan equations into simpler blocks, one for each irreducible representation, but experience has indicated that on occasion there are "Hartree–Fock instabilities" that destroy the symmetry of the resultant MOs. However, the Fock operator can always be constrained to have the symmetry of the nuclei.

It is difficult to imagine the world of chemistry without MO theory, and the possibilities that the Roothaan paper opened up for its future. Of course, it is possible that if Roothaan had not developed this as he did, the time was ripe, and others would have. In fact, as suggested earlier in this perspective, in the paper by Hall [5] similar ideas were developed, but without the same impact.[1] Nevertheless, the Roothaan paper "New developments in molecular orbital theory" is probably the keystone paper that led to the dominance of MO theory over VB theory, and its widespread and easy usage today.

## References

1. Roothaan CCJ (1951) Rev Mod Phys 23: 69
2. Parr RG (1963) Quantum theory of molecular electronic structure. Benjamin, New York
3. Slater JC (1963) Quantum theory of molecules and solids, vol 1. McGraw-Hill, New York, see, in particular, Appendix 7
4. These references are from the original Roothaan paper, and they are interesting reading. (1) Heitler W, London F (1927) Z Phys 44: 455; (2) Pauling L, Wilson EB (1935) Introduction to quantum mechanics. McGraw-Hill, New York, pp 340–380; (3) Eyring H, Walter J, Kimball E (1944) Quantum chemistry. Wiley, New York (4) Hund F (1931) Z Phys 51: 759; Hund F (1931) 73: 1; etc.; Mulliken RS (1928) Phys Rev 32: 186; Mulliken RS (1928) 32: 761; Mulliken RS (1932) 41: 49; etc
5. Hall GG (1951) Proc Roy Soc Lond Scr A 205: 541
6. Slater JC (1929) Phys Rev 34: 1293
7. Roothaan CCJ (1960) Rev Mod Phys 32: 179

[1]It is difficult to say why this is so. Perhaps the immediate interest of the Chicago group in detailed computation is part of the reason, a development in which the two 1960 papers by Ransil that reported self-consistent-field calculations on diatomic molecules became quite important in pointing toward the future high reliance on computers and their role in quantum chemistry [Ransil B (1960) Rev Mod Phys 32: 239; Ransil B (1960) Rev Mod Phys 32: 245]

Theor Chem Acc (2000) 103:219–220
DOI 10.1007/s002149900031

Theoretical
Chemistry Accounts
© Springer-Verlag 2000

*Perspective*

# Perspective on "A molecular orbital theory of reactivity in aromatic hydrocarbons"

## Fukui K, Yonezawa T, Shingu H (1952) J Chem Phys 20: 722

**Shigeki Kato**

Department of Chemistry, Graduate School of Science, Kyoto University, Kitashirakawa, Sakyo-ku, Kyoto 606-85024, Japan

Received: 4 March 1999 / Accepted: 30 March 1999 / Published online: 28 June 1999

**Abstract.** The development of theories for interpreting the course of chemical reactions is one of the most important achievements of theoretical chemistry in the twentieth century. I selected the paper by Fukui et al. from 1952, proposing the frontier electron density as the reactivity index for the orientation of electrophilic substitution reactions. This paper may be regarded as a bridge between an older reactivity theory, the electronic theory of organic chemistry, and new ones predicting the stereochemical courses of reactions such as frontier orbital theory and the Woodward–Hoffmann rule.

Rationalization of the chemical reactivity for organic compounds based on quantum mechanics is one of the goals of theoretical chemistry. In the early stages of the twentieth century, the electronic theory of organic chemistry had been established by Lapworth, Robinson and Ingold as a unified guiding principle for understanding complex organic reactions. [1] This assumed that the reactivity is determined by the electron density at a particular position in an organic molecule. Several concepts such as inductive, mesomeric and inductomeric effects were advanced for interpreting the electron distributions in organic reagents. Employing the Hückel molecular orbital (HMO) method, Coulson and Longuet-Higgins [2] developed the general theory for the electron distributions in conjugated molecules and proposed the $\pi$ electron density $q_r$ and self-polarizability $\pi_{rr}$ as the reactivity indices correlating the activation energies at different reactive positions in a conjugated molecule. Considering the energy changes which occur when a charged reagent approaches a particular position in a molecule, they deduced that electrophilic reagents react at the positions with high values of $q_r$, while nucleophilic reactions are favored by low values of $q_r$. They also related $\pi_{rr}$ to the reactivity for nonsubstituted aromatic hydrocarbons such as naphthalene in which $q_r$ are the same at all positions. Thus, the theory of Coulson and Longuet-Higgins is regared as the quantum mechanical generalization of the organic electronic theory which was derived empirically based on a large amount of experimental evidence in organic chemistry.

In 1952, Fukui, Yonezawa and Shingu [3] proposed an alternative reactivity index for aromatic hydrocarbons, referred to as the frontier electron density, $f_r$, which was different from the previous ones coming from the static [2] and localization [4] approaches, focusing on the electrons in the highest occupied (HO) $\pi$ MO. They carried out HMO calculations for the $\pi$ electronic structure of 15 aromatic hydrocarbons whose $\pi$ electron densities $q_r$ are the same at all sites. They assumed that the two electrons occupying the HOMO, distinct from the other $\pi$ electrons, are essential to the reactivity of the system and referred to them as the frontier electrons. It was found that the positions of the highest density of the frontier electrons, $f_r$, completely agreed with the experimental results of the structures of the reaction products of oxidation, halogenation, nitration and sulfonation for nine aromatic hydrocarbons including naphthalene, anthracene, phenanthrene, chrysene, naphthacene, perylene, pentacene and picene. The only discrepancy with experiment was for 3,4-benzophenanthrene, where the position of the largest frontier electron density is sterically hindered and the second largest position is correlated to the experimental results for the oxidation reactions.

Although the title paper from 1952 was only for electrophilic substitution reactions. Fukui et al. [5] further developed the frontier electron theory to be applicable to nucleophilic and radical reagents. In the case of reactions with a nucleophilic reagent, they postulated that the more susceptible position to attack has the higher density of the two electrons assumed to occupy the lowest unoccupied (LU) MO, while the position with the higher density of the two electrons, one occupying the HOMO and the other the LUMO, is more reactive with a radical reagent. With this extension of the theory, they reached the concept of frontier orbitals, i.e.

HOMO and LUMO, as particularly important MOs to predict the reactivities in many types of reaction. They also postulated that the electron delocalization or transfer between the reagent and reactant in the vicinity of the transition state is essential in determining the reactivity of organic compounds and the frontier orbitals play a most important role in such an electron delocalization.

Fukui [6] wrote a review paper in 1964 where the Diels–Alder reactivity was discussed by examining the stabilization energy due to the interconjugation between the diene and dienophile near the transition state using perturbation theory. He pointed out that the symmetry relation between the HOMO of diene and the LUMO of dienophile is crucial in determining the magnitude of the stabilization energy and showed that the symmetries of the HOMO and the LUMO are the same for all the diene–dienophile pairs (six dienes and nine dienophiles). This discovery of the importance of the symmetry relation between the HOMO and the LUMO was essentially the same as the symmetry rule proposed by Woodward and Hoffmann [7] for concerted cycloaddition reactions.

Although the frontier orbital theory had not received much attention from experimental organic chemists when proposed, the HOMO–LUMO interaction became widely recognized to be an important clue in interpreting the course of chemical reactions after the establishment of the Woodward–Hoffmann symmetry rule. In the 1960s and 1970s, theories of chemical reactivity were advanced remarkably by many theoretical chemists and the symmetry relation of the frontier orbitals, which is the essence of frontier orbital theory and the Woodward–Hoffmann rule, played the central role in these developments. Note that the MO theories of chemical reactivity are now standard methods for organic chemists and are included in many textbooks of organic chemistry. Fukui won the Nobel Prize with Hoffmann in 1981 for their theories, developed independently, concerning the course of chemical reactions.

As already described the theory of chemical reactivity has been developed in two stages in the twentieth century. The first is the electronic theory of organic chemistry, which was generalized by Coulson and Longuet-Higgins based on quantum mechnics. The second stage is the establishment of the symmetry rule for the MOs in predicting the course of a reaction, i.e. frontier orbital theory and the Woodward–Hoffmann rule. The title paper from 1952 may be regarded as a bridge connecting these two stages because it proposed the reactivity index, $f_r$, for interpreting the orientation effect in chemical reactions, the main subject of the electronic theory of organic chemistry, and was the starting point of the second stage after the concept of frontier orbitals was first introduced and it became the key ingredient in the further development of the theory.

At the present time, ab initio theoretical methods for calculating potential-energy surfaces have been greatly developed, providing quantitative information on the energies and structures of reactive species. Accurate calculations of the energy surfaces have become available for complex chemical reactions such as catalytic reactions including transition-metal compounds and reactions in solution. Considering that frontier orbital theory and the Woodward–Hoffmann rule were derived based on the HMO method, the third stage of chemical reaction theory will be developed based on highly accurate electronic structure theories, which can quantitatively predict reaction rates even for complex chemical reactions in solution.

# References

1. Ingold CK (1953) Structure and mechanism in organic chemistry. Cornell University Press, Ithaca
2. (a) Coulson CA, Longuet-Higgins HC (1947) Proc R Soc Lond Ser A 191: 39; (b) Coulson CA, Longuet-Higgins (1947) Proc R Soc Lond Ser A 192: 16
3. Fukui K, Yonezawa T, Shingu H (1952) J Chem Phys 20: 722
4. Wheland GW (1942) J Am Chem Soc 64: 900
5. Fukui K, Yonezawa T, Nagata C, Shingu H (1954) J Chem Phys 22: 1433
6. Fukui K (1964) In: Löwdin P-O, Pullman B (eds) Molecular orbitals in chemistry, physics, and biology. Academic Press, New York, p 513
7. (a) Hoffmann R, Woodward RB (1965) J Am Chem Soc 87: 2046; (b) Hoffmann R, Woodward RB (1965) J Am Chem Soc 87: 4388

Theor Chem Acc (2000) 103:221–224
DOI 10.1007/s002149900058

Theoretical
Chemistry Accounts
© Springer-Verlag 2000

*Perspective*

# Perspective on "The spectra and electronic structure of the tetrahedral ions $MnO_4^-$, $CrO_4^-$, and $ClO_4^-$"

## Wolfsberg M, Helmholz L (1952) J Chem Phys 20:837–843

**Michael B. Hall**

Department of Chemistry, Texas A&M University, College Station, TX 77843, USA

Received: 4 March 1999 / Accepted: 18 March 1999 / Published online: 9 September 1999

**Abstract.** The paper by Wolfsberg and Helmholz represents the first molecular orbital calculation on a transition-metal complex. Published in the heyday of ligand-field theory, the paper was 10 years ahead of its time. Here, the present author provides an overview of the title paper: a brief description of the results, his perspective on the historical context of the work and its influence on subsequent developments, personal reminiscences about the paper by Max Wolfsberg, and biographical sketches of the authors.

**Key words:** Transition metal – Molecular orbital – Extended Hückel – Metal oxides – Tetrahedral ions

## 1 Introduction

The title paper by Wolfsberg and Helmholz represents the beginning of a paradigm shift in the way chemists approached the electronic structure of transition-metal complexes. Prior to this work the main emphasis by chemists had been on the development of valence bond methods to explain the bonding and magnetic behavior, while physicists, less concerned about the bonding, preferred crystal-field theory to explain the magnetic and spectroscopic properties of the complexes. The paper by Wolfsberg and Helmholz represents the first molecular orbital calculation on a transition-metal complex. Published in the heyday of ligand-field theory, the paper was 10 years ahead of its time.

## 2 Summary of the work

After mentioning previous successes of both valence bond and molecular orbital methods in organic chemistry [1], the authors set two principal goals. First, the nature of the X=O bonds needed clarification; their relatively, short length had been attributed to resonance of double-bonded Lewis structures [2] and, alternatively, to a change in the hybridization on oxygen [3]. Second, while ions such as $ClO_4^-$ show no absorption in the visible or near ultraviolet (only a single band in the ultraviolet), the valence isoelectronic transition-metal species such as $MnO_4^-$ and $CrO_4^{2-}$ show two strong characteristic bands in this region. Permanganate, in particular, had been thoroughly studied in dilute single crystals [4].

The authors built a linear combination of atomic orbitals model of the ions $MnO_4^-$, $CrO_4^{2-}$, and $ClO_4^-$ by using only the valence orbitals of the central atom ($3d$, $4s$, $4p$ for Mn and Cr; $3s$ and $3p$ for Cl) and the $2p$ orbitals of O; thus, all complexes had 24 valence electrons. The inner-shell electrons were treated as part of the core. Using group theory [5], the authors determined the irreducible representations of the valence orbitals, the corresponding normalized linear combination of the oxygen orbitals (oxygen–oxygen overlap was neglected), and the group overlaps, $G$. The authors then set-up and solved the secular equation

$$|H_{ij} - G_{ij}\varepsilon| = 0$$

to obtain the eigenvalues. The $H_{ii}$ terms were taken as parameters whose range was restricted by known valence-shell ionization energies [6]. For O the authors gave $H_{\sigma\sigma}$ a more stable value than $H_{\pi\pi}$ in accord with earlier calculations [7]. The authors found that calculating the overlap integrals [8] with the usual Slater functions [9] gave values too small to yield reasonable interactions. To remedy the problem the authors created their own set of analytic atomic functions that more closely approximated known atomic self-consistent field functions [10].

The off-diagonal terms in the secular equation were approximated by the now famous "Wolfsberg–Helmholz formula".

$$H_{ij} = F_x G_{ij}(H_{ii} + H_{jj})/2 \ .$$

The adjustable parameter $F_x$ was given different values for the $\sigma$ and $\pi$ interactions.

Solution of the secular equation yielded a set of occupied orbitals $(1e)^4 (1t_2)^6 (1a_1)^2 (2t_2)^6 (t_1)^6$ for the 24 valence electrons. The authors used the bonding nature of the occupied orbitals to provide a qualitative description of the multiple-bond character of the X=O bonds. The calculation also gave the $3t_2$ orbital as the lowest-lying virtual orbital. Using the orbital energy differences, the authors assigned the lowest energy band in all the spectra to the allowed $^1A_1 \rightarrow {}^1T_2$ transition, arising from exciting an electron from the nonbonding oxygen $t_1$ molecular orbital to this $3t_2$ orbital. The second band, which was observed only for Cr and Mn, was assigned to the allowed $^1A_1 \rightarrow {}^1T_2$ transition arising from the $2t_2$ to $3t_2$ excitation. The calculations provided qualitative agreement with the experimentally observed band positions, given that more than qualitative comparison would require knowledge of several two-electron integrals, as described in the paper's appendix.

The authors also obtained qualitative agreement with the observed intensities and with the spectra of $MnO_4^-$ diluted into $KClO_4$ and $NaClO_4$ single crystals. The authors were able to account for many of the weaker transitions which were now allowed by the lower site symmetry.

## 3 Perspectives on subsequent developments

The importance of this work was not the accuracy of the assignments, but the fact that it was the first molecular orbital treatment of a transition-metal complex. Until then bonding had usually been described with Pauling's valence bond approach [2]. The explanation of magnetic properties had proponents in the valence bond school [2] and the crystal-field school [11], while spectral analysis tended to be the province of physics and was treated by the crystal-field approach. Earlier, Van Vleck [12] had argued that both valence bond and crystal-field theory were subsets of molecular orbital theory [12]. The paper by Wolfsberg and Helmholz was the first to apply quantitative molecular orbital theory to inorganic complexes. An early review states that "Modern quantum chemistry of coordination compounds started actually in 1952 when Wolfsberg and Helmholz carried out calculations for $MnO_4^-$ and $CrO_4^{2-}$ ..." [13].

The timing of this paper was also important. Inorganic chemists were just beginning to take a serious interest in spectroscopy, which had been the realm of physicists and chemical physicists, and they needed a method that would give a unified explanation of the bonding, magnetic properties, and spectra. Inorganic chemistry was also about to undergo a renaissance with the discovery of ferrocene in 1951 [14]. The subsequent development of the entire field of organometallic chemistry needed a robust molecular orbital approach; the Wolfsberg–Helmholz paper presaged this need.

Some early applications of the method were made by Japanese workers [15], but the 1950s were the "heyday" of crystal- and ligand-field theory [16]. It was not until the early 1960s that the molecular orbital approach for transition-metal complexes gained momentum. This interest was driven by a combination of spectroscopists'

interest in charge-transfer (non-ligand-field) transitions, as found in $MnO_4^-$ and $CrO_4^{2-}$, and chemists' desire to understand the bonding in organometallic complexes. The early 1960s found Harry Gray on a postdoctoral assignment with Carl Ballhausen. Together they developed the self-consistent charge and configuration method, an interative Wolfsberg–Helmholz approach [17]. The method was exploited by Gray and his coworkers to study the spectral properties of a wide variety of complexes [18].

Richard Fenske arrived as Assistant Professor at the University of Wisconsin in 1961 after having completed his PhD, with Donald Martin at Iowa State University on applying crystal-field theory to square-planar platinum complexes [19]. Fenske was interested in developing a method more closely tied to the ab initio molecular orbital method described so beautifully by Roothaan [20]. Building on some previous suggestions [21], he and his first students, especially Ken Caulton and Doug Radtke, developed an approximate self-consistent field method that had no empirical or adjustable parameters. With some later refinements by this author, the method became widely known as the Fenske–Hall method [22], and in this form it is still being used today [23].

Approximate molecular orbital theory continues to play an important role, especially as an adjunct to experimental work. However, by the late 1960s, as computers became faster and computer codes improved, theoreticians began applying ab initio methods to metal complexes [24].

## 4 Reminiscences by Max Wolfsberg

As a freshman at Washington University, I asked my instructor why the color of permanganate ion is purple. He answered "because God made it so". I remember saying to one of my friends that I would still like to find out "why". I started graduate study at Washington University in 1948 although the department did not think that this was very wise because (1) one should change universities between undergraduate school and graduate school and (2) I had declared an interest in theoretical chemistry and there was no one in the department who was considered to be a theoretical chemist. During my first year, Lindsay Helmholz consented to be my PhD mentor. I had already told him, when I was doing undergraduate study with him, of my interest in permanganate ion. He shared this interest and indeed he showed me notes on permanganate ion that Linus Pauling had written when he visited St. Louis. Lindsay's interest in permanganate had been fired by two beautiful papers by the German physicist J. Teltow on low temperature absorption spectroscopy of single crystals of dilute solutions of permanganate ion imbedded in perchlorate ion lattices with different site symmetries.

Since I had difficulties writing down mathematical expressions for the various Lewis multi-bonded resonance structures in Pauling's notes, I decided to try the molecular orbital approach, which in 1949 was not popular among chemists. I consequently spent the summer of 1949 studying papers on molecular orbital theory, especially those of R.S. Mulliken in *The Journal of Chemical Physics*. I then proceeded to set up the symmetry molecular orbital combinations for $RX_n$ compounds of various symmetries (tetrahedral, square-planar, etc.). Lindsay and I then started discussing permanganate ion spectroscopy in terms of molecular orbital theory. S.I. Weissman participated in many of these discussions. Central to these discussions was our "understanding" that the interactions between the orbitals on the manganese atom and the orbitals on the oxygen atoms are proportional to the overlap integral between these orbitals. While I deified Mulliken, who also emphasized the importance of overlap considerations, I must state that my own feeling about the importance of overlap came from Pauling's classic "The nature of the chemical bond". In the fall of 1949, I opened a new issue of *The Journal of Chemical Physics* and found the paper on the calculation of overlap integrals by Mulliken, Rieke, Orloff, and Orloff (*J Chem Phys* (1949) 17: 1248). We then realized that we could make quantitative statements about spectroscopy (rather than waving our arms) if we combined calculated overlap integrals and observed ionization potential data in a Hückel type of molecular orbital theory. I proceeded to calculate overlap integrals for Slater type *d* orbitals with Slater type *p* and *s* orbitals using the methodology of Mulliken et al. (see above). After comparing the single exponential Slater orbitals (using the Slater recipes for the screening constants) with Hartree Fock orbitals in the literature, I concluded that single exponential atomic orbitals were inadequate and fitted all (most) known theoretical data on Hartree Fock atomic orbitals in terms of linear combinations of single Slater functions with exponents which included the nuclear charges and appropriate new screening terms. This material is in my thesis, but we never wrote it up for publication. Other parts of the calculations are also given in more detail in my thesis than they are in the 1952 paper by Wolfsberg and Helmholz.

In the Spring of 1951, Lindsay wrote his PhD mentor Joe Mayer about our calculations, hoping he would show this communication to his wife Maria Goeppert Mayer. If I remember correctly, Maria wrote back a very encouraging letter and somehow she communicated with Mulliken. Mulliken invited Lindsay to give a seminar to his group, and Lindsay took me along to Chicago. It is typical of Lindsay that he paid for this trip for me out of his own pocket, as well as the cost of a trip to the Ohio State Symposium a couple of months later, where I presented our calculations at a session chaired by G. Herzberg. At the time when we visited Chicago, C. Roothaan was a postdoc of Mulliken; before the seminar, Mulliken warned that Roothaan would be unhappy about what we had done but not to be discouraged by that.

## 5 Biographical sketch of Max Wolfsberg

Max Wolfsberg was born on May 28, 1928 in Hamburg, Germany, and came to the US with his parents and brother, arriving via Sweden on September 22, 1939. He attended Washington University on an Honor Scholarship and graduated with an A.B. degree in 1948. He remained at Washington University for graduate work and obtained his Ph.D. degree under Prof. Lindsay Helmholz in 1951. He was an AEC predoctoral fellow during his final year at Washington.

Following graduation, he joined the staff of Brookhaven National Laboratory where he rose through the ranks to Senior Chemist. While at Brookhaven, he was awarded a National Science Foundation Senior Postdoctoral Fellowship (with R.P. Bell at Oxford, with C. Ballhausen at Copenhagen, and with P.O. Loewdin at Uppsala), and was visiting Professor of Chemistry at Cornell University in the spring of 1963 and at Indiana University in spring 1965. He was also Professor of Chemistry at the State University of New York at Stony Brook from 1966 to 1969.

Following a winter semester at the University of California, Irvine, as Regents' Lecturer, he joined the faculty as Professor of Chemistry in 1969. While at Irvine, Wolfsberg was awarded Alexander von Humbolt Awards in 1977, 1984, and 1993. He was Lecturer, Troisieme Cycle, at EPF Lausanne and the University of Berne in 1978, Deutsche Forschungs Gemeinschaft Guest Professor at the University of Ulm in 1986, National Academy Exchange Scholar at Leipzig in 1989, and Forchheimer Visiting Professor at Hebrew University in 1993. He also served as Chair of the Department of Chemistry from 1974 to 1980.

Wolfsberg's theoretical researches have focused on a variety of topics including quantum chemistry, isotope effects on thermodynamic properties and on chemical reaction rates, mass-spectrometric fragmentation patterns, translational–vibrational energy transfer, molecular dynamics calculations on condensed phases, and rotational–vibrational spectroscopy.

## 6 Biographical sketch of Lindsay Helmholz

Lindsay Helmholz was born in Chicago on November 11, 1909. From 1926 until 1928 he attended Cornell University and in 1933 received his Ph.D. degree from John Hopkins University, where he was an early student of Joseph E. Mayer. He was one of the few Mayer students to do experimental work and he worked on the Born–(Mayer)–Haber cycle to determine the electron affinities of F. Helmholz came to the California Institute of Technology first as a National Research Fellow (1934–1936) then as an Instructor in Chemistry (1936–1941). At the Institute Helmholz worked with Linus Pauling and became a crystallographer. In 1941 he moved to Dartmouth College as an Assistant Professor and set up his own X-ray diffraction apparatus.

During the Second World War, Helmholz was a chemist on the Manhattan Project in Los Alamos. In 1946 he received a Guggenheim Fellowship and later that year joined the Department of Chemistry at Washington University in St. Louis as an Assistant Professor. Helmholz and several others from the Manhattan Project joined the faculty of Washington University because Joseph W. Kennedy, previously Head of Chemistry at Los Alamos, moved there after the war and became Chairman of the Department in 1945. At Washington University Helmholz embarked on research programs in X-ray diffraction and spectroscopy; the diffractometer that he had set up at Dartmouth was moved to Washington University. He was promoted to Associate Professor in 1948 and to Professor in 1957. While at Washington University, he served two terms as Acting Chair (1963–1964 and 1976–1978). He retired in 1978 and died on March 17, 1993.

## References

1. Mulliken RS (1949) J Chim Phys 46: 497
2. Pauling L (1940) The nature of the chemical bond. Cornell University Press, Ithaca
3. Pitzer KS (1948) J Am Chem Soc 90: 2140
4. (a) Teltow J (1939) Z Phys Chem B 43: 198; (b) Teltow J (1938) Z Phys Chem B 40: 397
5. Wigner E (1931) Gruppen Theorie und ihre Anwendung auf die Quantenmechanik der Atomspektren. Vieweg, Brunswick, Germany
6. Mulliken RS (1934) J Chem Phys 2: 782
7. Mulligan JF (1951) J Chem Phys 19: 347

224

8. Mulliken RS, Rieke CA, Orloff D, Orloff H (1949) J Chem Phys 17: 1248
9. Slater JC (1930) Phys Rev 36: 57
10. Wolfsberg M (1951) PhD thesis. Washington University, St. Louis
11. Van Vleck JH (1932) Theory of magnetic and electric susceptibilities. Oxford University Press, Oxford
12. Van Vleck JH (1935) J Chem Phys 3: 803
13. Dyatkina ME, Klimenko NM, Rosenberg EL (1974) Pure Appl Chem 38: 391
14. (a) Kealy TJ, Pauson PL (1951) Nature 168: 1039; (b) Miller SA, Tebboth JA, Tremainer JF (1952) J Chem Soc 632
15. (a) Kuroda Y, Ito K (1955) J Chem Soc Jpn 76: 766; (b) Yamatera H (1956) J Inst Polytech Osaka City Univ 5: 163
16. (a) Bleaney B, Stevens KWH (1953) Rep Prog Phys 16: 108; (b) Bowers KD, Owen J (1955) Rep Prog Phys 18: 304; (c) Moffitt W, Ballhausen CJ (1956) Ann Rev Phys Chem 7: 107
17. (a) Ballhausen CJ, Gray HB (1962) Inorg Chem 1: 111; (b) Gray HB, Ballhausen CJ (1963) J Am Chem Soc 85: 260; (c) Ballhausen CJ, Gray HB (1964) Molecular orbital theory. Benjamin, New York. A typographical error in Table II in the title paper is corrected on page 269 of this book
18. Gray HB, Beach NA (1963) J Am Chem Soc 85: 2922; (b) Basch H, Viste A, Gray HB (1966) J Chem Phys 44: 9; (c) Gray HB (1968) In: Rich A and Davidson N (ed) Structural chemistry and molecular biology. Freeman, (San Francisco pp 783–805; Ballhausen CJ, Gray HB (1971) In: Martell AE (ed) Coordination chemistry, vol I. Van Nostrand Reinhold, New York, pp 3–83
19. Fenske RF, Martin DS, Reudenberg K (1962) Inorg Chem 1: 441
20. Roothaan CCJ (1951) Rev Mod Phys 23: 69
21. Richardson JW, Rundle RE (1956) US Atomic Energy Report ISC-830. Ames Laboratory, Iowa State
22. (a) Fenske RF, Caulton KG, Radtke DD, Sweeny CC (1966) Inorg Chem 5: 951; (b) Fenske RF, Caulton KG, Radtke DD, Sweeny CC (1966) Inorg Chem 5: 960; (c) Fenske RF, Radtke DD (1968) Inorg Chem 7: 479; (d) Caulton KG, Fenske RF (1968) Inorg Chem 7: 1273; (e) Hall MB, Fenske RF (1972) Inorg Chem 11: 768
23. (a) Graham JP, Wojcicki A, Bursten BE (1999) Organometallics 18: 837; (b) Weller AS, Shang MY, Fehlner TP (1999) Organometallics 18: 853; (c) Yam VWW, Qi GZ, Cheung KK (1998) Organometallics 17: 5448; (d) O'Brien TA, O'Brien JF (1998) J Coord (1998) Chem 44; 91; (e) Bahn CS, Tan A, Harris S (1998) Inorg Chem 37: 2770; (f) Yam VWW, Qi GZ, Cheung KK (1998) J Chem Soc Dalton 1819; (g) Tan A, Harris S (1998) Inorg Chem 37: 2205; (h) Tan A, Harris S (1998) Inorg Chem 37: 2215; (i) Fehlner TP (1998) J Organomet Chem 550: 21; (j) Chisholm MH, Lynn MA (1998) J Organomet Chem 550: 141; (k) Lin C, Ren T, Valente EJ, Zubkowski JD (1998) J Chem Soc Dalton Trans 571
24. (a) Veillard A (1969) Chem Commun 18: 1022; (b) Veillard A (1969) Chem Commun 23: 1427; (c) Hillier IH, Saunders VR (1970) Proc R Soc Lond Ser A 320: 161; (d) Moskowitz JW, Hollister C, Hornback CJ, Basch H (1970) J Chem Phys 53: 2570; (e) Guest MF, Hall MB, Hillier IH (1973) Mol Phys 25: 629

Theor Chem Acc (2000) 103:225–227
DOI 10.1007/s002149900053

Theoretical
Chemistry Accounts
© Springer-Verlag 2000

*Perspective*

# Perspective on "Equation of state calculations by fast computing machines"

## Metropolis N, Rosenbluth AE, Rosenbluth MN, Teller AH, Teller E (1953) J Chem Phys 21: 1087–1092

**William L. Jorgensen**

Department of Chemistry, Yale University, New Haven, CT 06520–8107, USA

Received: 6 April 1999 / Accepted: 14 April 1999 / Published online: 14 July 1999

**Abstract.** An overview is given of the paper by Metropolis et al. that has formed the basis for Monte Carlo statistical mechanics simulations of atomic and molecular systems.

**Key words:** Monte Carlo – Statistical mechanics – Metropolis sampling

## 1 Background

For many-particle systems in classical statistical mechanics, the key numerical problem is the solution of the configurational integral that appears in the averages for a property $Q$ (Eqs. 1, 2).

$$\langle Q \rangle = Q_k + \int Q(\vec{X}) P(\vec{X}) \, d\vec{X} \tag{1}$$

$$P(\vec{X}) = \exp[-\beta E(\vec{X})] \Big/ \int \exp[-\beta E(\vec{X})] \, d\vec{X}. \tag{2}$$

These equations are for the canonical NVT ensemble where $P(\vec{X})$ is the Boltzmann factor, $E(\vec{X})$ is the total potential energy, $\beta = 1/kT$, and the integrals are taken over all possible geometrical configurations, $\vec{X}$, of the system. $Q_k$ represents the contribution from the kinetic energy, which is taken as separable from the configurational contribution. For $N$ particles, $\vec{X}$ has about $3N$ dimensions or coordinates. In addition, at liquid or solid densities most arbitrary choices of $\vec{X}$ would have overlapping particles with low probability, $P(\vec{X})$, and, therefore, they would contribute little to the average. Consequently, a brute-force Monte Carlo solution of the configurational integral by random selection of configurations, $\vec{X}$, becomes impractical for more than a few particles with typical potential-energy functions.

## 2 Key contribution

In their classic paper, Metropolis et al. [1] recognized that a practical solution for the configurational integrals could be obtained by a modified Monte Carlo procedure where "instead of choosing configurations randomly, then weighting them with $\exp(-E/kT)$, we choose configurations with a probability $\exp(-E/kT)$ and weight them evenly." With this procedure and converting the integral to a sum over discrete configurations, Eqs. (1 and 2) are simplified to Eq. (3), where $L$ is the number of configurations considered and $\vec{X}_M$ indicates a Metropolis-sampled configuration.

$$\langle Q \rangle = Q_k + 1/L \sum_i^L Q_i(\vec{X}_M) \tag{3}$$

The Metropolis algorithm involves generation of a chain of configurations. When the system is at configuration $i$, an attempted move to another configuration $j$ is considered. The attempted move can be accepted and $j$ is the next configuration in the chain, or the move is rejected, $i$ repeats in the chain and another attempted move is tried. The decision to move from $i$ to $j$ is determined from a probability $p = \pi_j/\pi_i$, where for the NVT ensemble the Boltzmann factor (Eq. 4)

$$\pi_j = \exp(-\beta E_j) \tag{4}$$

is appropriate. Then, if $p \geq 1$, which for the NVT ensemble means $E_j \leq E_i$ (the potential energy went down), the move is accepted. If $p < 1$, $p$ is compared to a random number $x$ between 0 and 1, and if $p \geq x$, the move is still accepted, otherwise $j$ is rejected and $i$ repeats. In summary, configuration $j$ is accepted with a probability $\min[1, \pi_j/\pi_i]$. The chain of configurations generated in this way provides the Boltzmann-weighted configurations that are needed in Eq. (3). It is apparent that the Metropolis sampling, while allowing the energy to go up, focuses on low-energy configurations, which contribute the most to the Boltzmann averages. Nevertheless, care is required to guarantee convergence of the calculations and different computed properties converge at different rates [2, 3].

One could imagine that the Metropolis sampling algorithm could have been conceived by inspection from considering a simple one-particle system with two states, $r$ and $s$. If we let $\varepsilon_r - \varepsilon_s = kT \ln 3$, for example, then the Boltzmann factor $\pi_r/\pi_s = 1/3$ and the algorithm needs to sample state $s$ 3 times on average for every time it samples state $r$. This could be achieved by a walk where when in state $r$, the next step is back to state $s$, and when in state $s$, state $s$ needs to repeat an average of 2 times before a transition to $r$. For example, the following walk would work.

State $r$: $X$    $X$    $X$    $X$    $X$    $X$    $X$    .....
State $s$: $XXX$ $XXX$ $XXX$ $XXX$ $XXX$ $XXX$ $XXX$

However, to make the process more general and random, the decision on whether to repeat s should be based on a random number that is compared to $\pi_r/\pi_s$. A transition from a lower-energy state, $s$, to a higher-energy one, $r$, needs to occur for $\pi_r/\pi_s$ of the attempted moves.

## 3 Refinements, sampling and some details

Another issue is the size, $N$, of a sample that would be adequate to represent a liquid. To this end, Metropolis et al. [1] also introduced "periodic boundary conditions", which lead to simulating a liquid by explicitly considering a relatively small number of molecules, about 100–1000. Though they only treated hard disks in two dimensions in their paper, "extension to three dimensions is obvious." In three dimensions, the molecules are typically in a cubic or orthorhombic cell, which is surrounded by 26 images of itself. If on moving a molecule to create a new configuration it passes through a wall, then an image of it reenters the central cell through the opposite face. Thus, one just needs to keep track of the contents of the central cell.

With rigid molecules, a move normally consists of picking one molecule at random, translating it randomly in all three Cartesian directions and randomly rotating it about one randomly chosen Cartesian axis. For flexible molecules, both the external and internal degrees of freedom need to be sampled. The external ones are handled in the same way as for rigid molecules. The internal ones are usually treated by representing the molecule by a Z matrix (r, $v$, $\varphi$ internal coordinates). Upon an attempted move, the variable entries in the Z matrix are changed randomly within specified ranges and the molecule is rebuilt [4]. The use of internal coordinates also facilitates the enforcement of constraints, i.e., specification of bond lengths, bond angles and dihedral angles that are fixed, and avoids the introduction of additional terms in Eq. (4) with space-fixed coordinate systems [3]. It is clear that the acceptance rate is affected by the choices of ranges for the molecular translations and rotations; reasonable convergence behavior is normally obtained by adjusting the ranges such that 20–50% of the attempted moves are accepted.

An advantage of Monte Carlo simulations is that modified sampling procedures can potentially be devised to enhance convergence. One example is the preferential sampling of solvent molecules near a solute by attempting to move them more frequently than more distant solvent molecules [5]. The only caveat is to be sure that the intrinsic probabilities, i.e., in the absence of the Boltzmann factor, of states $i$ and $j$ are properly reflected. If this is not the case, then $\pi_j$ in Eq. (4) needs to include an appropriate Jacobian term representing the phase-space volume for state $j$. For example, with an atomic liquid, if one wanted to sample in spherical rather than Cartesian coordinates, the required Jacobian is $r_j^2 \sin \varphi_j$, and the acceptance probability would be $\min[1, (r_j^2 \sin \varphi_j / r_i^2 \sin \varphi_i) \exp(-\beta (E_j - E_i))]$. Another attractive point is that it is straightforward to execute Monte Carlo simulations in the NPT ensemble by including volume terms in Eq. (4) to yield Eq. (5) and by allowing the volume of the central cell to vary [2, 3].

$$\pi_j = \exp(-\beta H_j)V_j^N = \exp[-\beta(E_j + PV_j - Nk_{\mathrm{B}}T \ln V_j)]$$

(5)

The principal potential pitfall is that there is no ultimate guarantee that convergence of a property or of the simulation as a whole has been reached. Pathological cases can be constructed. For example, one could set up a simulation for a liquid secondary amide at 25°C with all of the monomers initially in the cis conformation. Since the rotational barrier is about 20 kcal/mol for conversion of cis to trans, under normal Monte Carlo conditions a conversion would be an exceedingly rare event. The system could be equilibrated and the energy seemingly converged, but the simulated liquid is metastable since with an energy difference of about 2.5 kcal/mol, 98.5% of the monomers should be trans. However, in general, for typical organic liquids with standard force fields, convergence of most key properties such as heat of vaporization, density, and radial distribution functions is not problematic [6]. In fact, Monte Carlo simulations have been shown to be more efficient than molecular dynamics for liquid hexane [7]. Such Monte Carlo calculations are trivial to set up with modern software [4] and can be executed for one state point in a few hours on Pentium-II-based personal computers [8]. For dilute solutions of a single flexible solute in a solvent, convergence of the conformational properties of the solute can understandably be more taxing as statistics are only being accumulated on one molecule rather than on $N$.

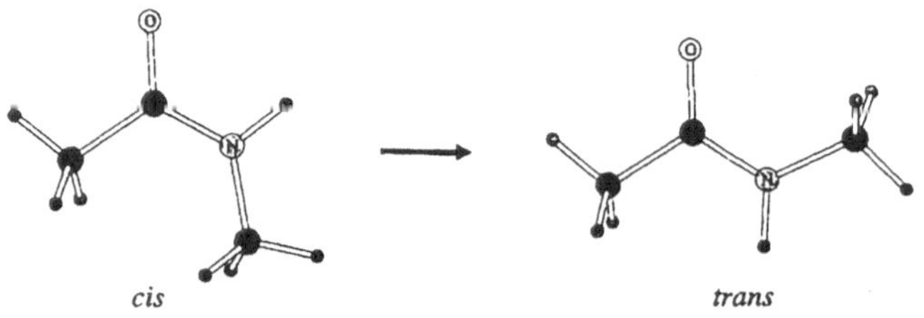

*cis*          *trans*

Figure 1.

Overall, Metropolis et al. introduced the sampling method and periodic boundary conditions that remain at the heart of Monte Carlo statistical mechanics simulations of fluids. This was one of the major contributions to theoretical chemistry of the twentieth century.

## References

1. Metropolis N, Rosenbluth AE, Rosenbluth MN, Teller AH, Teller E (1953) J Chem Phys 21: 1087–1092
2. (a) Barker JA, Henderson D (1976) Rev Mod Phys 48: 587–671; (b) Jorgensen WL (1983) J Phys Chem 87: 5304–5314; (c) Kalos MH, Whitlock PA (1986) Monte Carlo methods. Wiley, New York; (d) Levesque D, Weiss JJ (1995) Top Appl Phys 71: 121–204; (e) Jorgensen WL (1998) In: Schleyer PvR (ed) Encyclopedia of computational chemistry, vol 3. Wiley, New York, pp 1754–1763
3. Allen MP, Tildesley DJ (1987) Computer simulations of liquids. Clarendon, Oxford
4. Jorgensen WL (1999) BOSS, version 4.1. Yale University
5. Owicki JC, Scheraga HA (1977) Chem Phys Lett 47: 600–602
6. Jorgensen WL, Maxwell DS, Tirado-Rives J (1996) J Am Chem Soc 118: 11225–11236
7. Jorgensen WL, Tirado-Rives J (1996) J Phys Chem 100: 14508–14513
8. Tirado-Rives J, Jorgensen WL (1996) J Comput Chem 17: 1385–1386

Theor Chem Acc (2000) 103:228–230
DOI 10.1007/s002149900076

Theoretical
Chemistry Accounts
© Springer-Verlag 2000

*Perspective*

# Perspective on "Quantum theory of many-particle systems I, II, and III"

## Löwdin P-O (1955) Phys Rev 97:1474–1520

**Björn O. Roos**

Department of Theoretical Chemistry, Chemical Center, P.O. Box 124, S-221 00 Lund, Sweden

Received: 26 February 1999 / Accepted: 1 July 1999 / Published online: 4 October 1999

**Abstract.** The title paper is reviewed in perspective of the later development of computational methods in quantum chemistry. The importance of the concept of natural spin orbitals is discussed and its implication for the future development of multiconfigurational self-consistent-field theories is illustrated.

**Key words:** Natural orbitals – Configuration interaction – Multiconfigurational self-consistent-field

## 1 Introduction

The material we see first, the first books we read, and the first articles we encounter, will probably have a strong and may be decisive influence on how we develop our attitude to the scientific field we are working in. One of the first things I did as a young student was to travel the 70 km to Uppsala and take part in the 1961 summer school in quantum chemistry organized by Per-Olov Löwdin and the Uppsala Quantum Chemistry Group. It was in many ways an overwhelming experience. Löwdin was an excellent lecturer and to listen to his explanations of the foundations of quantum chemistry made you believe that it was all very simple, logical, and straightforward. When I therefore half a year later saw the green reprint of the title article(s) on the desk of my supervisor in Stockholm, I borrowed it from him. I have not yet returned it.

What Löwdin published in 1955 (the manuscripts were received in July that year) was a series of three articles under a common heading: *Quantum theory of many-particle systems*. The three parts had the following subtitles: I. Physical interpretation by means of density matrices, natural spin-orbitals, and convergence problems in the method of configuration interaction; II. Study of the ordinary Hartree–Fock approximation; and III. Extension of the Hartree–Fock scheme to include

degenerate systems and correlation effects. Without doubt the first article has had the greatest impact, but also in the two others is it possible to find little pearls that act as precursors for later developments of computational strategies in quantum chemistry. These articles had a profound influence on my own development, especially on the way I came to think about the molecular orbital (MO) concept in conjunction with accurate wave functions and configuration interaction (CI).

## 2 Natural spin orbitals

The single concept, which is defined and discussed in these articles, that without doubt has had the most penetrating impact on the whole field of quantum chemistry, is the concept of the natural spin orbitals. This concept is, in principle, very simple: the set of MOs that makes the first-order density matrix diagonal. Löwdin starts out by defining a hierarchy of reduced density matrices for a general wave function:

$$\Gamma^{(p)}(x_1' x_2' \ldots x_p' | x_1 x_2 \ldots x_p)$$
$$= \binom{N}{p} \int \Psi(1'2' \ldots p' \ldots N) \Psi(12 \ldots p \ldots N) dx_{1'2'(1)} .$$

He goes on to show that for the description of the energy it is sufficient to know the second-order reduced density matrix $\Gamma(x_1' x_2' | x_1 x_2)$. This was a favorite subject when he was lecturing and led to speculations about the possibility to compute the second-order density matrix directly, and discussions of the so called $N$-representability problem. In spite of several attempts, this way of attacking the quantum chemical many-particle problem has so far been unsuccessful. Of special interest was the first-order reduced density matrix, $\gamma(x_1' | x_1)$, which when expanded in a complete one-electron basis, $\psi$, is obtained as

$$\gamma(x_1' | x_1) = \sum_{k,l} \psi_k^*(x_1') \psi_l(x_1) \gamma(l|k) , \qquad (2)$$

where we use Löwdin's own notation. The matrix $\gamma(l|k)$ was identified as the charge and bond order matrix used earlier by Coulson and Longuet-Higgins (I will not repeat all the references given by Löwdin. They can be found in the article). The diagonal elements were interpreted as a measure of the occupation of a given spin orbital and it was shown that this always lies between 0 and 1, or more generally

$$0 \leq \gamma(k|k) \leq 1; \quad Tr(\gamma) = N \ . \tag{3}$$

### 2.1 Configuration interaction

Löwdin now makes a detour to discuss the method of CI. He expands the wave function in the set of Slater determinants that can be formed from the one-electron basis, $\psi$. He introduces the word complete CI to emphasize that the expansion is, in principle, exact. We have since learned to use the expression full CI in the case where the one-electron basis is finite (Löwdin made a sarcastic comment about our use of the word complete in the complete active space (CAS) concept at a Sanibel Conference in 1980). He computes the matrix elements between the Slater determinants without assuming orthogonality between the MOs. These are the famous Löwdin matrix elements, which generalized the Slater formulae to the nonorthogonal case.

The method of CI is outlined in all detail. Of course, he was not the first person to do this. Small CI calculations had been performed in the 1930s and he himself refers to the work of Boys and others and notes "that have overcome the numerical, difficulties by aid of electronic computers," but for me it was the first time I saw the CI method outlined in a systematic and detailed fashion. He also states that "one can solve secular equations of comparatively high orders by means of the modern electronic computers and one can expect a steady development of the methods of programming, etc." One might recall that at this time such a view of the future was not uncontroversial. He finally points out that there is a need for a simple physical interpretation of the complicated CI wave function. The tool are the natural orbitals (NOs).

### 2.2 The natural spin orbitals

The orbitals that diagonalize the first-order reduced density matrix are the natural spin orbitals $\chi$:

$$\gamma(x_1'|x_1) = \sum_k \chi_k^*(x_1')\chi_k(x_1)\eta_k \ , \tag{4}$$

where $\eta_k$ is the occupation number of natural spin orbital $k$ with the property

$$0 \leq \eta_k \leq 1; \quad \sum_k \eta_k = N \ . \tag{5}$$

He notices that the relation $\gamma^2 = \gamma$ (all occupation numbers are 1 or 0) is a sufficient condition for the possibility to reduce the wave function to a single Slater determinant. However, if this is not the case, but only a finite number of the occupation numbers are essentially different from zero, the wave function can be expressed as a sum of comparatively few terms and this would provide a simple solution to the CI convergence problem. Today we know that this was an overoptimistic view, but in this statement lies the basic idea behind multiconfigurational self-consistent-field (MCSCF) theory and in particular CASSCF.

The importance of the NO concept has been exactly these two things: analysis of correlation effects and the inspiration to develop methods to determine the most important NOs, MCSCF theory.

### 2.3 Extended Hartree–Fock

Actually Löwdin goes quite a long way to develop the MCSCF formalism. He defines a full CI on a limited one-electron basis set and points out that since the expansion is no longer complete, it is now important not only to determine the CI coefficients but also the MOs in order to obtain a solution "which is as accurate as possible." When the number of spin orbitals, $M$ equals the number of electrons, $N$, this becomes the ordinary Hartree–Fock equations, but when $M > N$ we obtain what he calls the extended Hartree-Fock equations. Twenty-five years later it became known as the CASSCF method. He writes the condition for optimum orbitals in the form

$$\sum_l \hat{F}_{kl}\psi_l = \sum_l \epsilon_{kl}\psi_l \ , \tag{6}$$

where

$$\hat{F}_{kl} = \hat{h}\gamma(k|l) + \sum_{m,n} \int \psi_m^*(2)\frac{1}{r_{12}}\psi_n(2)dx_2\Gamma(ln|km) \ , \tag{7}$$

and $\epsilon_{kl}$ are the Lagrangian multipliers, which form an Hermitian matrix. Integrating, the condition for optimal orbitals becomes equivalent to the condition that the matrix

$$F_{kl} = \sum_m h_{km}\gamma(m|l) + \sum_{m,n,o} (mn|ok)\Gamma(on|lm) \tag{8}$$

is Hermitian. This is nothing but what was later to become the extended Brillouin condition and the matrix above is what we today call the MCSCF Fock matrix. May be it should be emphasized that Löwdin did these formal derivations, but he did not suggest any effective procedures which could be used to obtain the CI coefficients and the MCSCF orbitals. This was left to the "programmers" of the next generation.

### 2.4 NOs for H₂O

As an illustration of the interpretative power of the NOs, we show here the eight most important orbitals for the water molecule. Löwdin was correct in assuming that only a rather small number of orbitals will have appreciable occupation numbers. He was less right in presuming that this would lead to a quickly converging CI expansion. Today we know that the contrary is true

230

**Fig. 1.** Eight natural orbitals for the water molecule. They were obtained from a complete active spare self-consistent-field calculation using a basis set of the size (O4s3p2d/H3s2p)

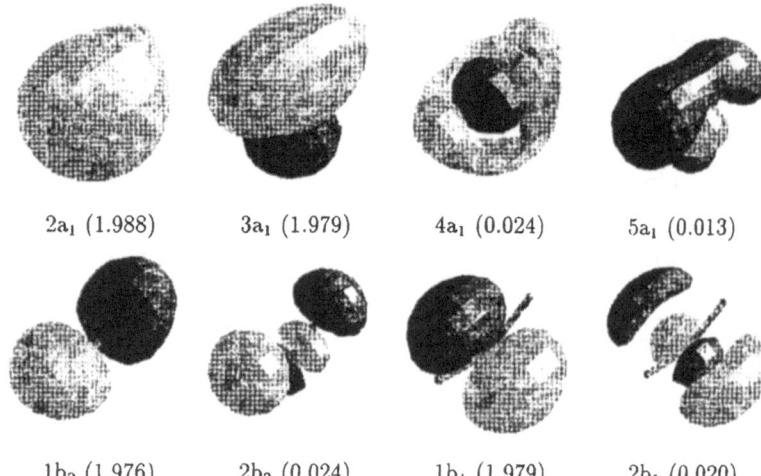

2a₁ (1.988)   3a₁ (1.979)   4a₁ (0.024)   5a₁ (0.013)

1b₂ (1.976)   2b₂ (0.024)   1b₁ (1.979)   2b₁ (0.020)

and that the reason for the slow convergence is the two-electron cusp, which is not easily described in an MO framework.

Typically, for a closed-shell molecule at equilibrium geometry there will be a set of NOs with occupation numbers close to 2 and a second set (usually valence in character) of about the same size with occupation numbers in the range 0.01–0.05. The rest of the NOs will have slowly decreasing occupation numbers without much structure.

The MOs with occupation numbers above 0.01 will describe clearly defined correlation effects: horizontal (along a bond), radial (sometimes called in-out), and angular. The strongly occupied NOs will of course resemble closely the Hartree–Fock orbitals. All this is nicely illustrated for the water molecule in Fig. 1.

We can see the four strongly occupied valence orbitals $2a_1$, $3a_1$, $1b_2$, and $1b_1$ and notice that their occupation numbers vary in the range 1.97–1.99. For each of these orbitals there is a correlating orbital with an occupation number that lies in the range 0.01–0.03. This pairing seems to be a rather general feature of closed-shell molecules. The correlation orbital will describe horizontal correlation for bonding orbitals ($4a_1$ and $2b_2$) and radial correlation for nonbonded electrons ($5a_1$ and $2b_1$). The latter will be composed of extra valence basis functions. Still, they have occupation numbers in the same range as the correlating orbital we can construct from valence orbitals. If larger calculations are performed, the additional NOs will have occupation numbers well below 0.01. The occupation numbers of the strongly occupied orbitals will decrease somewhat, but those of the correlating orbitals shown in Fig. 1 will not change much. It is a general experience that occupation numbers of NOs are very stable against variation in the correlation treatment, changes of basis set, etc.

## 3 Final remarks

I have talked mostly about the contents of the first of the series of three articles. There is no doubt that it is the most important part. The second article studies the ordinary Hartree–Fock approximation. There is not much new in this treatment but Löwdin concludes that many systems cannot be treated by a single Slater determinant, in particular not for the open-shell case, since they are not eigenfunctions of the spin. In the third article he starts out to tackle this problem. He introduces spin projection operators and discusses the possibility of using different orbitals for different spins. The alternant orbital method would emerge, which was to be used with some success by other members of the Uppsala quantum chemistry group. Löwdin was, however, a little bit overoptimistic about the possibility to treat correlation effects by the use of different orbitals for different spins. The latter developments of theories for electron correlation was to take other routes. The work on open-shell Hartree–Fock theory was a prelude to the later developments of open-shell Hartree–Fock theory in the late 1950s.

Quantum chemistry has had many actors. Per-Olov Löwdin was one of the dominating leaders in the late 1950s and the 1960s. Many were educated at his summer schools. In a way he set the stage for much of the development that was to come during the period 1960–80, when most of the computational strategies were developed. His 1955 article is certainly one of the foundation stones of modern computational quantum chemistry.

*Acknowledgements.* It should now be clear to the reader that the articles reviewed have had a considerable impact on the scientific career of the present author. I would therefore like to take this opportunity to express my gratitude to Per-Olov Löwdin for having written the articles that became such an inspiration for my own work in quantum chemistry. I am sure that I am not the only quantum chemist of my generation that has been strongly influenced by this work.

Theor Chem Acc (2000) 103:231–233
DOI 10.1007/s002149900016

Theoretical
Chemistry Accounts
© Springer-Verlag 2000

*Perspective*

# Perspective on "On the theory of oxidation–reduction reactions involving electron transfer. I"

## Marcus RA (1956) J Chem Phys 24: 966–978

**José M. Lluch**

Departament de Química, Universitat Autònoma de Barcelona, E-08193 Bellaterra (Barcelona), Spain

Received: 17 February 1999 / Accepted: 1 March 1999 / Published online: 7 June 1999

**Abstract.** This paper provides a retrospective overview of the title paper written by Marcus around the middle of the twentieth century. A description of the history that led to this work, the basic features of the theory of electron-transfer reactions in solution developed in it, and a comment on its huge influence on succeeding developments are presented.

**Key words:** Electron-transfer reactions – Marcus theory – Solvent fluctuations – Free-energy barriers – Inverted region

Electron transfers are perhaps the simplest of all chemical reactions, at least when no chemical bonds are broken or formed. The availability of many radioactive isotopes due to nuclear developments in World War II made possible the early experiments in the electron-transfer field. Artificial radioactivity made it possible to measure the rates of a large number of isotopic-exchange (self-exchange) electron-transfer reactions in aqueous solution. Prior to about 1950, it was commonly believed that a self-exchange reaction such as [where the asterisk marks a radioactive isotope]

$$[*Fe(aq)]^{2+} + [Fe(aq)]^{3+} \rightleftharpoons [*Fe(aq)]^{3+} + [Fe(aq)]^{2+} , \tag{1}$$

could proceed very rapidly (even almost instantaneously), because an electron could jump from the ferrous ion to the ferric ion, with no change in energy, across long distances. However, as several systems were explored, it soon became apparent that self-exchange reactions of simple cations in aqueous solutions, such as $Fe^{2+}/Fe^{3+}$, $Eu^{2+}/Eu^{3+}$ and $Ce^{3+}/Ce^{4+}$, are generally slow, while electron transfers in most systems involving complex ions, such as $MnO_4^-/MnO_4^{2-}$ and $Fe(CN)_6^{3-}/Fe(CN)_6^{4-}$, were faster by several orders of magnitude. In 1951, an important symposium on electron-transfer processes was held at the University of Nôtre Dame, and the proceedings were reported in the 1952 symposium issue of *J Phys Chem*. In his paper in that issue, Libby [1] noted that the surprising results could be rationalized by considering the role of the solvent molecules. For instance, in the reaction depicted in Eq. 1 the direct electron jump is much faster than the nuclear motion of the solvent, the new $[*Fe(aq)]^{3+}$ and $[Fe(aq)]^{2+}$ being generated with the hydration configurations appropriate for the original $[*Fe(aq)]^{2+}$ and $[Fe(aq)]^{3+}$, respectively. This idea comes from the Franck–Condon principle of molecular spectroscopy. The large change in initial and final equilibrium solvations around each of the two ions would account for the large energy barrier to the electron transfer. In self-exchange reactions involving complex ions, the change in equilibrium solvation would be smaller owing to the larger radii, thus leading to a smaller energy barrier to electron-transfer reactions.

Meanwhile, Marcus joined the faculty of the Polytechnic Institute of Brooklyn (New York) in 1951. After writing the 1952 RRKM papers, he was looking for a new field of theoretical research in which to work. As a result of a question posed by a student in his statistical mechanics class, Marcus focused on polyelectrolytes and their electrostatic free energy. To this aim, he had to expand considerably his electrostatics background, which turned out to be very useful for him when, in 1955, he found by chance the above-mentioned 1952 symposium issue on electron-transfer reactions. Reading that issue, Marcus realized that Libby's proposal involved a vertical transition, which violated the energy-conservation principle unless absorption of radiation took place. As a result he intended to understand how thermal electron transfers in the dark could occur satisfying both the Frank–Condon and energy-conservation principles. A month later, Marcus found the solution to the problem and published his results in the title paper, in which the first quantitative microscopic description of a chemical reaction in solution was made.

The basic features of the theory of electron-transfer reactions in solutions developed by Marcus in that paper can be summarized as follows. The electron transfer between the two reacting molecules is described in terms of a diabatic two-state model consisting of a system having the electronic configuration of the reactants (precursor complex) and a system having the electronic configuration of the products (successor complex) after the electron transfer. Each state gives rise to the corresponding diabatic potential energy hypersurfaces $U_p$ (precursor) and $U_s$ (successor), respectively, which are functions of the ensemble of nuclear coordinates of the entire system. Both $U_p$ and $U_s$ have minima that correspond to the equilibrium configuration of the solvent around the reacting molecules and the reaction products, respectively. Assuming a classical frame, the radiationless electron transfer takes place at the intersection region (X) between $U_p$ and $U_s$, which constitutes the transition state of the reaction. This implies that reorganization of the solvent is required before the electron transfer takes place. Random thermal fluctuations in the solvent configurations of the precursor complex occur until that X region is reached, then the energies of both diabatic states become equal and the electron jump happens. The transfer at X occurs at fixed positions and momenta of the nuclei, the Franck–Condon principle being satisfied. The appearance of the proper fluctuations costs free energy. It is this free energy that determines the rate of the reaction. The electronic coupling between both diabatic states is supposed to be large enough to permit the conversion from reactants to products at the intersection region, but small enough to be neglected in order to ignore the splitting of the two hypersurfaces in the vicinity of their intersection.

In addition Marcus treated the solvent as a dielectric continuum in order to make the calculation of the free-energy barrier feasible. As a result of the slowness of the vibrations and orientation changes of the solvent molecules, the solvent electrical polarization at the transition state is not in equilibrium with the electric field produced by the ionic charges of the reacting molecules or reaction products. Marcus had to use a method to calculate the electrostatic free energy of states having nonequilibrium polarization that was developed by himself in the paper following the title paper in the same volume of *J Chem Phys* [2]. The title paper finally provided a simple expression for the free-energy barrier, which when introduced into transition-state theory leads to the reaction rate.

The title paper was enormously important by itself, but in addition it was the first step (and the cornerstone) in a long series of papers on electron-transfer reactions which were published by Marcus from 1956 to 1965. During those years he extended [3, 4] the theory to include, for instance, intramolecular vibrational effects, numerically calculated rates of self-exchange and cross reactions, electrochemical electron-transfer reactions (i.e. including electrodes), chemiluminescent electron transfers, the relation between nonequilibrium and equilibrium solvation free energies for arbitrary geometries, and spectral charge-transfer processes.

One of the most fundamental achievements of the seminal Marcus theory on electron-transfer reactions is the nowadays widely used quadratic driving force–activation free energy Marcus relationship

$$\Delta G^{\#} = \frac{(\Delta G^0 + \lambda)^2}{4\lambda} \tag{2}$$

that relates the activation free energy $\Delta G^{\#}$ to the standard reaction free energy $\Delta G^0$ and the reorganization free energy $\lambda$ (i.e., the free energy released when the system evolves from equilibrium configurations corresponding to reactants to those corresponding to products, while an electronic wave function, that can be directly related to products in a valence-bond structure sense, is maintained to describe the solute). This classical expression contains a most interesting prediction: as the driving force (negative of the standard reaction free energy) of the reaction increases, the reaction rate rises to a maximum when $\Delta G^0 = -\lambda$, but then unexpectedly falls off again. The initial decrease in $\Delta G^{\#}$ with increasingly negative $\Delta G^0$ is the expected trend in chemical reactions (similar, for instance, to the usual trend in Brønsted plots of acid- or base-catalyzed reactions and in Tafel plots of electrochemical reactions) and corresponds to the "normal" region. Instead, the prediction for the region where $\Delta G^0 < -\lambda$, the so-called "inverted region", was one of the more startling and controversial results of Marcus theory. As a matter of fact, the first unambiguous experimental demonstration [5] of the existence of such an inverted region in solution electron-transfer reactions was not made until 1984, almost 25 years after Marcus had predicted it.

The huge impact of the title paper and the 1956–1965 series of papers originating from it on the field of theoretical chemistry (and on many new experimental developments in chemistry) was recognized by the Royal Swedish Academy of Sciences, which decided to award the 1992 Nobel Prize in Chemistry to Marcus for his contributions to the theory of electron-transfer reactions in chemical systems. Nowadays, the practical consequences of his theory extend over all areas of chemistry and many areas of biochemistry. Marcus theory describes and makes predictions concerning an increasingly growing set of widely differing phenomena, such as natural and artificial photosynthesis, metabolism, enzyme-catalyzed redox reactions, photochemical production of fuel, chemiluminescence, the conductivity of electrically conducting polymers, long-range electron transfer in proteins, corrosion, the methodology of electrochemical synthesis and analysis, and more. Much theoretical (and experimental) work on all these topics remains to be done, and it is expected that impressive progress on all of them will be achieved during the next century using new developments based on the Marcus theory initiated in the title paper. In this sense, computer simulations (Monte Carlo and molecular dynamics) will become excellent tools to study very important biological problems associated with electron-transfer reactions, such as radical processes intervening in chemical carcinogenesis and cellular aging.

Finally, it has to be stressed that, being aware of the corresponding differences, some concepts of electron-transfer reactions have been extended to other types of chemical reactions, such as the transfer of methyl groups, atoms or, especially, protons. It is clear that, apart from electron transfer, proton transfer is the most important type of chemical reaction and, in particular, the most common enzyme-catalyzed reaction. Through proton transfers, the original ideas formulated by Marcus in the title paper for electron-transfer reactions have found new fields to influence and in which to be successfully applied.

## References

1. Libby WF (1952) J Phys Chem 56: 863
2. Marcus RA (1956) J Chem Phys 24: 979
3. Marcus RA (1986) J Phys Chem 90: 3460
4. Marcus RA (1993) Angew Chem Int Ed Engl 32: 1111
5. Miller JR, Calcaterra LT, Closs GL (1984) J Am Chem Soc 106: 3047

Theor Chem Acc (2000) 103:234–235
DOI 10.1007/s002149900019

Theoretical
Chemistry Accounts
© Springer-Verlag 2000

*Perspective*

# Perspective on "Statistical mechanical theory of irreversible processes. I. General theory and simple applications to magnetic and conduction problems."

## Kubo R (1957) J Phys Soc Jpn 12: 570

**Daan Frenkel**

FOM Institute for Atomic and Molecular Physics, Kruislaan 407, 1098 SJ Amsterdam, The Netherlands

Received: 24 February 1999 / Accepted: 29 March 1999 / Published online: 21 June 1999

**Abstract.** Kubo's paper on linear-response theory provided a unified language to describe a wide variety of transport phenomena, both quantum and classical, in a suitable "microscopic" language. The paper has been crucial for subsequent developments in numerical simulation.

**Key words:** Linear-response theory – Correlation function – Fluctuation-dissipation theorem – Reaction rates

It is sometimes difficult to indicate precisely which paper heralds an important change in a particular field. In retrospect, one can often observe precursors and parallel developments in other publications; however, Kubo's paper on linear-response theory, although it certainly did not appear in a scientific vacuum, clearly marks a watershed.

Kubo's linear-response theory provides the full, quantum-mechanical relation between the response of a system to external perturbations and the spontaneous decay of fluctuations in the unperturbed system. Of course, the paper had important predecessors: Nyquist's [1] paper on thermal noise in resistors and Onsager's [2] seminal paper on the relation between decay of macroscopic and microscopic fluctuations, to name but the earliest.

The linear-response approach has played an important role in the construction of most modern theories of transport processes (see, e.g. Ref. [3]). Moreover, it has had a profound impact on the development of classical molecular dynamics simulations – a field that was emerging at the same time. Linear-response theory showed how linear transport coefficients can be computed in a simulation, by studying the decay of fluctuations in equilibrium. More specifically, most transport coefficients can be expressed as time integrals of (auto)correlation functions of microscopic "fluxes" (e.g. the current density, in the case of electrical conductivity) – the famous "Green–Kubo" relations. The microscopic fluxes that appear in these relations are explicit functions of the particle coordinates and momenta and can, therefore, be computed in a standard molecular dynamics simulation.

In fact, Kubo's paper also provided much of the language for the subsequent development of nonequilibrium molecular dynamics simulations; however, nonequilibrium molecular dynamics adresses problems that could not be handled in the context of the original linear-response theory (see, e.g. Ref. [4]).

However, when I focus on the implications of Kubo's paper for theoretical chemistry, two topics spring to mind that were subsequently affected by this work: "spectroscopy" and "chemical kinetics".

Let us first consider spectroscopy. Linear-response theory, in particular the fluctuation dissipation theorem – which relates the absorption of an incident monochromatic field to the correlation function of (e.g. dipole) fluctuations in equilibrium – has changed our perspective on "spectroscopy" of dense media. It has moved away from a static "Schrödinger" picture – phrased in terms of transitions between immutable (but usually incomputable) quantum levels – to a dynamic "Heisenberg" picture, in which the spectral line shape is related by Fourier transform to a correlation function that describes the decay of fluctuations. Of course, any property that cannot be computed in the Schrödinger picture, cannot be computed in the Heisenberg picture either; however, correlation functions, unlike wavefunctions, have a clear meaning in the classical limit. This makes it much easier to come up with simple (semi)classical interpretations and approximations.

The Kubo approach has also proven to be extremely fruitful for the theory of chemical rate processes. In fact, chemical rate constants can also be expressed as an integral of a flux autocorrelation function. As was shown by Chandler [5], however, the "brute-force" approach does not work in this case. In fact, in the context of diffusion in solids, this had been noted by Bennett [6]. Chandler showed how the expression for the rate constant of a unimolecular reaction can be cast in a form that can be used efficiently in computer simulations.

As is clear from the above, linear-response theory has stimulated many important developments; yet Kubo's approach has not been uncontroversial. In particular, in a paper that cannot be found in most libraries, van Kampen [7] has criticized the assumption made in linear-response theory that, for sufficiently weak fields, the change in the phase-space density is linear in the applied perturbation. In fact, due to the exponential divergence of phase-space trajectories, even the weakest perturbation will, on a macroscopic time-scale, result in large changes in the phase-space density; however, as was shown in the subsequent numerical work by Ciccotti et al. [8], on the microscopic time-scales that are usually relevant for the dynamical response of a system to a sudden perturbation, the linear-response assumption holds quite well.

## References

1. Nyquist H (1928) Phys Rev 32: 110
2. (a) Onsager L (1930) Phys Rev 35: 666; (b) Onsager L (1931) Phys Rev 38: 2265
3. Hansen JP, McDonald IR (1986) Theory of simple liquids. Academic Press, London
4. Evans DJ, Morris GP (1990) Statistical mechanics of non-equilibrium liquids. Academic Press, London
5. Chandler D (1978) J Chem Phys 68: 2959
6. Bennett CH (1975) In: Burton JJ, Norwick AS (eds) Diffusion in solids: recent developments. Academic Press, New York
7. van Kampen NG (1971) Phys Norv 5: 279
8. Ciccotti G, Jacucci G, McDonald IR (1979) J Stat Phys 21: 1

Theor Chem Acc (2000) 103:236–237
DOI 10.1007/s002149900012

Theoretical
Chemistry Accounts
© Springer-Verlag 2000

*Perspective*

# Using classical mechanics in a quantum framework.
# Perspective on "Semiclassical description of scattering"

## Ford KW, Wheeler JA (1959) Ann Phys (NY) 7: 259

**William H. Miller**

Department of Chemistry, University of California, and Chemical Sciences Division, Lawrence Berkeley National Laboratory, Berkeley, CA 94720-1460, USA

Received: 3 February 1999 / Accepted: 15 February 1999 / Published online: 21 June 1999

**Abstract.** Ford and Wheeler's paper on elastic scattering was the first to fully analyze the semiclassical limit of quantum mechanics for a collision process and thus reveal the nature of quantum corrections to classical mechanics therein. This "perspective" discusses the historical setting, the content, and present day implications of this work.

**Key words:** Semiclassical – Scattering – Quantum effects – Interference – Initial value representation

I have interpreted the invitation to write about an especially influential paper in our field in a very personal way, choosing one that had a particularly strong impact on me at the formative stage of my early graduate school years.

By the early 1960s the development of crossed molecular beam experimental methods had given one the vision of being able to study chemical reactions at a fundamental molecular level, and many of us that were theoretically inclined leapt at the challenges offered by this new field of chemical dynamics. Though it was clear that quantum mechanics was the right theoretical approach – and, indeed, much quantum reactive scattering theory was initiated in that early period and has born fruit up to the present – it also became clear that classical (i.e., Newtonian) mechanics was a useful approximation for describing atomic and molecular dynamics and, most importantly, that it was much easier to carry out classical trajectory calculations for reactive collisions than to solve the corresponding Schrödinger equation. (The pioneering paper of Karplus et al. [1] is thus justly one of these presently being honored as especially influential.) Today, of course, classical molecular dynamics simulations are a major industry, being

applied not only to gas-phase reactive scattering but also to molecular processes in liquids, in (or on) solids, and particularly to the description of dynamical processes in large biologically relevant molecules.

One worries, however, about the neglect of quantum effects in these classical simulations, particularly when the motion of hydrogen atoms is involved. For example, the zero-point vibrational energy in CH and OH bonds is enormous (compared to normal thermal energies), and it can cause unphysical behavior in a classical simulation if not treated properly. (In many simulations these bonds are held rigid, in part to avoid any problems with zero-point energy, but this is clearly not always satisfactory, most obviously so if they are involved in chemical reactions.) Tunneling of hydrogen-atom motion can also be a significant correction to purely classical dynamics, and quantum coherence (interference) effects may also survive on the short time scales relevant to the dynamics of chemical reactions.

In seeking a way to include these quantum phenomena, while at the same time retaining the usefulness of classical trajectory approaches, Ford and Wheeler's paper [2] (to which I was introduced in a course given by Dudley Herschbach) was the Rosetta stone: it showed in a beautifully simple way how classical mechanics could be used in the framework of a rigorous quantum theory for the case of elastic scattering. This so-called semiclassical theory combined classical mechanics in a consistent way with the quantum principle of superposition (of probability amplitudes), thereby describing interference effects in the scattering cross section. (Bernstein's 1964 review [3] of semiclassical elastic scattering is still one of the best.) A companion paper by Ford et al. [4] showed how tunneling effects could also be incorporated in this semiclassical theory of elastic scattering, and a third paper [5] described several applications of the theory.

Ford and Wheeler [2] described four approximations to the quantum scattering amplitude that led to what we would today call the "primitive" semiclassical approximation for the differential scattering cross section (or angular distribution), which has the form

$$\sigma(0) = \left| \sum_k \sigma_k(0)^{1/2} \exp\left[i S_k(0)/\hbar\right] \right|^2 . \tag{1}$$

Here the index $k$ denotes different classical trajectories that lead to the same final scattering angle $0$; $\sigma_k(0)$ is the classical cross section associated with the $k$th trajectory, and $S_k(0)$ is a classical action integral along it. The only vestige of quantum mechanics in Eq. (1) is that $\hbar$ sets the units for measuring the classical action; everything else is from classical mechanics. Amazingly, however, this theory describes *all* the quantum effects in elastic scattering, at least qualitatively. (Near the rainbow angle – the boundary between a "classically allowed" and a "classically forbidden" region – the primitive version of the theory breaks down and more sophisticated treatments, uniform asymptotic methods, are needed to obtain a quantitative description.)

All of the quantum features seen in Ford and Wheeler's treatment of elastic scattering have found generalizations in inelastic and reactive scattering cross sections and other dynamical quantities. For example there are interference (and rainbow) effects in the rotational/vibrational state distributions following an inelastic/reactive collision (or after photodissociation), and generalized (or "dynamic") tunneling gives rise to "classically forbidden" processes, i.e., those for which there are no purely classical contributions. In these more general situations it has also been found that all quantum effects (also including symmetry-based selection rules and quantization of bounded motion itself) are ultimately a result of the superposition of probability amplitudes and are thus contained (at least qualitatively) in the semiclassical description.

Interestingly, if the fourth approximation in Ford and Wheeler's sequence (the stationary-phase approximation for the partial wave sum/integral) is not made, then one obtains a result for the differential cross section that would nowadays be called an initial value representation,

$$\sigma(0) = (2\pi\hbar \sin 0)^{-1}$$
$$\times \left| \int_{-\infty}^{\infty} db(pb)^{1/2} \exp\{i[S(b) + pb0]/\hbar\} \right|^2 , \tag{2}$$

where

$$p = \sqrt{2mE} .$$

In Eq. (2) $b$ is the initial impact parameter for the classical trajectory, and $S(b)$ the classical action for this trajectory; here there is no sum over multiple trajectories since the initial impact parameter determines a unique trajectory, but there is an integral over all initial impact parameters. Unlike the primitive semiclassical expression in Eq. (1), Eq. (2) provides a quantitative description of quantum effects in the cross section at all scattering angles (except *very* small angles, $0 \gtrsim \hbar/pa$, where $a$ is the length scale of the long-range part of the potential), describing interference effects on the "bright" side of the rainbow angle and (generalized) tunneling behavior on the "dark" side, and also the transition between them. Initial value representations are of great interest today as a practical way of applying semiclassical theory to complex molecular processes and thus describing quantum effects therein.

In summary, Ford and Wheeler's paper was important at the conceptual level, in showing the basic nature of how quantum theory alters the classical description (of elastic scattering), and also at the practical level, in lending credence to the use of classical mechanics for atomic and molecular processes and providing the understanding of where quantum effects are likely to be significant. Semiclassical theory is still useful today in both these contexts, and certain versions of it (e.g., the initial value representation) are leading to practical ways of including quantum effects in classical molecular dynamics simulations of complex processes.

*Acknowledgements.* This work has been supported by the Director, Office of Energy Research, Office of Basic Energy Sciences, Chemical Sciences Division of the U.S. Department of Energy under contract no. DE-AC03-76SF00098, Lawrence Berkeley National Laboratory, and also by the National Science Foundation under grant no. CHE97-32758.

## References

1. Karplus M, Porter RN, Sharma RD (1965) J Chem Phys 93: 3259
2. Ford K W, Wheeler J A (1959) Ann Phys (NY) 7: 259
3. Bernstein R B (1966) Adv Chem Phys 10: 75
4. Ford K W, Hill D L, Wakano M, Wheeler J A (1959) Ann Phys (NY) 7: 239
5. Ford K W, Wheeler J A (1959) Ann Phys (NY) 7: 287

Theor Chem Acc (2000) 103:238–241
DOI 10.1007/s002149900048

Theoretical
Chemistry Accounts
© Springer-Verlag 2000

*Perspective*

# Chemical building blocks in quantum chemical calculations. Perspective on "The density matrix in many-electron quantum mechanics I. Generalized product functions. Factorization and physical interpretation of the density matrices"

## McWeeny R (1959) Proc R Soc Lond Ser A 253: 242–259

János G. Ángyán

Laboratoire de Chimie théorique, UMR CNRS no. 7565, Institut Nancéien de Chimie Moléculaire,
Université Henri Poincaré, B.P.239, F-54506 Vandœuvre-lès-Nancy, France

Received: 26 February 1999 / Accepted: 7 April 1999 / Published online: 14 July 1999

**Abstract.** The group function theory described in the title paper of McWeeny is overviewed by pointing out its influence on different fields of theoretical chemistry, in particular its serving as a general framework for various forms of building blocks and local treatments of extended systems.

**Key words:** Generalized product function – Group function – Ab initio model potentials – Embedding – Spectator groups

## 1 Introduction

In the middle of the 1950s the laboratory of Professor Slater at MIT gathered several brilliant young theoretical chemists to work on the derivation of matrix elements of the Coulombic Hamiltonian with respect to arbitrary Slater determinants. This research activity led to some very important general results, connected essentially to the names P.-O. Löwdin and R. McWeeny, both of whom were working independently along similar lines. Löwdin published his results about density matrices, natural orbitals as well as on the matrix elements between nonorthogonal Slater determinants in an internal report [1] in 1954 and published it in *Physical Review* [2] in 1955. McWeeny's technical report, which summarized his lectures given in May 1954, appeared in May 1955, followed by a series of papers in the *Proceedings of the Royal Society* in 1956. However, the most significant publications on this subject followed only a few years later: the title article [3] in 1959, as well as an exhaustive presentation of the density matrix formalism [4] in 1960. As McWeeny noted in the preface of the 1955 technical note [5], the objective of this work was twofold. On the one hand, there was a "tendency towards elaboration and codification of existing methods in preparation for electronic digital computation", but on the other there was "an equal need to keep an underlying physical picture." Löwdin's contribution to density matrix theory has been proven to be essential, and this is the subject of another perspective in this New Century Issue. While Löwdin focussed mainly on the interpretation of the one-particle density matrix (e.g. natural orbitals), McWeeny gave a detailed interpretation of the physical meaning of two-particle density matrices. This analysis has been reiterated in McWeeny's excellent textbooks [6, 7], which have educated several generations of theoretical chemists.

In spite of the primordial importance of density matrix theory, in the present account I would like to concentrate on another closely related result of the title paper: the theory of generalized product functions. This theory beautifully demonstrates that fundamental chemical concepts are not necessarily contradictory to rigorous quantum mechanics. On the contrary, such concepts can be advantageously exploited for the design of powerful approximation schemes. In fact, McWeeny has always insisted on the importance of a conceptual approach to the problems of quantum chemistry, as witnessed, for example, by his work on two-particle density matrices [4] or on the interpretation of dispersion energies in terms of propagators [8].

Chemists have always refused to consider each individual molecule as a completely new object and have tried to rationalize physical and chemical properties in terms of basic building blocks. The experience of more than a century supported the existence of transferable structural units in molecules, such as bonds, functional groups, chromophores, etc. The idea of functional groups or chromophores also implies that only a small part of the whole molecule, which is responsible for the chemical and spectroscopic properties, is active, while the remaining parts of the molecule behave as spectator groups. Standard methods of quantum chemistry do not take into account this distinction and they usually treat all the electrons of the system on an equal footing.

McWeeny proposed a generalization of the usual antisymmetrized product of one-electron wave functions in terms of an antisymmetrized product of many-electron group functions. The extreme elegance of his formalism lies in the fact that it is able to encompass in a natural way the usual molecular orbital theory, the method of geminals, on the one hand, and on the other hand, it opened the way for different methods where chemically identified electron groups are treated separately.

## 2 Generalized product functions

A generalized product function, as defined in the title paper, is an antisymmetrized product of the $\Phi_{Aa}(1, \ldots, N_A)$ individually antisymmetric wave functions describing the groups $A$ in their electronic state $a$:

$$\Phi_{Aa,Bb,\ldots}(1, 2, \ldots, N)$$
$$= \hat{A}' \Phi_{Aa}(1, \ldots, N_A) \Phi_{Bb}(N_A + 1, \ldots, N_A + N_B) \ldots \ . \tag{1}$$

The intergroup antisymmetrizer, $\hat{A}'$ permutes electrons between different groups. The wave function (Eq. 1) is a straightforward generalization of a Slater determinant: the one-electron wave functions (spin-orbitals) are replaced here by $N_A$-electron wave functions.

As a generalization of the configuration interaction, the total wave function can be written as a linear combination of generalized products of the type (Eq. 1)

$$\Psi(1, 2, \ldots, N) = \sum_{a,b,\ldots} C_{a,b,\ldots} \Phi_{Aa,Bb,\ldots}(1, 2, \ldots, N) \ . \tag{2}$$

Although in a strict sense, this form is inappropriate for expanding the exact wave function, which should contain all the possible partitions of the electrons among the individual groups, the wave function (Eq. 2) seems to be a good initial approximation to describe systems composed of loosely coupled electron groups.

McWeeny assumed that the group functions are completely arbitrary, i.e. they can correspond to highly correlated accurate wave functions of the individual groups. The only restriction, imposed upon the group functions was the strong orthogonality condition:

$$\int d\tau_1 \Phi_{Rr}(1, \ldots, N_R) \Phi_{Ss}(1, \ldots, N_S) = \delta_{RS} \delta_{rs} \ , \tag{3}$$

i.e. it was required that the overlap integral, taken with respect to any of the electron (space and spin) coordinates, should vanish. The strong orthogonality condition simplifies tremendously the formalism. Although in several cases it is quite straightforward to respect it without losing accuracy, this restriction proved to be somewhat frustrating in several physical applications where it had to be relaxed [9].

By the virtue of the strong orthogonality condition the two-particle density matrix of the total system could

be expressed in terms of the one- and two-particle density matrices of the individual groups, which is one of the central results of the title paper [3]. This allowed the derivation of the analogs to the Slater rules for generalized product functions, and the matrix elements of the molecular Hamiltonian could be written in terms of intergroup Coulomb and exchange operators $\mathcal{J}^S(i)$ and $\mathcal{K}^S(i)$. The effective Hamiltonian of a group is

$$\mathcal{H}^R_{\text{eff}}(1, 2, \ldots, N_R)$$
$$= \sum_{i=1}^{N_R} \left[ h^R(i) + \sum_{S(\neq R)} [\mathcal{J}^S(i) - \mathcal{K}^S(i)] \right] + \sum_{i<j}^{N_R} g(i, j) \ . \tag{4}$$

The local Brillouin theorem of the usual self-consistent-field (SCF) theory was also generalized to the following form

$$\langle \Phi_{Aa} | \mathcal{H}^A_{\text{eff}} | \Phi_{Aa'} \rangle = 0 \ , \tag{5}$$

corresponding to the physical condition that the first-order polarization energy of the total system vanishes. The above statement is equivalent to a variational principle applied to the group energies, i.e. the expectation values of the effective group Hamiltonians (Eq. 4). Thus the electronic Schrödinger equation of the full extended system can be replaced by a set of effective equations of lower dimensionality for the group functions which are coupled by the intergroup Coulomb and exchange potentials.

The above ingredients of the theory would, in principle, allow us to identify various types of electron groups, relying on our physical or chemical intuition to design an efficient procedure for the treatment of large systems, paying special attention to those groups where the most important chemical events take place. Even if the optimization of the group functions according to the principles described above may seem to be straightforward, one should solve a crucial technical problem: how to maintain the (strong) orthogonality during their variation. The response to this question depends strongly on the nature of the electron groups.

## 3 Some specific applications of the group function method

The necessity of partitioning the electrons in a molecule into core and valence groups was recognized in the early days of quantum chemistry. A similar separation seemed to be necessary for planar molecules where the mobile $\pi$ electrons are responsible for most of the spectroscopic properties and chemical reactivity. It was only after several decades of the successful use of the Hückel method and with the appearance of more sophisticated SCF-type $\pi$-electron approximations that some attempts were made at establishing the fundamental theoretical basis of these methodologies relating them to ab initio calculations. In fact, the treatment of the $\sigma$-$\pi$ separation by Lykos and Parr [10] was a precursor to the group function idea. In this special case, the strong orthogonality condition is automatically fulfilled due to the different symmetries of these two groups of electrons.

In semiempirical molecular orbital theories one assumes a formal orthogonality of the underlying atomic orbital basis, which makes it relatively straightforward to implement group function theory. One can cite the PCILO (Perturbative configuration interaction with localized orbitals) method [11], which uses strictly localized two-electron bond orbitals to construct a zeroth-order wave function and perturbational corrections to describe the interaction between them. In the spirit of group function theory, many-electron groups are used to describe delocalized systems, such as $\pi$ fragments in the extended PCILO method [12]. Surján [13] developed an elegant formalism in terms of strictly localized two-electron wave functions, called geminals. The zeroth-order wave function, which is an antisymmetric product of strictly localized geminals, can be systematically improved by many-body perturbation techniques. Since each bond is described locally at a "full configuration interaction" level, well-localized chemical reactions (e.g. bond breaking) can be efficiently represented with such wave functions.

The separated electron pair concept, which was first proposed by Hurley et al. [14] and which was later referred to as antisymmetrized product of strongly orthogonal geminals (APSG) [15], is also a special case of the group function concept. This kind of wave function is qualitatively correct at all internuclear distances and it can be improved either perturbationally [16, 17] or variationally [18].

In the above-mentioned cases the strong orthogonality condition could be fulfilled. In fact the simplest way to ensure strong orthogonality is to construct the individual group functions from orbitals taken from different sets which are orthogonal to each other [19]. Although an orthogonalization of the basis set is always possible, it is not a solution for the physically most attractive situations, where one would like to bring together separate systems, each described by their own wave functions, to form a composite system and estimate their interaction energy. Similarly, there would be no advantage in optimizing the wave function of the active electrons in the effective field of the spectator groups if, as a consequence of the orthogonalization, we were constrained to work with the whole one-electron basis set.

The relaxation of the strong-orthogonality constraint was studied first by McWeeny and Sutcliffe [9], who derived density matrix expressions for the four-electron case. The general formulation for arbitrary closed shell groups [20] allowed the development of a general theory of intermolecular potentials at short and intermediary distances. In such weakly interacting systems it is not necessary to relax the electron groups in the field of their partners: the electrostatic, polarization and dispersion energy components are calculated from the isolated wave functions of the components.

In the case of more strongly interacting groups it is necessary to optimize the individual groups in the field of the others. This is typically the case for the valence electrons in the field of the atomic cores, for a functional group in the field of the substituents or a defect subsystem in an ionic or molecular crystal. In the course of such optimizations one should take precautions in order to avoid a variational collapse, i.e. the partial occupation of the subspace belonging to the partner electron group.

A possible solution to this problem was given by Kleiner and McWeeny [21] in the case of the core–valence separation. They derived an ab initio effective core potential, obtained directly from the core wave function, which can be brought to a form which is analogous to the Phillips–Kleinmann-type core pseudopotentials. This line of thought has been pursued in a systematic manner by several authors. For example, Huzinaga's building-block equations [22] allow one to obtain the Hartree–Fock orbitals for one subsystem, provided that the solutions for the other subsystems are known. Applying a series of approximations, ab initio model potentials can be derived for different situations, such as atomic cores [23, 24] spectator groups [25–28], or ions or molecules embedded in crystals [29–31].

Yet another way of circumventing the nonorthogonality problem is by replacing the strong-orthogonality condition by the strong-biorthogonality condition, i.e. using a biorthogonal basis to preserve the formal simplicity of the density matrix and effective group energy expressions for the composite system. Nevertheless, the group function variational principle cannot be applied in an identical fashion, and further constraints should be applied to restrict each group to its own variational subspace. Mehler [32] proposed imposing the condition that the intergroup overlap integrals remain constant throughout the variation of the group orbitals. The effective group equations he obtained are closely related to Huzinaga's equations as well as to the Adams–Gilbert equations used in obtaining a priori localized orbitals.

Group function theory may serve as a valuable guide in establishing a firm basis for quantum mechanical solvent effect theories. Microscopic solvent effect models can be introduced by arguing in terms of solute and solvent groups and by considering the solvent molecules as spectator groups [33]. The microscopic reaction field and some related models can also be derived from a perturbational ansatz applied to the coupled set of solute–solvent group function equations [34]. The status of some heuristic hypotheses used in the derivation of the energy derivative expressions of the solvent cavity model has been analyzed in light of the theory of generalized product functions [35].

## 4 Perspectives

There are essentially two possible ways to invoke chemical concepts such as atoms, functional groups, etc., in theoretical chemistry. The a posteriori way consists of performing an analysis of the wave function and extracting the properties associated with some rigorously defined objects corresponding to such subsystems. For example, the work of Bader [36], who succeeded in formulating the concept of the atoms in molecules in a quantum mechanically well-founded manner, exemplifies this kind of approach. The a priori way of using chemical concepts, i.e. to postulate the existence of such chemical fragments or building blocks, while still remaining on the solid ground of rigorous quantum mechanics, seems to be even more difficult. McWeeny's pioneering work on group functions demonstrated that such a plan might be conducted with success.

The ever-growing interest in accurate quantum chemical treatments of extended systems such as macromolecules and crystals would greatly benefit from models which are based on well-defined basic hypotheses and a controlled hierarchy of approximations. Group function theory offers such a framework and continues to be a reference in the design of new models ranging from various ab initio model potentials applied to core electrons, spectator groups, embedded atoms or solvent molecules, to local space treatments and even to mixed quantum–classical models. I am convinced that in the future the empirical ingredients of these various models will be replaced by parameters which are rigorously related to high-quality wave functions of the fragments constituting the complete system. There are some encouraging results in this respect [37], but still a great deal of work remains to be done.

*Acknowledgement.* I am indebted to P. Surján (Budapest) for the many interesting discussions on the subjects related to group function theory and for reading this manuscript.

# References

1. Löwdin P-O (1954) Q Prog Rep Solid-State Mol Theor Group MIT 4: 1
2. (a) Löwdin P-O (1955) Phys Rev 97: 1474; (b) Löwdin P-O (1955) Phys Rev 97: 1490; (c) Löwdin P-O (1955) Phys Rev 97: 1509
3. McWeeny R (1959) Proc R Soc Lond Ser A 253: 242
4. McWeeny R (1960) Rev Mod Phys 32: 335
5. McWeeny R (1955) Tech Rep Solid-State Mol Theor Group MIT 7: 1
6. McWeeny R, Sutcliffe BT (1969) Methods of molecular quantum mechanics, 1st edn. Academic Press, London
7. McWeeny R (1989) Methods of molecular quantum mechanics, 2nd edn. Academic Press, London
8. McWeeny R (1984) Croat Chem Acta 57: 865
9. McWeeny R, Sutcliffe B (1963) Proc R Soc Lond Ser A 273: 103
10. Lykos P, Parr R (1955) J Chem Phys 24: 1166
11. (a) Diner S, Malrieu J-P, Claverie P (1969) Theor Chim Acta 13: 1; (b) Diner S, Malrieu J-P, Claverie P (1969) Theor Chim Acta 13: 18
12. Boča R (1982) Theor Chim Acta 61: 179
13. Surján PR (1984) Phys Rev A 30: 43
14. Hurley A, Lennard-Jones J, Pople J (1953) Proc R Soc Lond Ser A 220: 446
15. Kutzelnigg W (1964) J Chem Phys 40: 3640
16. Kapuy E (1968) Theor Chim Acta 12: 397
17. Surján PR (1995) Int J Quantum Chem 55: 109
18. Kapuy E (1961) Acta Phys Acad Sci Hung 13: 461
19. Arai T (1960) J Chem Phys 33: 95
20. Amovilli C, McWeeny R (1990) Chem Phys 140: 343
21. Kleiner M, McWeeny R (1973) Chem Phys Lett 19: 476
22. Huzinaga S, McWilliam D, Cantu A (1973) Adv Quantum Chem 7: 187
23. Huzinaga S, Seijo L, Barandiarán Z, Klobukowski M (1987) J Chem Phys 86: 2132
24. Colle R, Fortunelli A, Salvetti O (1986) Mol Phys 57: 1305
25. Baxter C, Cook D (1996) Int J Quantum Chem 60: 173
26. von Arnim M, Peyerimhoff S (1993) Theor Chim Acta 87: 41
27. Ferenczy G (1995) Int J Quantum Chem 53: 485
28. Colle R, Curioni A, Salvetti O (1993) Theor Chim Acta 86: 451
29. Seijo L, Barandiarán Z (1996) Int J Quantum Chem 60: 617
30. Mejiás J, Oviedo J, Sanz J (1995) Chem Phys 191: 133
31. Luaña V, Pueyo L (1989) Phys Rev B 39: 1093
32. Mehler E (1977) J Chem Phys 67: 2728
33. Thole B, van Duijnen P (1980) Theor Chim Acta 55: 307
34. Ángyán J (1992) J Math Chem 10: 93
35. Bofill J (1996) J Mol Struct (Theochem) 371: 45
36. Bader R (1990) Atoms in molecules – a quantum theory. University of Oxford Press, Oxford
37. Tchougréeff AL (1999) Phys Chem Chem Phys 1: 1051

Theor Chem Acc (2000) 103:242–247
DOI 10.1007/s002149900061

Theoretical
Chemistry Accounts
© Springer-Verlag 2000

*Perspective*

# Perspective on "Some recent developments in the theory of molecular energy levels"

## Longuet-Higgins HC (1961) Adv Spectrosc 2:429–472. The geometric phase effect

**David R. Yarkony**

Department of Chemistry, Johns Hopkins University, Baltimore, MD 21218, USA

Received: 20 April 1999 / Accepted: 26 May 1999 / Published online: 9 September 1999

**Abstract.** Conical intersections of two potential-energy surfaces have the obvious, but important, effect of facilitating radiationless decay of the excited state. They. also have a less obvious, but potentially more general, impact on single and multisurface dynamics through the geometric phase effect. The geometric phase effect, the subject of this perspective, requires that the adiabatic electronic wavefunction, real-valued and continuous with respect to nuclear coordinates, change sign when transported along a closed loop – a pseudorotation path – surrounding a single point of conical intersection. This was discovered by Longuet-Higgins in 1958 and carefully described in papers between 1958 and 1963. In the title article Longuet-Higgins demonstrates, in the context of a theoretical exposition of the dynamic Jahn–Teller (and Renner–Teller) effects, the connection between conical intersections and the geometric phase effect, and establishes the consequences of the geometric phase effect in nuclear dynamics. Since that time appreciation of the importance of the geometric phase effect has increased enormously aided in no small measure by Berry's 1984 work that established the role of the geometric phase effect in general adiabatic processes. That work spurred research in areas well outside the realm of molecular spectroscopy/dynamics. However, recent work demonstrating the prevalence of conical intersections of two Born–Oppenheimer states of the same symmetry suggests that conical intersections and the geometric phase effect will be issues of significant importance in molecular/chemical dynamics in the next century.

**Key words:** Geometric phase effect – Conical intersection – Molecular Aharonov–Bohm effect – Berry's phase

# 1 The 'origin' of the geometric phase effect

This perspective is concerned with the geometric phase effect, a property of a real-valued adiabatic wavefunc-tion that results from a conical intersection involving the potential-energy surface of the state in question. It has profound and fundamental effects (modifying the basic Schrödinger equation) on nuclear motion. The geometric phase effect is known to be broadly applicable [1] as a result of Berry's highly influential 1984 work [2]; yet, the discovery of the geometric phase effect and the description of its potential impact, by Longuet-Higgins [3–5] 25 years earlier, are firmly rooted in the theoretical treatment of electronically nonadiabatic processes using the extension of the Born–Oppenheimer approximation, due to Born and Huang [6]. Interestingly although conical intersections were themselves known from the pioneering work of von Neumann and Wigner in 1929 [7] it was not until approximately 30 years later, long after the 1937 work of Jahn and Teller [8] on the effect bearing their names, that Longuet-Higgins [3–5] established the existence of the geometric phase effect in Jahn-Teller systems. The featured article, one of several papers Longuet-Higgins published between during the period 1958–1963 dealing with the geometric phase effect, presents a careful description of the geometric phase effect and its implications. Hence the choice of the article for this perspective.

## 1.1 Relation to the theory of nonadiabatic processes

Within the Born–Oppenheimer approximation the faster moving electrons create a potential-energy surface on which the nuclei move. This approximation forms the basis of our understanding of chemical bonding and molecular dynamics, see, for example, the article by Tully in this issue. Despite its central position in chemical theory, breakdowns of the Born–Oppenheimer approximation, electronically nonadiabatic processes, are ubiquitous. Nonadiabatic processes include charge-transfer [9] and electronic-quenching reactions, and many photochemical reactions are nonadiabatic [10], including some of nature's most basic processes: the initial radiationless energy transport step in photosynthesis [11] and the cis-trans isomerization that initiates the process of vision.

Most treatments of nonadiabatic processes do not abandon the Born–Oppenheimer idea of nuclear motion on potential-energy surfaces, instead in a nonadiabatic process the nuclei move on more than one Born–Oppenheimer potential-energy surface. Nonadiabatic transitions between potential-energy surfaces occur when the nuclei encounter a region where two potential-energy surfaces are in close proximity. Regions where the potential-energy surfaces intersect linearly, conical intersections, are of preeminent importance.

## 1.2 Conical intersections

A conical intersection of two potential-energy surfaces is depicted in Fig. 1 [12]. From this figure it is seen that for the excited state the conical intersection provides an efficient pathway for internal conversion: radiationless decay to the ground electronic state [10]. The effect of a conical intersection on ground-state dynamics is more subtle but is of enormous fundamental and practical importance. It might appear that for nuclear motion beginning on the ground-state potential-energy surface the conical intersection is of limited importance since it represents a mountain peak and therefore is comparatively inaccessible. In fact, for that reason nuclear motion on the ground-state potential-energy surface is not usually expected to sample the conical intersection directly; however the conical intersection can effect nuclear motion indirectly through the geometric phase effect [13]. In this case the particle need only traverse a closed path around the conical intersection. Proximity to the conical intersection is not the issue.

## 1.3 Geometric phase effect

The geometric phase effect, the signature property of a conical intersection, requires that the adiabatic electronic wavefunction (chosen to be real-valued and continuous with respect to nuclear coordinates) change sign when

transported along a closed loop – a pseudorotation path [14] – surrounding (only) a single point of conical intersection. The pseudorotation path for an $X_3$ molecule was presented in the featured work. Figure 2 presents a one-third of the pseudorotation path for $H_3$, following Ref. [15]. In the featured work Longuet-Higgins explains the mixing of the two wavefunctions, that are degenerate at the conical intersection point, along the pseudorotation path that leads to the sign change. This result can be understood using a simple Jahn–Teller model Hamiltonian. (see Sect. 3). Degeneracy is essential here. A nondegenerate wavefunction cannot exhibit the geometric phase effect since it must remain normalized, i.e., the leading coefficient in its perturbation expansion about a point $\mathbf{R}$ must always be large.

## 1.4 Implications: dynamic Jahn–Teller effect

Longuet-Higgins pointed out that the sign change in the electronic wavefunctions has profound implications for the associated nuclear Schrödinger equation. To see this note that since the electronic wavefunction changes sign along a circular path around the conical intersection, the vibrational wavefunction must compensate by also changing sign so that the overall wave function returns to itself after traversing the loop. Let $\theta$ denote the angle that transports nuclear configuration around the loop. Then a basis function for the nuclear wavefunction of the form $\Phi_m(\theta) = \exp im\theta$ is unacceptable since $\Phi_m(\theta + 2\pi) = \Phi_m(\theta)$. Rather functions of the form $\Phi_{m+1/2}(\theta) = \exp i(m + 1/2)\theta$ are required since $\Phi_{m+1/2}(\theta + 2\pi) = -\Phi_{m+1/2}(\theta)$. This is the origin of the well-known half-integer quantization in the dynamic Jahn–Teller effect [4].

## 2 Subsequent work

### 2.1 Molecular Aharonov–Bohm effect

The idea in the preceding section was subsequently used to consider to the single adiabatic state problem [16], where it is known as the molecular Aharonov–Bohm (MAB) effect [17] owing to its analogy with a magnetic effect discussed by Aharonov and Bohm [18]. In MAB theory the electronic wavefunction is altered to remove the sign change. It is multiplied by a phase factor so that it returns to itself after traversing a closed loop. The effect of this modification is to introduce into the nuclear Schrödinger equation a vector potential, analogous to the magnetic vector $\mathbf{A}$ in the standard semiclassical treatment of radiation [19]. This term must be included whether or not the system actually encounters the conical intersection. Thus through the geometric phase effect a conical intersection can exhibit a highly non local influence on nuclear dynamics.

Recently this idea has received much attention in the chemical physics community, brought about by accurate measurements [20–22] of rotational–vibrational energy transfer in collisions of H with $H_2$. Kuppermann [23–26]

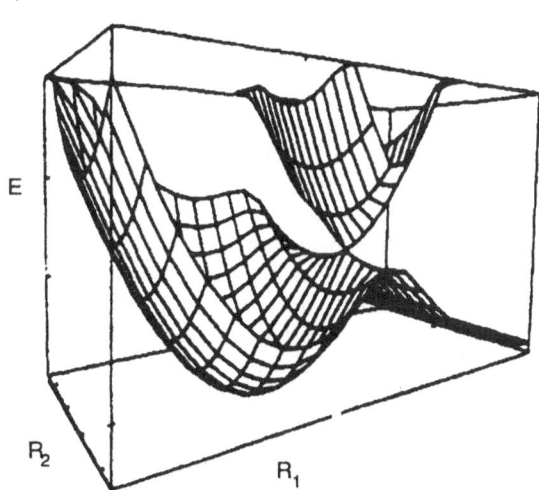

**Fig. 1.** A conical intersection of two potential-energy surfaces

**Fig. 2.** Symmetry-unique portion of the pseudorotation path for $H_3$, *outer plates*, around the $D_{3h}$ conical intersection point, *center plates*, as originally discussed by Longuet-Higgins [4]. Each point on the path is the constructed from ($\cos\theta\,Q_y + \sin\theta\,Q_x$). Present drawing from Ref. [15]. The third internal coordinate, not shown, is the $D_{3h}$-preserving ring breathing mode, $Q_s$

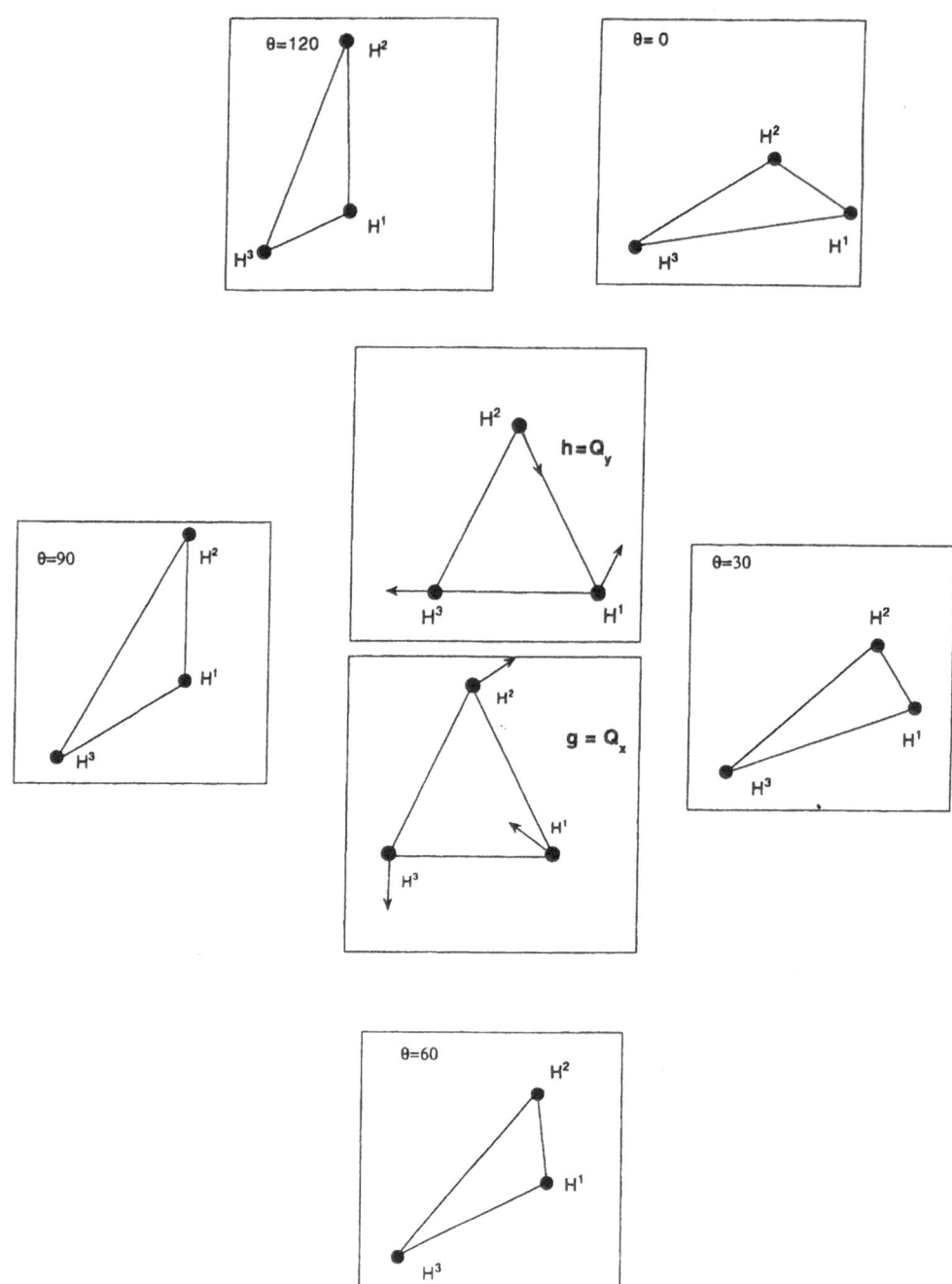

has argued that it is precisely the geometric phase effect that needs to be included in the adiabatic description of the reaction $H + H_2(v = 1, j) \rightarrow H + H_2(v = 0, j')$ to obtain agreement with experiment.

## 2.2 The quest for the geometric phase effect in optical spectroscopy

Berry's 1984 work led to manifestations of the geometric phase effect being detected outside the area of optical spectroscopy discussed by Longuet-Higgins. Observation of the geometric phase effect in molecular spectroscopy has proved surprisingly difficult. One reason for this difficulty is suggested in Sect. 3.

In 1986 the first apparent observation of a half-integer pseudorotation quantum number was reported [27]. The B band of $Na_3$ (interpreted as the $3^2E' - X^2E'$ transition) was observed through resonant two-photon ionization spectroscopy. A fit to the spectral data was obtained with a linear Jahn–Teller model, indicating the existence of the geometric phase effect; however, less than 2 years later ab initio calculations called this result into question, indicating that the transition in question was to the lowest, nondegenerate component, of a pseudo Jahn–Teller triple of electronic states [28]. Recent investigations at rotational resolution using resonant two-photon ionization and optical–optical double resonance spectroscopy [29, 30] have confirmed the absence of a geometric phase effect in this state. In fact it was not until

1998 that the first unambiguous demonstration of the geometric phase effect in Na$_3$ was reported on the basis of an analysis of the A$^2$E'' → X$^2$E' emission [31].

In the above-mentioned Na$_3$ studies the observation of the geometric phase effect was indirect, i.e., it is inferred by fitting the observed spectra to a detailed spectroscopic Hamiltonian that evinces the geometric phase effect. Preferable would be a direct observation, i.e., the observation of a 'spectral signature' of the geometric phase effect; however, this remains an elusive goal. It has been suggested the geometric phase effect could be "directly" observed using short-duration, coherent light pulses. In the experiment a nondegenerate ground state would be pumped to a degenerate excited state exhibiting the geometric phase effect and allowed to evolve until a second coherent pulse causes interference half a cycle later. The resulting interference would be evident in the fluorescence emission from the excited state [32–35]. This theoretical proposal of Cina has yet to be observed experimentally.

## 3 Current issues/mathematical formulation

Although the discovery of geometric phase effect is over 40 years old, it and related issues remain areas of active research. Below we illustrate in more precise terms the nature of the geometric phase effect as described above, and briefly outline some current issues in this field.

We begin by describing the origin of the geometric phase effect using the quadratic Jahn–Teller Hamiltonian of Longuet-Higgins [4]:

$$H^e(Q) = m(Q)I + K^{(2)}(Q)\sigma_z + \Lambda^{(2)}(Q)\sigma_x \qquad (1)$$

Here vectors (matrices) are written in boldface (bold italic), the $Q = (Q_x, Q_y, Q_s)$ are the three standard Jahn–Teller coordinates, see Fig. 2, $\sigma_w$ are the Pauli matrices, $K^{(l)}(Q) = \sum_{i=1}^l k_i(Q)$, $\Lambda^{(l)}(Q) = \sum_{i=1}^l \lambda_i(Q)$,

$$k_1(Q)/g_1(Q_s) = Q_x \equiv \rho\cos\theta$$
$$\lambda_1(Q)/g_1(Q_s) = Q_y \equiv \rho\sin\theta \qquad (2a)$$

$$k_2(Q)/g_2(Q_s) = Q_x^2 - Q_y^2 = \rho^2\cos 2\theta$$
$$\lambda_2(Q)/g_2(Q_s) = -2Q_xQ_y = -\rho^2\sin 2\theta \qquad (2b)$$

and $m(Q)$ is arbitrary in the present context. $g_1(Q_s)$ and $g_2(Q_s)$ are the $Q_s$-dependent parameters of the model. Note from Eq. (2a, b) that $H^e$ has a conical intersection for $Q_x = Q_y = 0$ ($\rho = 0$).

$H^e$ has eigenvalues [36]:

$$E_\pm(\rho, \theta, Q_s)$$
$$= m(Q) \pm \rho|g_1(Q_s)|$$
$$\times \sqrt{[1 - \rho/\rho_c(Q_s)]^2 \pm 2\rho/\rho_c(Q_s)(\cos 3\theta \pm 1)} . \qquad (3)$$

Its eigenfunctions, the adiabatic electronic states, are

$$[\Psi^+(r; Q), \Psi^-(r; Q)] = [(\psi_1(r; Q), \psi_2(r; Q))u(\Theta(Q))]^\dagger \qquad (4a)$$

where $r \equiv (r_1, \ldots, r_{N^e})$ are the coordinates of the $N^e$ electrons, $\psi_1$ and $\psi_2$ are the basis for $H^e$, and $u[\Theta(Q)]$ is given by

$$u[\Theta(Q)] = \begin{pmatrix} \cos\Theta & \sin\Theta \\ -\sin\Theta & \cos\Theta \end{pmatrix} \qquad (4b)$$

$$\tan 2\Theta(\rho, \theta, Q_s) = \frac{\sin\theta \mp (\rho/\rho_c)\sin 2\theta}{\cos\theta \pm (\rho/\rho_c)\cos 2\theta} . \qquad (4c)$$

Here $\rho_c(Q_s) \equiv |g_1(Q_s)/g_2(Q_s)|$, and the upper (lower) signs in the radical in Eq. (3) and in Eq. (4c) are used if $g_1g_2$ is $> (<)0$.

From Eq. (4c) for small $\rho$, $\Theta = \theta/2$. In this case after a closed loop has been traversed, $\theta \to \theta + 2\pi$ but $\Theta \to \Theta + \pi$; therefore, from Eq. (4b) we see that $[\Psi^+(r; \rho, \theta + 2\pi), \Psi^-(r; \rho, \theta + 2\pi)] = -[\Psi^+(r; \rho, \theta), \Psi^-(r; \rho, \theta)]$. This is the geometric phase effect!

The situation is different for large $\rho$ for which $\Theta = -\theta$ and so $[\Psi^+(r; \rho, \theta + 2\pi), \Psi^-(r; \rho, \theta + 2\pi)] = +[\Psi^+(r; \rho, \theta), \Psi^-(r; \rho, \theta)]$. The reason for this change is the existence of three more conical intersections at $\rho = \rho_c$ [36]. These additional conical intersections complicate the observation of the geometric phase effect. For further discussion of this point see Ref. [37].

As noted previously the key feature of the geometric phase effect is that it leads to a modification of the nuclear Schrödinger equation. This modification is described below.

The total wave function for a molecule can be expanded as

$$\Psi_k^T(r, R) = \sum_{I=1}^{N^a} \tilde{\chi}_I^k(R)\tilde{\Psi}_I^a(r; R) , \qquad (5)$$

where $R \equiv (R_1, \ldots, R_{N^{nuc}})$ are the coordinates of the $N^{nuc}$ nuclei, in a space fixed frame, $\tilde{\chi}_I^k(R)$ are the nuclear wave functions and $\tilde{\Psi}_I^a(r; R)$ are the adiabatic electronic wavefunctions which satisfy the standard (Coulombic) electronic Schrödinger equation

$$[H^e(r; R) - E_I^a(R)]\tilde{\Psi}_I^a(r; R) = 0 \qquad (6)$$

with the total Hamiltonian given by $H^T(r, R) = \sum_{\alpha=1}^{N^{nuc}} (\frac{1}{2M_\alpha})p_\alpha \, p_\alpha + H^e(r; R)$.

The difference between $\Psi_I^a(r; R)$ and $\tilde{\Psi}_I^a(r; R) \equiv \exp iA^I(R)\Psi_I^a(r; R)$ is crucial. As a consequence of the geometric phase effect, at each $R$ the sign of $\Psi_I^a(r; R)$ is arbitrary and path-dependent. Such a function is called double-valued, since it can be either $\pm\Psi_I^a(r; R)$; however, the total wave function must be single-valued and so as noted previously the geometric phase effect must be compensated for by either choosing the phase factor $\exp iA^I(R)$ such that $\tilde{\Psi}_I^a(r; R)$ is single-valued, or by the phase properties of $\tilde{\chi}_I^k(R)$. The only requirement on $A^I(R)$, sometimes referred to as the Longuet-Higgins phase [38], is that it change sign when transported in a closed loop around a conical intersection. The determination of $A^I(R)$ has been a topic of recent discussion [39, 40]. The second option is less frequently used since in general, although not for the Jahn–Teller effect, it is more difficult to implement.

Using Eq. (5) in the total Schrödinger equation $(H^T - E_k)\Psi^T_k = 0$ and taking $\langle\tilde{\Psi}^a_l(\mathbf{r};\mathbf{R})|$ the nuclear Schrödinger equation becomes

$$\left[\sum_{\alpha=1}^{N^{nuc}}\frac{1}{2M_\alpha}\left[(\mathbf{p}_\alpha + \mathbf{A}^l_\alpha)^2 - (\mathbf{p}_\alpha + \mathbf{A}^l_\alpha)\cdot i\mathbf{f}^{ll}_\alpha(\mathbf{R})\right.\right.$$
$$\left.- i\mathbf{f}^{ll}_\alpha(\mathbf{R})\cdot(\mathbf{p}_\alpha + \mathbf{A}^l_\alpha)\right] + \bar{E}_l(\mathbf{R}) - E_k\right]\tilde{\chi}^k_l(\mathbf{R})$$
$$= -\sum_{J(\neq l)}^{N^a}\exp[iA_{Jl}(\mathbf{R})]\sum_{\alpha=1}^{N^{nuc}}\left[\frac{1}{2M_\alpha}\{k^{lJ}_\alpha(\mathbf{R}) + (\mathbf{p}_\alpha + \mathbf{A}^J_\alpha)\right.$$
$$\left.\cdot i\mathbf{f}^{lJ}_\alpha(\mathbf{R}) + i\mathbf{f}^{lJ}_\alpha(\mathbf{R})\cdot(\mathbf{p}_\alpha + \mathbf{A}^J_\alpha)\}\right]\tilde{\chi}^k_J(\mathbf{R}) \ , \tag{7}$$

where $\mathbf{A}^l_\alpha(\mathbf{R}) = \nabla_\alpha A^l(\mathbf{R})$, $A_{Jl}(\mathbf{R}) = A_J(\mathbf{R}) - A_l(\mathbf{R})$, the first derivative couplings or simply the derivative couplings [41], $\mathbf{f}^{ll}_\alpha(\mathbf{R})$, are given by

$$\mathbf{f}^{lJ}_\alpha(\mathbf{R}) = \langle\Psi_J(\mathbf{r};\mathbf{R})|\nabla_\alpha\Psi_l(\mathbf{r};\mathbf{R})\rangle_r \tag{8}$$

$$\bar{E}_l(\mathbf{R}) = E_l(\mathbf{R}) + \sum_{\alpha=1}^{N^{nuc}}\frac{k^{ll}_\alpha}{2M_\alpha} \tag{9}$$

and

$$k^{lJ}_\alpha(\mathbf{R}) = \langle\nabla_\alpha\Psi_J(\mathbf{r};\mathbf{R})\,|\,\nabla_\alpha\Psi_l(\mathbf{r};\mathbf{R})\rangle_r$$
$$= \sum_{K=1}^{N^{CSF}}\mathbf{f}^{KJ}_\alpha(\mathbf{R})^*\cdot\mathbf{f}^{Kl}_\alpha(\mathbf{R}) \ . \tag{10}$$

If $\Psi^a_l$ is chosen to be real then $\mathbf{f}^{ll}_\alpha(\mathbf{R}) = 0$, otherwise it is purely imaginary.

Of particular interest is the adiabatic, or MAB, limit $N^a = 1$ where $\Psi^T_k = \exp[iA^l(\mathbf{R})]\Psi^a(\mathbf{r};\mathbf{R})\tilde{\chi}^k_l(\mathbf{R})$ and $\Psi^a$ is real-valued. The nuclear Schrödinger equation is

$$\left[\sum_{\alpha=1}^{N^{nuc}}\frac{+1}{2M_\alpha}\left[(\mathbf{p}_\alpha + \mathbf{A}^l_\alpha)^2\right] + E_l(\mathbf{R}) - E_k\right]\tilde{\chi}^k_l(\mathbf{R}) = 0 \ . \tag{11}$$

This clearly demonstrates that the effect of the conical intersection is to add a vector potential, analogous to the vector potential $\mathbf{A}$ in semiclassical electromagnetic theory, to the nuclear Schrödinger equation in the adiabatic limit. Calculations based on Eq. (11) have recently been reported [39, 42, 43].

For the Jahn–Teller problem $N^a = 2$ is required. The evaluation of $k^{lJ}_\alpha(\mathbf{R})$ in Eq. (7) is quite costly [44, 45]. It is therefore convenient to approximate $k^{lJ}_\alpha(\mathbf{R})$ by the second term in Eq. (10) with $N^{CSF} = N^a$. A somewhat stronger version of this assumption is found in the featured work. In this case for real $\Psi^a_l(\mathbf{r};\mathbf{R})$ and assuming $\mathbf{A}^J = \mathbf{A}^l = \mathbf{A}$, Eq. (7) becomes

Equation (12) illustrates the standard interpretation of $\mathbf{f}^{lJ}_\alpha$ as coupling the two electronic states in question. It is therefore interesting to observe the effect of the following transformation to a complex electronic basis used by Longuet-Higgins in the featured article (see also Ref. [46]).

$$\Psi^T \equiv \tilde{\Psi}_+\tilde{\chi}_- + \tilde{\Psi}_-\tilde{\chi}_+ \ , \tag{13a}$$

where

$$\tilde{\Psi}_+ = (1/\sqrt{2})(\tilde{\Psi}^a_l + i\tilde{\Psi}^a_J) \quad \tilde{\Psi}_- = (1/\sqrt{2})(\tilde{\Psi}^a_l - i\tilde{\Psi}^a_J)$$
$$\tilde{\chi}_- = (1/\sqrt{2})(\tilde{\chi}_l - i\tilde{\chi}_J) \quad \tilde{\chi}_+ = (1/\sqrt{2})(\tilde{\chi}_l + i\tilde{\chi}_J) \ . \tag{13b}$$

Then $\tilde{\chi}_\pm$ satisfy (see also Ref. [46])

$$\begin{pmatrix} h(\mathbf{p}_\alpha,\mathbf{A}_\alpha,\mathbf{f}^{lJ}_\alpha) + \bar{E}^{avg}_{lJ}(\mathbf{R}) - E_k & (E^a_l - E^a_J)/2 \\ (E^a_l - E^a_J)/2 & h(\mathbf{p}_\alpha,\mathbf{A}_\alpha,-\mathbf{f}^{lJ}_\alpha) + \bar{E}^{avg}_{lJ}(\mathbf{R}) - E_k \end{pmatrix}$$
$$\times \begin{pmatrix} \tilde{\chi}^k_-(\mathbf{R}) \\ \tilde{\chi}^k_+(\mathbf{R}) \end{pmatrix} = \begin{pmatrix} 0 \\ 0 \end{pmatrix} \ , \tag{14a}$$

where $\bar{E}^{avg}_{lJ}(\mathbf{R}) = [\bar{E}^a_l(\mathbf{R}) + \bar{E}^a_J(\mathbf{R})]/2$ and

$$h(\mathbf{p}_\alpha,\mathbf{A}_\alpha,\mathbf{f}^{lJ}_\alpha) = \sum_{\alpha=1}^{N^{nuc}}\frac{1}{2M_\alpha}\left[(\mathbf{p}_\alpha + \mathbf{A}_\alpha)^2 - (p_\alpha + \mathbf{A}_\alpha)\cdot\right.$$
$$\left.i\mathbf{f}^{lJ}_\alpha(\mathbf{R}) - i\mathbf{f}^{lJ}_\alpha(\mathbf{R})\cdot(\mathbf{p}_\alpha + \mathbf{A}_\alpha)\right] \ .$$

Using the replacement $\tilde{\chi} = \chi\exp(-iA)$ the total wavefunction can be rewritten in terms of double-valued functions as $\Psi^T_k \equiv \Psi_+\chi^k_- + \Psi_-\chi^k_+$. In terms of $\chi_\pm$ Eq. (14a) becomes

$$\left[\sum_{\alpha=1}^{N^{nuc}}\frac{1}{2M_\alpha}(\mathbf{p}_\alpha + \mathbf{f}^{lJ}_\alpha)^2 + E_l(\mathbf{R}) - E_k\right]\chi^k_-$$
$$+ i\left[(E^a_l - E^a_J)/\sqrt{2}\right]\chi^k_2 = 0$$
$$- i\left[(E^a_l - E^a_J)/\sqrt{2}\right]\chi^k_2 + \left[\sum_{\alpha=1}^{N^{nuc}}\frac{1}{2M_\alpha}(\mathbf{p}_\alpha - \mathbf{f}^{lJ}_\alpha)^2\right.$$
$$\left.+ E_J(\mathbf{R}) - E\right]\chi^k_- = 0. \tag{14b}$$

The most interesting aspect of Eq. (14a, b) is that the derivative coupling is transferred from the off-diagonal term to the diagonal term and the coupling is purely of the potential form.

Note the similarity between the term involving $(\mathbf{p}_\alpha + \mathbf{f}^{lJ}_\alpha)^2$ in Eq. (14b) and that involving $(\mathbf{p}_\alpha + \mathbf{A}^l_\alpha)^2$ in Eq. (11). Thus it is interesting to ask whether the replacement $\mathbf{A}^l_\alpha \to \mathbf{f}^{lJ}_\alpha$ is appropriate. In a sense it is, but

$$\begin{pmatrix} \sum_{\alpha=1}^{N^{nuc}}\frac{1}{2M_\alpha}(\mathbf{p}_\alpha + \mathbf{A}_\alpha)^2 + \bar{E}_l(\mathbf{R}) - E_k & \sum_{\alpha=1}^{N^{nuc}}\frac{-1}{2M_\alpha}\left[(\mathbf{p}_\alpha + \mathbf{A}_\alpha)\cdot i\mathbf{f}^{lJ}_\alpha(\mathbf{R}) + i\mathbf{f}^{lJ}_\alpha(\mathbf{R})\cdot(\mathbf{p}_\alpha + \mathbf{A}_\alpha)\right] \\ \sum_{\alpha=1}^{N^{nuc}}\frac{-1}{2M_\alpha}\left[(\mathbf{p}_\alpha + \mathbf{A}_\alpha)\cdot i\mathbf{f}^{Jl}_\alpha(\mathbf{R}) + i\mathbf{f}^{Jl}_\alpha(\mathbf{R})\cdot(\mathbf{p}_\alpha + \mathbf{A}_\alpha)\right] & \sum_{\alpha=1}^{N^{nuc}}\frac{1}{2M_\alpha}(\mathbf{p}_\alpha + \mathbf{A}_\alpha)^2 + \bar{E}_J(\mathbf{R}) - E_k \end{pmatrix}\begin{pmatrix} \tilde{\chi}^k_l(\mathbf{R}) \\ \tilde{\chi}^k_J(\mathbf{R}) \end{pmatrix}$$
$$= \begin{pmatrix} 0 \\ 0 \end{pmatrix} \ . \tag{12}$$

ONLY when certain approximations are valid. To see this define a candidate $\bar{A}^I$ by

$$\bar{A}^I[\alpha, C_\rho(\mathbf{R}), \mathbf{f}^{IJ}] = \int\limits_{\alpha, C_\mu} \mathbf{f}^{IJ}(\mathbf{R}') \cdot d\mathbf{R}' \;, \qquad (15)$$

where $C_\rho(\mathbf{R})$ is a circle of radius $\rho$ centered at $\mathbf{R}$, $[\alpha, C_\rho(\mathbf{R})]$ denotes an arc of $\alpha^\circ$ on $C_\rho(\mathbf{R})$ and $\mathbf{f}^{IJ} = (\mathbf{f}_1^{IJ}, \ldots, \mathbf{f}_{N^{nuc}}^{IJ})$. Then we have the following [47]:

$$\bar{A}^I[2\pi, C_\rho(\mathbf{Q}), \mathbf{f}^{IJ}] \xrightarrow[\rho \to 0]{} \pi \quad \text{if } \mathbf{Q} = \mathbf{R}_x \qquad (16a)$$

$$\bar{A}^I[2\pi, C_\rho(\mathbf{Q}), \mathbf{f}^{IJ}] \xrightarrow[\rho \to 0]{} 0 \quad \text{if } \mathbf{Q} \neq \mathbf{R}_x \;, \qquad (16b)$$

where $\mathbf{R}_x$ is a point of conical intersection of states $I$ and $J$. This is the requisite property for an $\mathbf{A}^I$. However, it must hold for circles of arbitrary $\rho$. For finite $\rho$ Eq. (16a, b) is only an approximation since $\mathbf{f}^{IJ}$ has a nonvanishing curl [48]. Within the two-state approximation, $N^{CSF} = N^a$, the curl does in fact vanish [49]. Thus in this case the identification of $\bar{A}^I$ with $\mathbf{A}^I$ is appropriate; although, for $N^{CSF} > N^a$ it is only an approximation [50].

One way around this dilemma is to use an approximate derivative coupling that has the same singularity as $\mathbf{f}^{IJ}$, so that Eq. (16a, b) is satisfied, but is explicitly the gradient of a scalar so that its curl vanishes. The details of such a construction – applicable in favorable situations – can be found in Ref. [40].

## 4 The future

The geometric phase effect described in the title work results from a symmetry-required conical intersection, a comparatively rare occurrence in molecular systems. However the recent development of efficient algorithms [51, 52] for locating accidental conical intersections of two states of the same symmetry, an issue in which Longuet-Higgins also played a seminal role [53], has shown that conical intersections, rather than being rare occurrences, are a common phenomenon. The absence of point group symmetry as a prerequisite for these intersections and the recent discovery of unexpected loci for the seams of conical intersection [54] suggest that in the future conical intersections in general and the geometric phase effect in particular will be encountered often and in unexpected situations.

*Acknowledgements.* D.R.Y. is pleased to acknowledge the support provided by grants from AFOSR, NSF and DOE that made this work possible.

## References

1. Shapere A, Wilczek F (eds) (1989) Geometric phases in physics. World Scientific, Singapore
2. Berry MV (1984) Proc R Soc Lond Ser A 392: 45
3. Longuet-Higgins HC, Öpik U, Pryce MHL, Sack RA (1958) Proc Roy Soc Lond Ser A 244: 1
4. Longuet-Higgins HC (1961) Adv Spectrosc 2: 429
5. Herzberg G, Longuet-Higgins HC (1963) Discuss Faraday Soc 35: 77
6. Born M, Huang K (1954) Dynamical theory of crystal lattices. Oxford University Press, Oxford
7. von Neumann J, Wigner E (1929) Phys Z 30: 467
8. Jahn HA, Teller E (1937) Proc R Soc Lond Ser A 161: 220
9. Baer M, Ng CY (eds) (1991) State-selected and state-to-state ion-molecule reaction dynamics: parts 1 and 2, Vol 82. Wiley, New York
10. Michl J, Bonacic-Koutecky V (1990) Electronic aspects of organic photochemistry. Wiley, New York
11. Hu X, Schulten K (1997) Phys Today 50: 28
12. Kash PW, Waschewsky GCG, Morss RE, Butler LJ, Francl MM (1994) J Chem Phys 100: 3463
13. Mead CA (1980) J Chem Phys 72: 3839
14. Berry RS (1960) J Chem Phys 32: 933
15. Yarkony DR (1996) J Phys Chem 100: 18612
16. Mead CA, Truhlar DG (1979) J Chem Phys 70: 2284
17. Mead CA (1980) Chem Phys 49: 23
18. Aharonov Y, Bohm B (1959) Phys Rev 115: 485
19. Schiff LI (1960) Quantum mechanics. McGraw-Hill, New York
20. Kliner DAV, Rinnen K-D, Zare RN (1990) Chem Phys Lett 166: 107
21. Kliner DAV, Adelman DE, Zare RN (1991) J Chem Phys 95: 1648
22. Neuhauser D, Judson RS, Kouri DJ, Adelman DE, Shafer NE, Kliner DAV, Zare RN (1992) Science 257: 519
23. Kuppermann A, Wu Y-SM (1993) Chem Phys Lett 205: 577
24. Wu Y-SM, Kuppermann A (1993) Chem Phys Lett 201: 178
25. Wu Y-SM, Kuppermann A (1995) Chem Phys Lett 235: 105
26. Kuppermann A (1996) In: Wyatt RE, Zhang JZH (eds) Dynamics of molecules and chemical reactions. Dekker, New York, pp 411–472
27. Delacrétaz G, Grant ER, Whetten RL, Wöste L, Zwanziger JW (1986) Phys Rev Lett 56: 2598
28. Cocchini F, Upton TH, Andreoni W (1988) J Chem Phys 88: 6068
29. Ernst WE, Rakowsky S (1995) Phys Rev Lett 74: 58
30. Ernst WE, Rakowsky S (1995) Ber Bunsenges Phys Chem 99: 441
31. Busch HV, Dev V, Eckel H-A, Kasahara S, Wang J, Demtröder W, Sebald P, Meyer W (1998) Phys Rev Lett 81: 4584
32. Romero-Rochín V, Cina JA (1989) J Chem Phys 91: 6103
33. Cina JA, Romero-Rochin V (1990) J Chem Phys 93: 3844
34. Cina JA (1991) Phys Rev Lett 66: 1146
35. Cina JA, Smith TJ, Romero-Rochín V (1993) In: Prigogine I, Rice SA (eds) Advances in Chemical Physics, Vol 83. Wiley, p 1
36. Zwanziger JW, Grant ER (1987) J Chem Phys 87: 2954
37. Sadygov RG, Yarkony DR (1999) J Chem Phys 110: 3639
38. Baer M (1997) J Chem Phys 107: 2694
39. Kendrick B, Pack RT (1996) J Chem Phys 104: 7502
40. Yarkony DR (1996) J Chem Phys 110: 701
41. Lengsfield BH, Yarkony DR (1992) In: Baer M, Ng C-Y (eds) State-selected and state-to-state ion-molecule reaction dynamics: part 2 theory, Vol 82. Wiley, New York, pp 1–71
42. Kendrick B, Pack RT (1996) J Chem Phys 104: 7475
43. Kendrick B (1997) Phys Rev Lett 79: 2431
44. Saxe P, Yarkony DR (1987) J Chem Phys 86: 321
45. Jensen JO, Yarkony DR (1988) J Chem Phys 89: 3853
46. Baer M (1997) J Chem Phys 107: 10662
47. Yarkony DR (1997) J Phys Chem A 101: 4263
48. Mead CA, Truhlar DG (1982) J Chem Phys 77: 6090
49. Baer M (1975) Chem Phys Lett 35: 112
50. Mead CA (1983) J Chem Phys 78: 807
51. Radazos IN, Robb MA, Bernardi MA, Olivucci M (1992) Chem Phys Lett 197: 217
52. Yarkony DR (1996) Rev Mod Phys 68: 985
53. Longuet-Higgins HC (1975) Proc R Soc Lond Ser A 344: 147
54. Yarkony DR (1998) Acc Chem Res 31: 511

Theor Chem Acc (2000) 103:248–251
DOI 10.1007/s002149900034

Theoretical
Chemistry Accounts
© Springer-Verlag 2000

*Perspective*

# Perspective on "The physical nature of the chemical bond"

## Ruedenberg K (1962) Rev Mod Phys 34: 326–376

**Mark S. Gordon**[1], **Jan H. Jensen**[2]

[1] Department of Chemistry, Iowa State University, Ames, IA 50011, USA
[2] Department of Chemistry, University of Iowa, Iowa City, IA 52242, USA

Received: 25 February 1999 / Accepted: 25 March 1999 / Published online: 14 July 1999

**Key words:** Covalent bonding – Kinetic-energy
pressure – Potential-energy section – Interference density

## 1 Introduction

"I believe that a good theory combines mathematical rigor with physical interpretation." With these words, Klaus Ruedenberg began a summary of the highlights of his scientific career in 1996. Many seminal papers by this giant in the field of quantum chemistry serve to illustrate this philosophy; none accomplish the task with more beauty and insight than "The physical nature of the chemical bond" [1]. This paper and those that followed it [2–4] represent the first rigorous, first-principles analysis of the fundamental quantum mechanical origins of covalent bonding which, as Mulliken [5] noted in 1977, "are commonly misunderstood". With the insight that has become his trademark, Ruedenberg combined four fundamental precepts – the virial theorem, wave–particle duality, the variational principle and the decomposition of the total energy into kinetic and potential components – to devise a model that is both simple and broadly applicable. This paper clearly establishes the important point that it is "possible to extract from a rigorous wavefunction (or a bonafide approximation to it), in a quantitative fashion, a partitioning of the energy which justifies conceptual interpretations."

The guiding principle in the "chemical bond" paper is the endeavor to isolate the energy-lowering that is associated with bond formation and then to further isolate that contribution to the energy-lowering that is fundamentally quantum mechanical in nature. This is accomplished by proposing a "reaction mechanism" that is constructed in such a way that the virial theorem and the variational principle combine to force the energy-lowering, thereby enabling the decomposition into kinetic and potential-energy contributions to emerge in a natural and easily understandable manner. This focus on the energy-lowering upon bond formation

is critical, since it is precisely with this lowering that we, as chemists, identify bond formation. Additionally, we associate the strength of a bond with the magnitude of the energy decrease. The use of the virial theorem allows for a direct connection with classical physical concepts as is illustrated beautifully in the 1970 paper by Feinberg et al. [2] in the section entitled "The paradox, analogy to space travel."

## 2 Discussion

The underlying basis for the analysis of the chemical bond may be understood by considering the simplest chemical bond – that which is formed when a hydrogen atom and a proton combine to make $H_2^+$, the hydrogen molecule ion [3]:

$$H + H^+ \rightarrow H_2^+ \qquad (1)$$

In the simplest possible case, the two nuclear centers each support one $1s$ (atomic) wavefunction with which to accommodate electronic motion, and one imagines the formation of a molecular wavefunction $\Psi$ from the two atomic wavefunctions $1s_a$ and $1s_b$ as the two atoms come together,

$$\Psi = N(1s_a \pm 1s_b) , \qquad (2)$$

where $N$ is a normalization constant. The + and − in Eq. (2) correspond to the bonding and antibonding molecular wavefunctions, respectively, since they correspond to energy-lowering and energy-raising, respectively. For the bonding wavefunction,

$$N = [2(1 + S_{ab})]^{-1/2} , \qquad (3)$$

where $S_{ab}$ is the overlap of the two atomic wavefunctions. Then, the electron density for the wavefunction in Eq. (2) is

$$\rho_{qm} = \Psi^*\Psi = [2(1 + S_{ab})]^{-1}(1s_a + 1s_b)^2 . \qquad (4)$$

A fundamental point regarding $\rho_{qm}$ is to note that to obtain a quantum mechanical probability density, one first adds the appropriate wavefunctions. In contrast,

*Correspondence to:* M.S. Gordon

classically, one would simply add the corresponding densities to obtain the (normalized) composite probability density,

$$\rho_{cl} = 1/2(\rho_a + \rho_b) = 1/2(1s_a^2 + 1s_b^2) , \qquad (5)$$

Therefore, the inherently quantum mechanical contribution to the density is the difference between $\rho_{qm}$ and $\rho_{cl}$. Since this difference ultimately may be traced to the wave nature of our treatment of electrons, and, in particular to the interference of the waves that we refer to here as atomic wavefunctions, Ruedenberg termed this difference density the interference density, $\rho_I$:

$$\rho_I = \rho_{qm} - \rho_{cl} = [1 + S_{ab}]^{-1}$$
$$\times (1s_a 1s_b - 1/2 S_{ab}[1s_a^2 + 1s_b^2]) . \qquad (6)$$

Since both the classical and quantum mechanical densities must integrate to the correct number of electrons in the system (in this case, 1), the interference density must integrate to zero over all space. So, the interference density represents the rearrangement of electron density, with respect to the classical density of Eq. (5), that occurs when the two atoms approach each other. A plot of this rearrangement density along the internuclear axis for the bonding wavefunction at the equilibrium internuclear distance exhibits the well-known buildup of electron density in the bond region between the nuclei, at the expense of a concomitant depletion of electron density around the nuclei. The reverse would, of course, be true for the higher energy antibonding wavefunction.

This decomposition of the total density in terms of a classical and an inherently quantum mechanical interference contribution leads to an analogous resolution of the energy. By considering the energy in a similar light, one may approach the quantum mechanical origins of the energy-lowering responsible for covalent binding. In considering the energy changes due to formation of a covalent bond, Ruedenberg notes that the application of the variational principle to minimize the energy of the newly formed molecule necessarily (due to the virial theorem) requires the balance of two competing factors. This competition arises as a result of the alteration of the electron probability density from a compact distribution around two separated atoms to a more diffuse distribution of the electron density on the two, now bonded, atoms. Due to the nature of the kinetic-energy operator,

$$\mathscr{T} = -1/2 \nabla^2 \qquad (7)$$

this relaxation of electron density relieves the kinetic-energy pressure: i.e., it reduces the kinetic energy relative to the atoms. On the other hand, the same relaxation of the electron density distribution reduces the magnitude of the nuclear suction; i.e., the potential due to electron–nuclear attraction is attenuated since the electron density is less concentrated around the nuclei. It is the balance between these two competing phenomena that is critical in determining the nature of the chemical bond.

To place the foregoing on a more quantitative footing, consider the formation of $H_2^+$ from $H + H^+$ to occur in two steps. We begin with H described by the exact $1s$ wavefunction (exponent $= 1.0$), with an identical $1s$ wavefunction centered on the infinitely separated proton. The electron density may then be described as either $\rho = \rho_a$ or $\rho = \rho_b$, depending whether a or b is the proton. An energetically equivalent description is $\rho = 1/2(\rho_a + \rho_b)$. In the first step, H and $H^+$ are brought from infinity to their equilibrium internuclear separation of 2.0 bohr in such a manner that the exponent on the $1s$ wavefunctions remains 1.0. The energy change due to this step is almost certainly negative (exoergic), due to the formation of the covalent bond. The second step is simply to permit the molecular wavefunction, given by Eq. (2), to relax its exponent (and thereby the radial distribution) from 1.0 to the optimum 1.24 and thus re-establish the virial relation $V = -2T$. An energy decrease for this step is guaranteed by the variational principle. One may then calculate the changes in the total ($E$), kinetic ($T$) and potential ($V$, including nuclear repulsion) energies for each step and for the overall process. The results are presented in Table 1. In the first step, bond formation, the energy decrease is driven by release of the kinetic-energy pressure, due to the spreading out of the electron density over the two nuclear centers. In the second step, the electron density is contracted (the wavefunction becomes more compact) as the exponent of the wavefunction increases from 1.0 to 1.24. This causes an increase in the kinetic-energy pressure (illustrated by the positive value of $\Delta T$ for this step) and a concomitant decrease in the potential-energy suction, as the electron density contracts around the nuclei; however, the total energy-lowering is smaller than for step 1.

Note that the energies quoted in Table 1 do not correspond to the exact $H_2^+$ bond energy. This is because there is an additional step, in which polarization is introduced into the wavefunction. This is discussed in detail in the original papers by Ruedenberg and coworkers.

The foregoing analysis may be more directly connected with the decomposition of the electron density into classical and interference components by considering an alternative "mechanism" for the formation of $H_2^+$. In this second two-step mechanism, we first contract the H atom wavefunction by increasing its orbital exponent from the optimum value of 1.0 to the final $H_2^+$ value of 1.24. One may think of this as a contractive promotion step, analogous to the hybridization of the orbitals of a carbon atom in preparation for the formation of methane. This step costs energy, but it is energy that will be recovered upon bond formation.

We next define the total molecular potential energy, $V^{mol}$, as

**Table 1.** Energetics (hartree) for two-step formation of $H_2^+$

|        | $\Delta T$ | $\Delta V$ | $\Delta E$ |
|--------|-----------|-----------|-----------|
| Step 1 | −0.1138   | +0.0600   | −0.0538   |
| Step 2 | +0.2009   | −0.2336   | −0.0327   |
| Total  | +0.0871   | −0.1736   | −0.0865   |

$$V^{mol} = -\langle \Psi | Z/r_a + Z/r_b | \Psi \rangle + Z/R_{ab}$$
$$= -[2(1 + S_{ab})]^{-1}$$
$$\times \langle (1s_a + 1s_b) | Z/r_a + Z/r_b | (1s_a + 1s_b) \rangle + Z/R_{ab},$$
$$(8)$$

where $Z$ is the nuclear charge, $r_a$ is the distance from an electron on atom A from its nucleus, and $R_{ab}$ is the internuclear distance. $V^{mol}$ contains (a) the attractions of the electron density on an atom for its own nucleus, (b) the attractions of electron density on one atom for the other nucleus, and (c) the attraction of the interference density for both nuclei. To separate these contributions, we define $V_A$ as the atomic attraction, $V_{QC}$ as the two-center quasiclassical (QC) attraction, and $V_I$ as the attraction due to interference:

$$V_A = -\langle 1s_a | Z/r_a | 1s_a \rangle$$
$$= -1/2[\langle 1s_a | Z/r_a | 1s_a \rangle + \langle 1s_b | Z/r_b | 1s_b \rangle] \quad (9)$$

$$V_{QC} = -1/2[\langle 1s_a | Z/r_b | 1s_a \rangle + \langle 1s_b | Z/r_a | 1s_b \rangle]$$
$$+ 1/R_{ab} \quad (10)$$

$$V_I = (1 + S_{ab})^{-1}[\langle 1s_a | -Z/r_a - Z/r_b | 1s_b \rangle$$
$$- 1/2 S_{ab}(\langle 1s_a | -Z/r_a - Z/r_b | 1s_a \rangle$$
$$+ \langle 1s_b | -Z/r_a - Z/r_b | 1s_b \rangle)] \quad (11)$$

$$V^{mol} = V_A + V_{QC} + V_I . \quad (12)$$

For the ground-state hydrogen atom,

$$V_H = -\langle 1s_a | Z/r_a | 1s_a \rangle . \quad (13)$$

For an exponent of 1.0, $V_A = V_H$, so upon bond formation at a fixed exponent of 1.0, the change in potential energy is given by

$$\Delta V = V_{QC} + V_I . \quad (14)$$

In general,

$$\Delta V = (V_A - V_H) + V_{QC} + V_I = V_{CP} + V_{QC} + V_I , \quad (15)$$

where $V_{CP}$ is the energy increase due to the contractive promotion that occurs when the exponent is increased to 1.24. The kinetic energy can be partitioned in a similar manner, except, of course, that there is no analog for the two-center classical (QC) term:

$$\Delta T = T_{CP} + T_I . \quad (16)$$

These energy differences are summarized in Table 2. It is clear from Table 2 that

1. The driving force for covalent bonding in $H_2^+$ is the constructive interference contribution to the kinetic energy.

2. The wavefunction contraction, and thus the lower final potential energy, is not directly due to the sharing of electrons as was generally believed prior to the publication of this paper. Rather, it is "atomic" in nature.

An examination of the antibonding wavefunction, at the ground-state equilibrium internuclear distance of 2.0 bohr, reveals an analogous energy decomposition: the energy increase arises from the destructive interference contribution to the kinetic energy. Similar conclusions were arrived at by the Ruedenberg group for studies on more complex molecules, and also by Kutzelnigg [6] and Mulliken [5].

The important role of the electronic kinetic energy, first established by Ruedenberg for covalent bonding has been demonstrated for many other important chemical phenomena such as steric repulsion [7], ionization potentials [8], three-center/two-electron bonding [9], and hydrogen bonding [10]. Thus, this very general conceptual framework serves to highlight the similarities and differences between these diverse situations.

## 3 Concluding remarks

The essence of this model of the chemical bond may be summarized as follows

1. The origins of covalent chemical bonding lie in the kinetic-energy lowering caused by interference effects that arise from the fundamentally quantum mechanical wave–particle duality. "...the interference energy owes its binding effect entirely to a lowering of the kinetic energy...the interference process is unfavorable as regards the potential energy. The ubiquitous statement that overlap accumulation of electrons in a bond leads to a lowering of the potential energy is based on fallacious reasoning."

2. "The wave mechanical kinetic behavior, which differs typically from the classical behavior...is a fundamentally essential element of covalent bonding. Any explanation of chemical binding based essentially on an electrostatic, or any other nonkinetic concept, misses the very reason why quantum mechanics can explain chemical binding, whereas classical mechanics cannot."

3. While the formulation of the model is most commonly made in terms of atomic orbital components, it does not rely on such a formulation and is quite general.

**Table 2.** Energetics (hartree) for bond formation in $H_2^+$

| | $H_2^+$ exponent = 1.0 | | | $H_2^+$ exponent = 1.24 | | |
| --- | --- | --- | --- | --- | --- | --- |
| | $\Delta T$ | $\Delta V$ | $\Delta E$ | $\Delta T$ | $\Delta V$ | $\Delta E$ |
| Contractive | 0.0 | 0.0 | 0.0 | 0.2672 | -0.2387 | 0.0285 |
| Promotion two-center | 0.0 | 0.0275 | 0.0275 | 0.0 | 0.0123 | 0.0123 |
| Classical interference | -0.1138 | 0.0325 | -0.0813 | -0.1801 | 0.0528 | -0.1273 |
| Total binding | -0.1138 | 0.0600 | -0.0538 | 0.0871 | -0.1736 | -0.0865 |

4. While any energy decomposition is necessarily arbitrary, such pictures are useful, since "they allow us to visualize and predict similarities and differences in the solution of the Schrödinger equation for different molecules without continuous appeal to an electronic computer." This comment, expressed originally in 1962, is even more relevant today.

*Acknowledgements.* The authors are grateful to Professors Ruedenberg and Rioux for reading and commenting on this manuscript. Their suggestions have been most helpful.

## References

1. Ruedenberg K (1962) Rev Mod Phys 34: 326–376
2. Feinberg M, Ruedenberg K, Mehler EL (1970) In: Lowdin PO (ed) Advances in quantum chemistry, vol 5. Academic Press, New York, pp 27–98
3. Feinberg MJ, Ruedenberg K (1971) J Chem Phys 54: 1495–1511
4. Ruedenberg K (1997) In: Chalvet O, Daudel R, Malrieu JP (eds) Localization and delocalization in quantum chemistry. Reidel, Dordrecht, pp 223–245
5. Mulliken RS, Ermler WC (1977) In: Diatomic molecules. Academic Press, New York
6. Kutzelnigg W (1978) In: Einführung in die Theoretische Chemie, vol 2. VCH, Weinheim
7. Baerends EJ (1992) In: Pacchione G, Bagus PS, Parmigiani F (eds) Cluster models for surface and bulk phenomena. Plenum, New York, pp 189–207
8. Rioux F, DeKock RL (1988) J Chem Educ 75: 537–539
9. Webb SP, Gordon MS (1988) J Am Chem Soc 120: 3846
10. (a) Jensen JH, Gordon MS (1995) J Phys Chem 99: 8091–8101; 10(b) Minikis RM, Jensen JH, Int J Quantum Chem (in press)

Theor Chem Acc (2000) 103:252–256
DOI 10.1007/s002149900064

Theoretical
Chemistry Accounts
© Springer-Verlag 2000

*Perspective*

# Perspective on "An extended Hückel theory. I. Hydrocarbons"

## Hoffmann R (1963)  J Chem Phys 39:1397–1412

**Myung-Hwan Whangbo**

Department of Chemistry, North Carolina State University, Raleigh, NC 27695-8204, USA

Received: 26 February 1999 / Accepted: 14 June 1999 / Published online: 9 September 1999

**Abstract.** An overview is given on how the title paper by Hoffmann has influenced theoretical studies of molecules and solids over the years. The strengths and weaknesses of the extended Hückel theory are also discussed.

**Key words:** Extended Hückel theory, advantages and disadvantages – Structure–property correlation – Extended Hückel theory, history and impact – Roald Hoffmann

## 1 Introduction

Physical properties of molecules and solids are intimately related to their atom arrangements. To understand this structure–property relationship, it is necessary to examine how the electronic structure of a system depends on its atom arrangement by performing electronic structure calculations at a certain level of approximation. The level of calculation one needs to employ depends on the nature of the answer one hopes to find from the calculations. An important role of an electronic structure theory is to provide a conceptual framework in which to think, to organize experimental knowledge [1]. In this role, theoretical predictions need not be quantitative but should provide a bias for correct thinking [1]. For the past three and a half decades, the extended Hückel (EH) theory [2] devised by Hoffmann has provided such a role. In this paper, we briefly review how this semiempirical theory has influenced our understanding of physical and electronic properties of discrete molecules and extended solids. Then we discuss the strengths and weaknesses of the EH theory to help avoid misusing EH calculations.

## 2 EH theory

The genesis of the EH method [2] was the linear combination of atomic orbitals/molecular orbital (MO) study of polyhedral molecules by Hoffmann and

Lipscomb [3]. The title paper published by Hoffmann in 1963 [2] was featured as a citation classic in *Current Contents* in 1989 [4], where Hoffmann described the background of the EH method in detail. The firm foundation of the EH method was given in the title paper [2] and in the subsequent three papers published under the general title of "*Extended Hückel theory*" in 1964 [5–7]. In the EH method the MOs $\psi_i$ ($i = 1, 2, \ldots, m$) of a system are constructed using a set of valence atomic orbitals $\{\chi_1, \chi_2, \chi_3, \ldots, \chi_m\}$. The effective one-electron Hamiltonian, $H^{\mathrm{eff}}$, determining the energies $e_i$ of the MOs is not specified, but its matrix representation, $H_{\mu\nu} = \langle \chi_\mu | H^{\mathrm{eff}} | \chi_\nu \rangle$, in an atomic orbital basis is defined semiempirically; namely, the diagonal element $H_{\mu\mu}$ is approximated by the valence state ionization potential (VSIP) of the atomic orbital $\chi_\mu$ [8, 9]

$$H_{\mu\mu} = -\mathrm{VSIP} \tag{1}$$

and the off-diagonal element $H_{\mu\nu}$ is approximated by the Wolfsberg–Helmholz formula [8],

$$H_{\mu\nu} = K S_{\mu\nu}(H_{\mu\mu} + H_{\nu\nu})/2 \ , \tag{2}$$

where $S_{\mu\nu}$ is the overlap integral $S_{\mu\nu} = \langle \chi_\mu | \chi_\nu \rangle$ and $K = 1.75$. In the weighted Wolfsberg–Helmholz approximation [10], the coefficient $K$ of Eq. (2) is replaced with another coefficient $K'$, which is given by

$$K' = K + \Delta^2 + \Delta^4(1 - K) \ , \tag{3}$$

where $\Delta = (H_{\mu\mu} - H_{\nu\nu})/(H_{\mu\mu} + H_{\nu\nu})$. Equation (3) is used to reduce the extent of counterintuitive orbital mixing [10, 11]. The variational theorem leads to a set of simultaneous equations:

$$\sum_{\mu=1}^{m} \left( H_{\mu i} - e_i S_{\mu i} \right) C_{\mu i} = 0 \quad (i = 1, 2, \ldots, m) \ . \tag{4}$$

The solution of these equations provides the energies $e_i$ of the MOs $\psi_i$.

The approximations of the EH theory in leading up to Eq. (4) are very crude, and so it has serious drawbacks that first-principles electronic structure theories do not. In this age of powerful computers and commercial program packages of first-principles electronic structure theories, one may ask if there is any room for the use of the semi empirical EH theory in chemistry and physics. The answer is an emphatic yes, because the EH theory possesses a few advantages that no first-principles theory can ever match (see Sect. 4). For these reasons, EH calculations have been extensively used to study structure–property relationships in all kinds of materials, from molecules to solids, from organic to inorganic compounds.

## 3 Brief history of the use of EH calculations

During the years 1963–1973, EH calculations were largely applied to organic molecules. The concept of orbital symmetry conservation, introduced by Woodward and Hoffmann [12], provided a strong influence on what qualitative features to find from electronic structure calculations. To arrive at the concept of orbital symmetry conservation, it was necessary to know the frontier MOs of organic molecules and how their nodal properties vary as the molecular geometry changes. It must be noted that Hoffmann carried out EH calculations for a large number of organic molecules [2, 5–7] before his celebrated collaboration with Woodward, which led to the series of five communications [13–17] establishing the concept of orbital symmetry conservation. During the period 1963–1972, the concept of orbital interaction was firmly established. The most significant review resulting from EH calculations in this period is "Interaction of orbitals through space and through bonds" by Hoffmann [18].

Since 1974 the focus of EH calculations has shifted from organic molecules to compounds of transition-metal elements. This shift began with the work of Rösch and Hoffmann [19] and that of Hoffmann et al. [20]. It is interesting to note that around the time of this shift an efficient program package of first-principles electronic structure theory (i.e., Gaussian 70) became available and enabled one to study the structural details of organic molecules on a quantitative basis. Representative studies on polyene conformational preferences in organometallic compounds were carried out by Albright and coworkers [21, 22]. These studies established that rotational barriers are quite well reproduced by EH calculations. Chemical reactivities of transition-metal compounds were examined in a number of papers. Representative studies on this topic include the work of Tatsumi et al. [23] on reductive elimination of $d^8$ organotransition-metal complexes, that of Eisenstein and Hoffmann [24] on reactivity of transition-metal complexed olefins towards a nucleophile, and that of Saillard and Hoffmann [25] on C—H and H—H activation in transition-metal complexes. Studies by Hoffmann's group during the years 1974–1982 made it possible to describe the structure and bonding of transition-metal compounds in terms of fragment MOs (FMOs). These studies eventually led to the development of the isolobal analogy between organic and inorganic compounds [26, 27].

Studies of the structures and bonding in extended solids based on the EH theory started with the work of Whangbo and Hoffmann [28] in 1978 and that of Whangbo et al. [29] in 1979. These papers showed that electronic structures of solid-state materials can also be described using the concepts of perturbation and orbital interaction, thus laying down the foundation for the electronic structure studies of solids and surfaces by S. Alvarez, J.K. Burdett, E. Canadell, J.-F. Hallet, R. Hoffmann, T.R. Hughbanks, S. Lee, G.J. Miller, M.-H. Whangbo and their coworkers in the 1980s and 1990s. These research efforts led to a number of review articles and monographs on the structures and bonding in solids [30–39]. Representative studies of chemical reactivity and bonding on surfaces include the work of Saillard and Hoffmann [25] on C—H and H—H activation on surfaces and that of Sung and Hoffmann [40] on bonding of carbon monoxide to metal surfaces.

Beginning with the work of Parkinson et al. [41] in 1991, the EH theory has been extensively used to interpret atomic- and molecular-scale images of scanning tunneling microscopy (STM) and atomic force microscopy (AFM). These studies showed that an STM image of a surface is well described by the electron density map resulting from the frontier orbitals of the surface, and an AFM image by that from all the occupied orbitals of the surface. These STM and AFM studies for a variety of layered compounds and overlayers were reviewed in the book by Magonov and Whangbo (42) in 1996.

## 4 Strengths and weaknesses of the EH theory

In predicting the optimum structure of a system (in particular, bond-length optimization), EH calculations are not reliable if trusted blindly as one might with the program packages of first-principles electronic structure theories. For systems of known geometry, however, EH calculations have been invaluable as evidenced by numerous studies over the years.

The energy levels and orbitals generated by EH calculations for a system do not depend on the number of electrons the system has, because the EH theory does not consider electron–electron repulsion explicitly. Consequently, the EH method does not provide a correct way of describing the relative energies of different electronic states available for a given system. For example, consider a molecular system that has two electrons to fill its two energy levels, for example the highest occupied MO (HOMO) and the lowest unoccupied MO (LUMO). The ground state of this system can be a singlet state in which the HOMO is doubly occupied or a triplet state in which the HOMO and LUMO are each singly occupied with the same spin. In EH calculations the total energy of a system is given by the sum of the energies of its occupied orbitals, thus predicting that the singlet state is always more stable than the triplet state, in disagreement with experiment. However, it is important to recognize that EH calculations do provide information concerning when such a failure is likely to occur. The essential effect of electron–electron repulsion is to make the double occupancy of an orbital energetically unfavorable; therefore, the triplet state becomes more stable than the singlet state if the HOMO–LUMO energy difference is small enough [43]. Thus, when EH calculations for a molecule lead to a small HOMO–LUMO gap, $\Delta e$, one must note that the system may adopt a triplet state as the ground state.

It is important to note the solid-state counterpart [44] of this observation. When the unit cell of a solid contains an odd number of electrons, the highest-occupied band of this system becomes half filled. If electron–electron repulsion is neglected, the levels of the bottom half of the band are each doubly filled, and those of the top half are empty, thereby leading to a metallic state. Thus EH calculations predict that a system with a partially filled band is always metallic, in conflict with experiment. This is again a serious failure, if results of EH calculations are accepted blindly. However, EH calculations do provide information concerning when such a failure is likely to occur. For a system with a partially filled band, a magnetic insulating state may become more stable than the metallic state when the width of the partially filled band is narrow. The metallic and magnetic insulating states are similar in that they possess a partially filled band, but they differ in the way the band levels are occupied. In a magnetic-insulating state, a partially filled band has all its band levels singly filled [44]; therefore, when EH calculations for a system lead to narrow partially filled bands, one must consider the possibility that its ground state is magnetic-insulating rather than metallic.

The fact that EH calculations do not depend on the number of electrons in a system gives rise to advantages that no first-principles theory can ever provide. EH calculations are simple and fast. Consequently, they can be used to study those molecular and extended solids that contain so many atoms per unit cell that first-principles calculations are impossible or difficult to apply. It is intellectually challenging and satisfying to find correct explanations and correct predictions for such problems by employing an imperfect method such as the EH theory. In addition, our research life may be too short to leave the problems untackled until the arrival of a computer powerful enough to treat the problems at the level of first-principles theories. The simplicity of the EH method does not necessarily mean that the electronic structures it generates are unreliable. The usefulness of any calculation, be it first-principles or semiempirical, rests ultimately on the test of whether the calculated results are consistent with experimental observations and provide insight into experimental problems under question. EH calculations carried out for a variety of materials manifest that the EH theory is useful if one asks the kinds of questions that it can answer.

EH calculations for a complex system can be approximated by those for the relevant part of the system. For example, in the organic conducting salts (BEDT-TTF)$_2X$ with a mononegative anion $X^-$, where BEDT-TTF refers to bis(ethylenedithio)tetrathiafulvalene, the layers of the (BEDT-TTF)$_2^+$ cations alternate with layers of the $X^-$, anions [34, 45]. It is the cation layers that are largely responsible for the transport properties of the salt, and the cation layers are separated by the anion layers. Therefore, for the purpose of studying the transport properties of a (BEDT-TTF)$_2X$ salt, the electronic band structure of the salt can be well approximated by that of an isolated cation layer. Such an approximation greatly simplifies the computational task and has been widely applied to a variety of organic conducting salts [34, 45]. In a similar manner, for the purpose of studying the transport properties of the hexagonal alkali tungsten bronze $A_xWO_3$ ($A$ = K, Rb, Cs; $x < 1/3$), the electronic band structure of this bronze can be approximated by that of the $WO_3^{x-}$ lattice [46]. In EH calculations, these kinds of approximations are valid, and as a consequence simplify the task of calculations enormously.

In EH calculations it is easy to express the orbitals of a composite system A-B in terms of the orbitals of its fragments A and B even if the fragments are functional groups. This FMO analysis is useful and meaningful in EH calculations because the orbitals of a fragment are transferable from one molecule to another as long as its geometry remains the same. For instance, consider a composite molecule A-B in which the fragments A and B are joined by a single bond. In principle, this molecule can be divided into A$^+$ and B$^-$, A$^-$ and B$^+$, A$^-$ and B$^-$, etc. In EH calculations, the energies and the nodal properties of fragment orbitals do not depend on the number of electrons a fragment has. However, in any theory taking electron–electron repulsion into consideration, the number of electrons in a fragment should be assigned to calculate its orbitals, and the energies and the nodal properties of fragment orbitals depend on the number of electrons a fragment has, and so it is complicated to perform an FMO analysis [47, 48]. Nevertheless, a meaningful FMO analysis can be carried out within the framework of self-consistent-field MO (SCF-MO) theory if fragment orbitals are defined by the Fock matrix partitioning method [47, 48] rather than by SCF-MO calculations on molecular fragments. In this case, the orbitals of a fragment become "environment-adjusted" fragment orbitals [47, 48], and so the transferability of fragment orbitals from one molecule to another is not so good as in EH calculations.

## 5 Use of EH calculations

In EH calculations, one should pay attention to the parameters of the atomic orbitals used to construct the matrix elements $H_{\mu\nu}$ and $S_{\mu\nu}$. In the EH method valence atomic orbitals are approximated by Slater-type orbitals (STOs). A single-zeta STO, $\chi_\mu$, is defined by

$$\chi_\mu(r, \theta, \phi) \propto r^{n-1} \exp(-\zeta r) Y(\theta, \phi) , \tag{5}$$

where $n$ is the principal quantum number, $\zeta$ is the exponent, and $Y(\theta, \phi)$ is the spherical harmonic. In a double-zeta STO, a linear combination of two exponential functions is used as

$$\chi_\mu(r, \theta, \phi) \propto r^{n-1}[c_1 \exp(-\zeta_1 r) + c_2 \exp(-\zeta_2 r)] Y(\theta, \phi) . \tag{6}$$

The $\zeta$ values for single-zeta STOs, the $\zeta_1$, $\zeta_2$, $c_1$ and $c_2$ values for double-zeta STOs, and the VSIP values can be taken from results of atomic electronic structure calculations using the Hartree–Fock method [49, 50]. There are two program packages for EH theory that allow one to perform EH calculations for molecules and solids. One is YAeHMOP [51], which is freely available; the other is CAESAR [52], which is a commercial program designed to run on personal computers. Detailed descriptions on how to carry out EH calculations using the program package CAESAR have been given by Ren et al. [52]. The freely available program package CA-CAO, originally written by Mealli and Proserpio [53], is designed to do EH calculations for molecules. This package allows one to generate Walsh diagrams and orbital interaction diagrams as well as to plot perspective views of MOs.

When results of EH calculations for a molecule or a solid of known structure are not consistent with the physical properties of the molecule or the solid, it is necessary consider two possible sources leading to this disagreement:

1. The failure is caused by the assumptions inherent in any electronic structure theory. For example, because of electron–electron repulsion, the lowest possible spin state of a molecule may not be the ground state, and the metallic state of a solid may be less stable than the corresponding magnetic insulating state. In this case, one should consider an alternative way of filling the calculated energy levels with electrons to generate a high-spin state for a molecule and a magnetic insulating state for a solid. Then one should attempt to extract, from results of EH calculations, useful information needed to go beyond a one-electron electronic structure theory.

2. The failure is not caused by the assumptions of electronic structure theory but originates from the use of EH approximations. In this case, it is important to analyze the source of the failure from the viewpoint of the atomic parameters employed, modify the parameters appropriately, and repeat the calculations. This task is not difficult, if one becomes familiar with the concepts of perturbation and orbital interaction. In spirit, this process is not different from what one does with first-principles calculations. For example, when a chosen basis set or correlation level does not give correct results, one tries another basis set or correlation level. First-principles methods are based on rigorous theoretical and mathematical formulations, but their actual calculations do include approximations.

## 6 Concluding remarks

Approximate electronic structures calculated by the EH method may not provide quantitative predictions, but they are mostly adequate for qualitative structure–property correlation analysis. The relevance of the EH method lies in this role of facilitating the search for structure–property correlations. When used in conjunction with the concepts of perturbation and orbital interaction, EH calculations have been instrumental in discovering a number of qualitative concepts useful for thinking about the structures and properties of various organic and inorganic materials. To name a few, examples include FMOs [26, 27], orbital correlation diagrams [12, 27], through-space and through-bond interactions [18, 27], and the isolobal analogy [26, 27] for molecular systems; a band orbital picture of magnetic-insulating states [30, 31, 44], an orbital-mixing view of electronic phase transitions in metals [30, 37, 54], and hidden Fermi surface nesting [37, 55, 56] for solid-state materials; and tip-force induced local stiffness variation [42, 57] in STM and AFM.

In closing, it is refreshing to read what Hoffmann had to say about theory 25 years ago [1]:

In my mind, the most important role of theory in chemistry is to provide a framework in which to think, to organize experimental knowledge. I picture chemistry as a living organism – an ever-expanding amorphous sphere with extensions along frontier areas, capable of movement, progress, that may be lightning fast and excruciatingly slow. I see theory as a kind of primitive nervous system of this organism, needed to organize the frontier information, to connect it to the accumulated store of knowledge, to communicate among the myriad advancing areas. The cues for further action that this nerve system gives need not be 100% correct – how could it be? Only a slight bias for the correct option, when a million decisions are to be made, endows the organism with the statistical equivalent of an intuition.

*Acknowledgements.* This work was supported by the Office of Basic Energy Sciences, Division of Materials Sciences, US Department of Energy, under grant DE-FG05-86ER45259.

## References

1. Hoffmann R (1974) Chem Eng News 52(30): 32
2. Hoffmann R (1963) J Chem Phys 39: 1397
3. Hoffmann R, Lipscomb WN (1962) J Chem Phys 36: 2179
4. Hoffmann R (1989) Curr Contents 20(19)
5. Hoffmann R (1964) J Chem Phys 40: 2745
6. Hoffmann R (1964) J Chem Phys 40: 2474
7. Hoffmann R (1964) J Chem Phys 40: 2480
8. Wolfsberg M, Helmholz L (1952) J Chem Phys 20: 837
9. Ballhausen CJ, Gray HB (1965) Molecular orbital theory. Benjamin, New York
10. Ammeter JH, Bürgi H-B, Thibeault J, Hoffmann R (1978) J Am Chem Soc 100: 3686
11. Whangbo M-H, Hoffmann R (1978) J Chem Phys 68: 5498
12. Woodward RB, Hoffmann R (1970) The conservation of orbital symmetry. Verlag Chemie and Academic Press, New York
13. Woodward RB, Hoffmann R (1965) J Am Chem Soc 87: 395
14. Hoffmann R, Woodward R (1965) J Am Chem Soc 87: 2046
15. Woodward RB, Hoffmann R (1965) J Am Chem Soc 87: 2511
16. Hoffmann R, Woodward R (1965) J Am Chem Soc 87: 4388
17. Hoffmann R, Woodward R (1965) J Am Chem Soc 87: 4389
18. Hoffmann R (1971) Acc Chem Res 4: 1
19. Rösch N, Hoffmann R (1974) Inorg Chem 13: 2656
20. Hoffmann R, Chen MM-L, Elian M, Rossi AR, Mingos DMP (1974) Inorg Chem 13: 2666
21. Albright TA, Hoffmann R, Thibeault JC, Thorn DL (1979) J Am Chem Soc 101: 3801
22. Albright TA (1982) Acc Chem Res 15: 149
23. Tatsumi K, Hoffmann R, Yamamoto A, Stille JK (1981) Bull Chem Soc Jpn 54: 1857
24. Eisenstein O, Hoffmann R (1981) J Am Chem Soc 103: 4308
25. Saillard JY, Hoffmann R (1984) J Am Chem Soc 106: 2006
26. Hoffmann R (1982) Angew Chem Int Ed Engl 21: 711
27. Albright TA, Burdett JK, Whangbo M-H (1985) Orbital interactions in chemistry. Wiley, New York
28. Whangbo M-H, Hoffmann R (1978) J Am Chem Soc 100: 6093
29. Whangbo M-H, Hoffmann R, Woodward RB (1979) Proc R Soc Lond Ser A 366: 23
30. Whangbo M-H (1982) In: Miller JS (ed) Extended linear chain compounds, vol 2. Plenum, New York, p 127

256

31. Whangbo M-H (1983) Acc Chem Res 16: 95
32. Burdett JK (1984) Prog Solid-state Chem 15: 173
33. Whangbo M-H (1986) In: Rouxel J (ed) Crystal chemistry and properties of materials with quasi-one dimensional structures. Reidel, Dordrecht p 27
34. Williams JM, Wang HH, Emge TJ, Geiser U, Beno MA, Leung PCW, Carlson KD, Thorn RJ, Schultz AJ, Whangbo M-H (1987) Prog Inorg Chem 35: 51
35. Burdett JK (1988) Acc Chem Res 21: 189
36. Hoffmann R (1988) Solids and surfaces: a chemist's view of bonding in extended structures. VCH, New York
37. Canadell E, Whangbo M-H (1991) Chem Rev 91: 965
38. Burdett JK (1995) Chemical bonding in solids. Oxford University Press, New York
39. Iung C, Canadell E (1997) Description orbitalaire de la structure électronique des solids. 1. De la molécule aux composés 1D. Ediscience Paris
40. Sung SS, Hoffmann R (1985) J Am Chem Soc 107: 578
41. Parkinson BA, Ren J, Whangbo M-H (1991) J Am Chem Soc 113: 7833
42. Magonov SN, Whangbo M-H (1996) Surface analysis with STM and AFM. VCH, Weinheim
43. Hay PJ, Thibeault JC, Hoffmann R (1975) J Am Chem Soc 97: 4884
44. Whangbo M-H (1979) J Chem Phys 70: 4963
45. Williams JM, Ferraro JR, Thorn RJ, Carlson KD, Geiser U, Wang HH, Kini AM, Whangbo M-H (1992) Organic superconductors. Prentice Hall, New York

46. Lee K-S, Seo D-K, Whangbo M-H (1997) J Am Chem Soc 119: 4043
47. Whangbo M-H, Schlegel HB, Wolfe S (1977) J Am Chem Soc 99: 1296
48. Whangbo M-H (1981) In: Csizmadia IG, Daudel R (eds) Computational theoretical organic chemistry. Reidel, Dordrecht p 233
49. Clementi E, Roetti C (1974) At Data Nucl Data Tables 14: 177
50. McLean AD, McLean RS (1981) At Data Nucl Data Tables 26: 197
51. Landrum GA (1997) Yet another extended Hückel molecular orbital package (YAeHMOP). Cornell University Ithara. http://overlap.chem.cornell.edu:8080/yaehmop.html
52. Ren J, Liang W, Whangbo M-H (1988) Crystal and electronic structure analysis using CAESAR. PrimeColor Software Raleigh. This book can be downloaded free of charge from the Web site http://www.primec.com
53. Mealli C, Proserpio DM (1990) J Chem Ed 67: 399
    A PC version of the computer aided composition of atomic orbitals (CACAO) program package can be downloaded from the Web site http://apamac.ch.adfa.oz.au/OzChemNet/Programs/Windows/WinSoftware.html/
54. Whangbo M-H (1989) Adv Chem Ser 226: 269
55. Whangbo M-H, Canadell E, Foury P, Pouget J-P (1991) Science 252: 96
56. Whangbo M-H, Canadell E (1992) J Am Chem Soc 114: 9587
57. Bengel H, Cantow H-J, Evain M, Magonov SN, Whangbo M-H (1996) Surf Sci 365: 461

Theor Chem Acc (2000) 103:257–258
DOI 10.1007/s002149900022

Theoretical
Chemistry Accounts
© Springer-Verlag 2000

*Perspective*

# Perspective on "Stereochemistry of polypeptide chain conformations"

## Ramachandran GN, Ramakrishnan C, Sasisekharan V (1963) J Mol Biol 7: 95–9

**Richard Lavery**

Laboratoire de Biochimie Théorique CNRS UPR 9080, Institut de Biologie Physico-Chimique,
13 rue Pierre et Marie Curie, F-75005 Paris, France

Received: 26 February 1999 / Accepted: 16 March 1999 / Published online: 21 June 1999

**Abstract.** Despite its apparent simplicity the "Ramachandran map" has been an enormously successful tool for describing and understanding protein structure. Thirty-five years after its invention, it is still used daily for checking the quality of experimental and modeled protein structures. It is, moreover, founded on a rational, reduced-coordinate model of the polypeptide chain which continues to be useful in computational attempts at predicting protein folding.

**Key words:** Ramachandran map – Protein structure – Molecular modeling – Conformational analysis

Living in the midst of an explosion in structural molecular biology, it is difficult to remember that at the beginning of the 1960s only a handful of protein structures had been solved by X-ray crystallography and little was known of the range of the folding patterns that the polypeptide chain could adopt. The only simplifying feature of these patterns was the existence of regular secondary structures, the α-helix and the β-sheet, which were the result of stereochemical insight on the part of Pauling [1, 2], rapidly confirmed by the fiber diffraction work carried out by Perutz [3]. The first protein structures solved, hemoglobin [4] and myoglobin [5], although integrating these fundamental building blocks exhibited considerably more complex forms, whose analysis clearly required new mathematical and graphical tools. However, in the early 1960s, computers were only just beginning to enter the lives of chemists and biologists. Force fields as we know them today were already evolving from their spectroscopic roots [6], but computations on biopolymers were still in the future (this did not exclude some courageous attempts at fold prediction [7]). Similarly, only the most rudimentary graphic systems were available, although the importance of visualization in structural biology would soon become clear [8].

It was in this setting that G.N. Ramachandran (more commonly addressed, with the southern Indian penchant for shortening names, as "GNR") and his colleagues proposed to represent peptide chain conformations on a two-dimensional map. The map was defined by the dihedrals $\phi$ and $\psi$, which describe the bonds on either side of an amino acid α-carbon ($\phi$: C'-N-Cα-C', $\psi$: N-Cα-C'-N). This was an inspired use of reduced coordinates. By choosing to ignore side-chain conformations, rotations around the partially conjugated peptide linkage and deformations in bonded geometry, protein conformations were reduced to a problem of $2N$ variables for $N$ amino acids (gaining roughly a factor of 50 over Cartesian coordinates and even a factor of 3 over a complete dihedral angle description).

Simple steric calculations (using minimum contact distances between classed atom pairs) enabled the "Ramachandran map" to be divided up into allowed and forbidden regions (the latter representing roughly 75% of the map!). Although obtained for a dipeptide, this result was equally applicable to polypeptides with either regular or complex folded conformations. Indeed, the original paper already showed that helical parameters could be mapped onto $\phi/\psi$ space (see also Ref. [9]), helping to rationalize the nature and the relationships of not only the α-helix and the β-sheet, but also the other helical forms which had recently been identified (2.2,-ribbon, π- and γ-helices, and Ramachandran's own, triple helical collagen structure [10]). The map also played an important role in understanding the structural role of specific amino acids such as proline or glycine, which could either constrain or expand the normally allowed domains.

Thirty-five years on, the Ramachandran map has been calculated and recalculated with increasingly sophisticated treatments of both molecular [11] and quantum mechanics [12], but its basic content remains unchanged. It has become an indispensable tool for all those dealing with protein structures as the touchstone for judging the quality of both experimental and simulated protein conformations [13], and it is also present in

258

most of the latest models of protein folding as a means of correctly biasing peptide conformational sampling (see, for example, Ref. [14]).

I arrived at the Indian Institute of Science in Bangalore (founded by J.N. Tata, at a site chosen by Sir William Ramsay) in 1976 to begin my theoretical studies of biopolymers. During my postdoctoral research with GNR, I did not expect that the famous map would have such a profound influence, but, looking back, I see two factors which have influenced my work and that of many others in our domain – the power of simple models and the importance of appropriate analytical tools. Reduced-coordinate models, on which the Ramachandran map is based, continue to play a major role in describing the static and dynamic conformations of both proteins [15] and nucleic acids [16]. With their help, considerable progress has been made in understanding the governing principles behind protein folding, even if reliable predictions on all but the smallest proteins are still out of reach. A new challenge faces us today in the case of the RNAs, where the passage from two-dimensional base pairing schemes to three-dimensional conformations is far from trivial [17].

Without replacing the Ramachandran map, protein conformational analysis tools have progressed, notably by extending a simple description of the polypeptide chain into fold classifications which help to reveal the principles and the evolution of protein architecture [18]. With the appearance of molecular dynamics simulations it has also become necessary to develop tools for analyzing protein conformational fluctuations [19, 20]. Similar analysis problems are posed by nucleic acids. While it is now possible to describe the subtle variations in the geometry of the double helix [21], the number of significant variables per nucleotide step makes it difficult to achieve the simplicity of the Ramachandran maps, although attempts continue to be made in this direction [22]. DNAs have recently been seen to adopt a surprising range of conformations [23, 24], but it is once again RNAs and their protein complexes which represent the hardest challenge for future theoretical studies.

In all these areas, I think we would do well to remember that simple techniques, handled intelligently, can be as powerful, and sometimes have a further-reaching impact, than those relying on the latest feats of technical prowess.

## References

1. Pauling L, Corey RB, Branson HR (1951) Proc Natl Acad Sci USA 37: 205
2. Pauling L, Corey RB (1951) Proc Natl Acad Sci USA 37: 729
3. Perutz MF (1951) Nature 167: 1053
4. Perutz MF, Rossman MG, Cullis AF, Muirhead H, Will G, North ACT (1960) Nature 185: 416
5. Kendrew JC, Dickerson RE, Strandberg BE, Hart RG, Davies DR (1960) Nature 185: 422
6. (a) Shimanouchi T, Mizushima SI (1955) J Chem Phys 23: 707; (b) Lippincott ER, Schroeder R (1955) J Chem Phys 23: 1099; (c) Lifson S, Oppenheim I (1960) J Chem Phys 33: 109; (d) de Santis P, Giglio E, Liquori AM, Ripamonti A (1963) J Polym Sci Part A Chem 1: 1383
7. Scheraga HA (1960) J Am Chem Soc 82: 3847
8. Levinthal C (1966) Sci Am 214: 42
9. Sklenar H, Etchebest C, Lavery R (1989) Proteins Struct Funct Genet 6: 46
10. Ramachandran GN, Kartha G (1955) Nature 176: 593
11. Lee CH, Zimmerman SS (1995) J Biomol Struct Dyn 13: 201
12. Perczel A, Farkas O, Csizmadia IG (1996) J Comput Chem 17: 821
13. EU 3-D validation network (1998) J Mol Biol 276: 417
14. Rooman MJ, Kocher JPA, Wodak SJ (1991) J Mol Biol 221: 961
15. Skolnick J, Kolinski A, Ortiz AR (1998) J Biomol Struct Dyn 16: 381
16. Lafontaine I, Lavery R (1999) Curr Opin Struct Biol 9: 170
17. Westhof E (1993) J Mol Struct (Theochem) 286: 203
18. Swindells MB, Orengo CA, Jones DT, Hutchinson EG, Thornton JM (1998) BioEssays 20: 884
19. Swaminathan S, Ravishankar G, Beveridge DL, Lavery R, Etchebest C, Sklenar H (1990) Proteins Struct Funct Genet 8: 179
20. Ech-Cherif El-Kettani MA, Zakrzewska K, Durup J, Lavery R (1993) Proteins Struct Funct Genet 16: 393
21. Lavery R, Sklenar H (1989) J Biomol Struct Dyn 6: 655
22. Beckers MLM, Buydens LMC (1998) J Comput Chem 19: 695
23. Lebrun A, Lavery R (1997) Curr Opin Struct Biol 7: 348
24. Allemand JF, Bensimon D, Lavery R, Croquette V (1998) Proc Natl Acad Sci USA 95: 14152

Theor Chem Acc (2000) 103:259–262
DOI 10.1007/s002149900030

Theoretical
Chemistry Accounts
© Springer-Verlag 2000

# *Perspective*

# Perspective on "Inhomogeneous electron gas"

## Hohenberg P, Kohn W (1964) Phys Rev 136: B864

**Matthias Ernzerhof, Gustavo E. Scuseria**

Department of Chemistry, Rice University, P.O. Box 1892, Houston, TX 77005, USA

Received: 16 March 1999 / Accepted: 30 March 1999 / Published online: 28 June 1999

**Abstract.** The Hohenberg–Kohn theorems in "Inhomogeneous electron gas" established a whole new perspective for the study of electronic structure theory and marked the birth of modern density functional theory (DFT). In our view, DFT and wavefunction theories complement each other. Starting with the invention of the Kohn–Sham method a fruitful synthesis of DFT and wavefunction theories took place and the most powerful computational tools currently available are combinations of both methods. The Hohenberg–Kohn theorems inspire the quest for simple density functionals of increased accuracy. We believe that the synthesis of accurate density functionals and computationally efficient wavefunction methods will continue to dominate electronic structure theory.

**Key words:** Density functional theory – Hohenberg–Kohn theorems

## 1 Introduction

The electronic structure of molecules and atoms has long been at the focus of chemists' interests. Soon after the electron was discovered in 1897, electron theories of valence were developed [1, 2]. However, at that time, nothing was known about the driving force behind the formation of a chemical bond or an ion. The discovery of quantum mechanics in 1925 made it possible to address the fundamental questions of chemistry. Schrödinger's equation [3] became the central point in electronic structure theory. In 1927, Condon [4] gave a quantum mechanical explanation of the bond in $H_2$ and initiated molecular orbital theory. In the same year, Heitler and London [5] developed the valence-bond description of $H_2$.

With the development of computers and efficient algorithms, remarkable advances have been made in the calculation of wavefunctions. The starting point in conventional wave-mechanical treatments of electronic structure is the Hartree–Fock approximation, which is a molecular orbital type of approximation. In Hartree–Fock theory, the exchange energy is calculated exactly, but electron correlation is completely neglected. We understand now that this treatment often leads to an unbalanced description. A typical example is the rupture of a covalent bond. The correlation energy is larger in magnitude in the molecule compared to the atoms, and neglecting it introduces significant errors in the dissociation energy. Unfortunately, accurate calculations of the correlation energy using wavefunction techniques are still severely limited because of the steep scaling of the computational effort with molecular size.

Modern density functional theory (DFT) was born with the title paper and was developed in parallel with wavefunction methods. The Kohn–Sham method [6] is an example of a unification between pure DFT methods and wavefunction theory, a synthesis that has lead to some of the most powerful tools in computational quantum chemistry.

## 2 Early work in DFT

Shortly after the Schrödinger equation had been formulated it became obvious that finding its solution for the most simple atoms and molecules is a challenging problem. Back in the early days of quantum mechanics even the independent-particle problem posed to be a difficult task. In 1927, Thomas [7] and Fermi [8] independently introduced an approximation which avoids the calculation of single-particle orbitals. In the Thomas–Fermi approach, the electron density appears as the variational degree of freedom in an equation for the ground-state energy. For a given external potential $v(\mathbf{r})$, Thomas and Fermi introduced the minimization problem

$$E_v = \min_{\rho} \left[ \frac{3}{10} (3\pi^2)^{2/3} \int d^3r \rho^{5/3}(\mathbf{r}) \right.$$

$$\left. + \int d^3r v(\mathbf{r})\rho(\mathbf{r}) + U[\rho] - \mu \left( \int d^3r \rho(\mathbf{r}) - N \right) \right] . \quad (1)$$

---

*Correspondence to*: G. E. Scuseria

$\mu$ denotes the chemical potential which is adjusted such that the electron density $\rho$ integrates to the desired number of electrons (N).

$$U[\rho] = \frac{1}{2}\int \mathrm{d}^3r\mathrm{d}^3r'\rho(\mathbf{r})\rho(\mathbf{r}')/|\mathbf{r} - \mathbf{r}'| \qquad (2)$$

approximates the electron–electron interaction energy $V_{\mathrm{ee}}$. The remarkable step taken by Thomas and Fermi was to approximate the kinetic energy of a Slater determinant ($T_S$) by a functional of the electron density, i.e.,

$$T_S[\rho] = \frac{3}{10}(3\pi^2)^{2/3}\int \mathrm{d}^3r\rho^{5/3}(\mathbf{r}) . \qquad (3)$$

This expression gives the exact kinetic energy for a homogeneous electron gas of noninteracting electrons.

In 1930, Dirac [9] proposed that a density functional for exchange be added to the Thomas–Fermi energy expression (Eq. 1). Dirac's exchange functional

$$E_X = \frac{-3}{4}\left(\frac{3}{\pi}\right)^{1/3}\int \mathrm{d}^3r\rho^{4/3}(\mathbf{r}) \qquad (4)$$

is identical to the local density approximation (LDA) [6] for exchange.

These early steps in DFT (for a review see Ref. [10]) were too crude for chemical applications since, as later shown (see Ref. [11] and references therein), no chemical binding can be obtained within Thomas–Fermi theory. In 1951, Slater [12] derived a local approximation to the nonlocal exchange operator in the Hartree–Fock equations. His single-particle equations, in which the kinetic energy is evaluated exactly, are precursors of the Kohn–Sham equations [6]. A method which has an even greater similarity with the widely used Kohn–Sham scheme was proposed by Gáspár [13]. Gáspár used the LDA for the exchange energy in the Hartree–Fock approach and derived the corresponding self-consistent equation.

### 3 The paper on the inhomogeneous electron gas

The early density functional approximations were not deduced from Schrödinger's equation by a series of well-understood approximations. Rather they were based on an intuitive physical picture of electronic systems. As such, these approximations cannot be systematically improved, and it was not clear at the time if density functionals could in principle be exact.

The lack of a theoretical framework certainly did not promote the development of density functionals. This situation changed radically in 1964 with the paper of Pierre Hohenberg and Walter Kohn. Hohenberg and Kohn established a one-to-one correspondence between electron densities of nondegenerate ground states and external local potentials, $v(\mathbf{r})$, which differ by more than a constant. All physical properties obtainable with $v$ can therefore be expressed in terms of the electron density. It was thus established that, for example, the Thomas–Fermi approximation to the kinetic energy can in principle be refined to yield arbitrary precision. Hohenberg and Kohn defined the density functional $F[\rho]$

$$F[\rho] = \langle\psi[\rho]|\hat{T} + \hat{V}_{\mathrm{ee}}|\psi[\rho]\rangle , \qquad (5)$$

where $\psi$ denotes the nondegenerate ground-state wavefunction which yields $\rho$, and $\hat{V}_{\mathrm{ee}}$ is the electron repulsion operator. Having introduced the functional $F[\rho]$, Hohenberg and Kohn showed that for a given external potential, $v$, the ground-state density minimizes the energy functional

$$E_v[\tilde{\rho}] = \int \mathrm{d}^3r\tilde{\rho}(\mathbf{r})v(\mathbf{r}) + F[\tilde{\rho}] . \qquad (6)$$

$\tilde{\rho}$ denotes an appropriate trial density which integrates to the correct particle number. The problem of finding the ground-state energy for a given external potential has hereby been completely reformulated in terms of the electron density which is a function of three variables regardless of the number of electrons. There is, however, no systematic way of generating practical approximations to $F[\rho]$. Nevertheless, the knowledge that such a functional exists has greatly motivated physicists and chemists ever since.

For certain limiting cases, Hohenberg and Kohn were able to construct the functional $F[\rho]$. They showed that $F[\rho]$ can be expressed in terms of the electronic polarizability for densities of the form $\rho(\mathbf{r}) = \rho_0 + \tilde{\rho}(\mathbf{r})$ with $\tilde{\rho}(\mathbf{r})/\rho_0 \ll 1$. Furthermore, they formulated the gradient expansion, valid in the slowly varying limit, and presented a scheme by which the gradient coefficient can be obtained from the $n$th-order ($n = 1, 2, \ldots, \infty$) polarizabilities of the system. These two construction schemes for $F[\rho]$ have inspired subsequent approaches to developing approximate density functionals.

An important extension of the original Hohenberg–Kohn approach has been proposed by Levy [14, 15] based on earlier work by Percus [16]. The functional $F[\rho]$ of Hohenberg and Kohn is defined only for densities which are obtained from a nondegenerate ground-state wavefunction corresponding to an external local potential. Levy introduced a functional $\mathscr{F}$

$$\mathscr{F}[\rho] = \min_{\psi\to\rho}\langle\psi|\hat{T} + \hat{V}_{\mathrm{ee}}|\psi\rangle , \qquad (7)$$

where $\psi \to \rho$ means that the search extends over all antisymmetric wavefunctions yielding $\rho$. Obviously, this functional is defined for any density which can be obtained from an antisymmetric wavefunction. $\mathscr{F}$ can then be used in Eq. (6) to obtain the ground-state energy even if the ground state is degenerate.

### 4 The Kohn–Sham equations: a synthesis of wavefunction theory and DFT

Following the Hohenberg–Kohn paper, a successful synthesis was made between DFT and wavefunction theory. Useful approximations to the kinetic energy as a functional of the electron density are difficult to obtain. The kinetic energy is of the same order of magnitude as the total energy, and even small errors in the absolute value of the kinetic energy can lead to unacceptably large energy errors upon reactions or other chemical transformations.

The Hohenberg–Kohn theorems find a very important application in the derivation of the Kohn–Sham equations, in which the problem of approximating the noninteracting kinetic energy ($T_S$) is eliminated by introducing single-particle orbitals $\varphi_i$. The exact electron density is written as the electron density of a Slater determinant,

$$\rho(\mathbf{r}) = \sum_i^{occ} \varphi_i(\mathbf{r})\varphi_i(\mathbf{r}) \ . \tag{8}$$

The ground-state energy is split up according to

$$E = T_S + \int d^3 r\, v(\mathbf{r})\rho(\mathbf{r}) + U[\rho] + E_{XC}[\rho] \ , \tag{9}$$

where $E_{XC} = T - T_S + V_{ee} - U$ is the exchange–correlation energy, and $T_S$ is the kinetic energy of the determinant formed from the orbitals $\varphi_i$, i.e., $T_S = -1/2 \sum_i^{occ} \langle \varphi_i | \Delta | \varphi_i \rangle$. The variational equation (Eq. 6) is then simply solved by requiring that

$$\frac{\delta E}{\delta \varphi_i(\mathbf{r})} = 0 \quad \text{and} \quad \langle \varphi_i | \varphi_j \rangle = \delta_{ij} \ . \tag{10}$$

The resulting single-particle eigenvalue equations are the Kohn–Sham equations. The Hohenberg–Kohn theorems ensure that the exchange–correlation energy in Eq. (9) is a functional of the electron density.

## 5 Some developments in quantum chemistry initiated by the Hohenberg–Kohn paper

It is not practical to give a complete account of the impact of the Hohenberg–Kohn paper on quantum chemistry. Therefore, in this section, we mostly focus on a few aspects which influenced our own research. For a general overview we refer to Refs. [11, 17–19].

After the computational resources became available and reliable algorithms for the solution of the Kohn–Sham equations were in place, the LDA for exchange and correlation employed in the Kohn–Sham scheme turned out to be a powerful tool for electronic structure calculations [18]. Consequently, a large fraction of the early work in DFT focused on the question why LDA works so well even for very inhomogeneous systems. The exchange–correlation hole, which surrounds an electron, became the key in understanding the success of the LDA to $E_{XC}$. It turned out that many important constraints satisfied by the exact exchange correlation hole are also satisfied by its LDA (for an overview we refer to Refs. [11, 17, 18, 20–24]. Based on the understanding of the LDA improved approximations for exchange and correlation, known as generalized gradient approximations (GGAs) [25–31], were developed. GGAs reached a level of accuracy which made the Kohn–Sham method the most effective tool in quantum chemistry. An additional boost in accuracy was obtained with hybrid methods [32–39] in which a fraction of exact exchange (calculated from the Kohn–Sham wavefunction) is added to density functional approximations for exchange and correlation. This ensures that the long-range part of the exchange-correlation hole is accounted for. Recently, a new generation of kinetic energy density dependent functionals appeared in the literature [40–46]. In these methods, the kinetic energy density of the Kohn–Sham determinant and the electron density are employed to model $E_{XC}$. These functionals challenge the accuracy of hybrid schemes but avoid the computationally unfavorable evaluation of exact exchange.

In chemistry, excited states and time-dependent phenomena are often of interest. Although the Hohenberg–Kohn variational principle applies only to ground states, various extensions of DFT have been developed over the years which successfully model excitation energies and time-dependent phenomena (for an overview see Refs. [47–49]).

For us, a very exciting development initiated by the Hohenberg and Kohn paper is that insight into the problem of exchange and correlation became the essential ingredient in the development of approximations. Exchange and correlation effects are condensed into the relatively simple expressions of present days functionals and can sometimes be explained by a few general conditions for $E_{XC}$ [30, 46]. This is a significant advance in our understanding of electronic structure provided by DFT.

It is often said that the lack of a systematic procedure to improve the accuracy of an approximate density functional is a disadvantage of DFT; however, at a high level of accuracy probably any systematic approximation scheme for the correlation energy produces a considerable amount of information which is not relevant for the problem at hand. The lack of a systematic procedure to construct density functionals and the knowledge of its existence motivates us to uncover the essentials of electronic structure. Often it turns out that difficult problems in wavefunction theory can be solved with simple density functionals and, in other cases, computationally inexpensive wavefunction methods increase the accuracy of approximate density functionals. We expect that in the foreseeable future electronic structure theory will benefit from new ideas emerging from the synthesis of DFT and wavefunction theory.

*Acknowledgements.* Our DFT work is supported by the National Science Foundation (grant no. CHE-9618323) and the Welch Foundation.

## References

1. Lewis GN (1916) J Am Chem Soc 38: 762
2. Pauling L (1948) The nature of the chemical bond. Cornell University Press, Ithaca
3. Schrödinger E (1926) Ann Phys 79: 361
4. Condon EU (1927) Proc Natl Acad Sci USA 13: 466
5. Heitler W, London F (1927) Z Phys 44: 455
6. Kohn W, Sham LJ (1965) Phys Rev 140: A1133
7. Thomas LH (1927) Proc Camb Philos Soc 23: 542
8. Fermi E (1928) Z Phys 48: 73
9. Dirac P (1930) Proc Camb Philos Soc 26: 376
10. March NH (1957) Adv Phys 6: 1
11. Parr RG, Yang W (1989) Density-functional theory of atoms and molecules. Oxford University Press, Oxford
12. Slater JC (1951) Phys Rev 81: 385
13 Gáspár R (1954) Acta Phys Acad Sci Hung 3: 263
14. Levy M (1979) Proc Natl Acad Sci USA 76: 6062
15. Levy M (1982) Phys Rev A 26: 1200

262

16. Percus JK (1978) Int J Quantum Chem 13: 89
17. Dreizler RM, Gross EKU (1990) Density functional theory. Springer, Berlin Heidelberg New York
18. Jones RO, Gunnarsson O (1989) Rev Mod Phys 61: 689
19. Joubert DP (ed) (1998) Density functionals: theory and applications. Lecture notes in physics, vol 500. Springer, Berlin Heidelberg New York
20. Gunnarsson O, Jonson M, Lundqvist BI (1979) Phys Rev B 20: 3136
21. Perdew JP, Kurth S (1998) In: Joubert DP (ed) Lecture notes in physics, vol 500. Density functionals: theory and applications. Springer, Berlin Heidelberg New York, pp 8–59
22. Ernzerhof M, Burke K, Perdew JP (1996) In: Seminario JM (ed) Recent developments and applications of modern density functional theory, theoretical and computational chemistry. Elsevier, Amsterdam, pp 207–238
23. Ernzerhof M, Perdew JP, Burke K (1996) In: Nalewajski R (ed) Topics in current chemistry, vol 180. Density functional theory I. Springer, Berlin Heidelberg New York, pp 1–30
24 Burke K, Perdew JP, Ernzerhof M (1998) J Chem Phys 109: 3760
25. Perdew JP, Wang Y (1986) Phys Rev B 33: 8800
26. Becke AD (1988) Phys Rev A 38: 3098
27. Lee C, Yang W, Parr RG (1988) Phys Rev B 37: 785
28. Perdew JP (1991) In: Ziesche P, Eschrig H (eds) Electronic structure of solids '91. Akademie, Berlin, pp 11–20
29. Perdew JP, Burke K, Wang Y (1996) Phys Rev B 54: 16533
30. (a) Perdew JP, Burke K, Ernzerhof M (1996) Phys Rev Lett 77: 3865; (b) Perdew JP, Burke K, Ernzerhof M (1997) Phys Rev Lett 78: 1396
31. (a) Perdew JP, Chevary JA, Vosko SH, Jackson KA, Pederson MR, Singh DJ, Fiolhais C (1992) Phys Rev B 46: 6671; (b) Perdew JP, Chevary JA, Vosko SH, Jackson KA, Pederson MR, Singh DJ, Fiolhais C (1993) Phys Rev B 48: 4978
32. Becke AD (1993) J Chem Phys 98: 1372
33. Becke AD (1993) J Chem Phys 98: 5648
34. Becke AD (1996) J Chem Phys 104: 1040
35. Perdew JP, Ernzerhof M, Burke K (1996) J Chem Phys 105: 9982
36. Ernzerhof M (1996) Chem Phys Lett 263: 499
37. Ernzerhof M, Perdew JP, Burke K (1997) Int J Quantum Chem 64: 285
38. Ernzerhof M (1998) In: Joubert DP (ed) Lecture notes in physics, vol 500. Density functionals: theory and applications. Springer, Berlin Heidelberg New York, pp 60–90
39. Ernzerhof M, Scuseria GE (1999) J Chem Phys 110: 5029
40. Becke AD, Roussel MR (1989) Phys Rev A 39: 3761
41. Becke AD (1994) Int J Quantum Chem Symp 28: 625
42. Van Voorhis T, Scuseria GE (1998) J Chem Phys 109: 400
43. Becke AD (1998) J Chem Phys 109: 2092
44. Schmider HL, Becke AD (1998) J Chem Phys 108: 9624
45. Perdew JP, Kurth S, Zupan A, Blaha P (1999) Phys Rev Lett 82: 2544
46. Ernzerhof M, Scuseria GE (1999) J Chem Phys (in press)
47. Gross EKU, Kohn W (1990) In: Trickey SB (ed) Advances in quantum chemistry, vol 21. Academic Press, New York, pp 255–291
48. Gross EKU, Dobson JF, Petersilka M (1996) In: Nalewajski R (ed) Topics in current chemistry. vol 181. Density functional theory II, Springer, Berlin Heidelberg New York, pp 81–160
49. Casida ME (1996) In: Seminario JM (ed) Recent developments and applications of modern density functional theory, theoretical and computational chemistry. Elsevier, Amsterdam, pp 391–434

Theor Chem Acc (2000) 103:263–264
DOI 10.1007/s002149900020

Theoretical
Chemistry Accounts
© Springer-Verlag 2000

*Perspective*

# Perspective on "Correlations in the motion of atoms in liquid argon"

## Rahman A (1964) Phys Rev 136: 405

Peter J. Rossky

Department of Chemistry and Biochemistry, University of Texas at Austin, Austin, TX 78712-1167, USA

Received: 19 February 1999 / Accepted: 25 March 1999 / Published online: 21 June 1999

**Abstract.** This article provides a discussion of the title paper by Aneesur Rahman. Here, the use of numerical integration of Newtonian equations of motion to simulate the classical dynamics of a liquid system with arbitrary continuous interatomic potentials was first introduced. Emphasis is placed on the author's motivations, the depth of his investigation, and the legacy that both the methodology and the style of investigation has left us.

**Key words:** Molecular dynamics simulation – Liquid dynamics

The terminology "molecular dynamics simulation" is no doubt a familiar one to essentially all graduate students in science. It is recognized as a description of a computational procedure for studying molecular motion at an atomic level, based (most frequently) on a defined model for the molecular-level potential energy. The dynamics is simulated via numerical integration of (typically) classical equations of motion. It has now become as common a term as "quantum chemistry", the latter term describing the use of approximate numerical solutions of the Schrödinger equation to describe molecular electronic structure. There is no question that the use of these dynamical methods to study the molecular dynamics of condensed matter systems, ranging from pure liquids to solutions to solids to biological assemblies, has exploded in the last 20 years. The technique has evolved to the point where implementations can be practically routine, allowing wide access outside the theoretical community. In addition, the use of molecular dynamics simulation, when explicitly coupled with supplemental, experimentally derived, structural constraints, has become a critical tool in the refinement of the structural content of X-ray crystallography and multidimensional NMR data on large systems.

The inauspicious first sentence of the abstract of the title paper summarizes the content of the paper suc-cinctly: "A system of 864 particles interacting with a Lennard–Jones potential and obeying classical equations of motion has been studied on a digital computer ... to simulate molecular dynamics in liquid argon ..." Monte Carlo simulation methods [1] had already been implemented in similar contexts, and the classical dynamical simulation of hard-sphere liquids had been implemented in the classic work of Alder and Wainwright [2, 3]; however, the methods used in Monte Carlo approaches did not generalize to Hamiltonian molecular dynamics, and the algorithms for hard-sphere fluids did not generalize to more realistic potentials. This generalization, in the use of a finite-difference numerical integrator, was a key step.

What cannot be seen in the abstract are the several themes of this short (seven-page) paper. These themes set standards both for the quality of simulation and the style of attack on complex systems. The paper discusses the methodology of numerical finite-difference integration of the dynamical equations of motion in detail in an appendix, with proper attention to numerical accuracy. The limitations of the simple pair-additive interatomic potential, of the cutoff range of the interaction, and of the periodic boundary conditions are also all noted. Validation of the underlying potential model is seriously considered via direct comparison with available experimental data for the atomic diffusion constant and for the interatomic-pair distribution function, from X-ray scattering.

Nevertheless, the paper is clearly focused not on the novel methodology, but rather on the physics of the liquid and, more importantly, on the connection of the physics to experimental observables. (In fact, according to one personal account by the author, the significance of the methodological step was not apparent to him at the time of publication.) The methodology was a tool, and the author's goal was to evaluate the physical content of inelastic neutron scattering (which addresses relative interatomic dynamics) and to evaluate the veracity of available analytical models of these dynamics.

In the words of the paper, "If neutron scattering data of unlimited accuracy and completeness was available, then the kind of work presented here would serve the useful but unexciting purpose of confirming the results already obtained with neutrons." While perhaps a bit exaggerated, the point of the author was to emphasize that the simulation served in two critical ways that remain essential in all current and future implementations. First, the results of the simulation could be used to interpret experimental observations. In the case at hand for Rahman, the aspect at issue was the physical picture of motion; for example, do the atoms exhibit hopping dynamics, with infrequent relatively large displacements, or small displacement random-walk-like dynamics? Do the statistical dynamical correlations exhibit remnants of solidlike behavior, such as the non-monotonic oscillating decay of velocity fluctuations, or monotonic, simple overdamped frictional decay? Second, recognizing that one cannot reasonably expect to answer all questions directly via simulation, the results could be used to test the validity of simpler and more analytical models, and, further, suggest ways for improvement of the veracity of such models. This emerged as well from the title paper, in the context of Brownian/Langevin dynamics, as well as in the representation of relative-pair motion. It is notable that the paper includes only six literature references, the majority of these being to papers reporting experimental investigations and those describing proposed theoretical models for liquid dynamics.

The capacity of theoretical and computational chemistry continues to expand, and nowhere is the new growth more evident than in combined "molecular dynamics simulation" and "quantum chemistry". The products, variously referred to as quantum, or mixed quantum–classical, molecular dynamics or as ab initio molecular dynamics, have already been widely used to investigate systems as complex as biological molecules and polymer films as well as electronic processes in solution, without complete dependence on molecular mechanics models. The role of these methods both as primary simulation tools and as methods for development of more reliable model potentials will certainly continue to grow. At the same time, there are substantial opportunities for new methods which can focus on the quantum aspects of nuclear motion, and not just on the quantum electronic structure, and for those which can address quantum state-to-state processes.

The title paper provides a template for such future investigations in a number of fundamental ways. I had the exceptional opportunity to first learn about molecular dynamics methodology by working with Anees for an extended summer workshop nearly 25 years ago. One critical element to his effectiveness that was immediately evident was the importance of connecting a theoretical framework to the physical insight that arose from the numerical output of such "computer experiments". Insight into the appropriate framework for an analysis is as important as any other element of molecular dynamics simulation – this feature is evident in the title paper, and it has lost no element of significance in the intervening years or for the future.

# References

1. Metropolis N, Rosenbluth AW, Rosenbluth MN, Teller AH, Teller E (1953) J Chem Phys 21: 1087
2. Alder BJ, Wainwright TE (1957) J Chem Phys 27: 1208
3. Alder BJ, Wainwright TE (1959) J Chem Phys 31: 459

Theor Chem Acc (2000) 103:265–269
DOI 10.1007/s002149900067

Theoretical
Chemistry Accounts
© Springer-Verlag 2000

*Perspective*

# Perspective on "Self-consistent equations including exchange and correlation effects"

## Kohn W, Sham LJ (1965) Phys Rev A 140:133–1138

**E.J. Baerends**

Scheikundig Laboratorium der Vrije Universiteit, De Boelelaan 1083, 1081 HV Amsterdam, The Netherlands

Received: 16 February 1999 / Accepted: 22 June 1999 / Published online: 9 September 1999

**Abstract.** The paper by Kohn and Sham (KS) is important for at least two reasons. First, it is the basis for practical methods for density functional calculations. Second, it has endowed chemistry and physics with an independent particle model with very appealing features. As expressed in the title of the KS paper, correlation effects are included at the level of one-electron equations, the practical advantages of which have often been stressed. An implication that has been less widely recognized is that the KS molecular orbital model is physically well-founded and has certain advantages over the Hartree–Fock model. It provides an excellent basis for molecular orbital theoretical interpretation and prediction in chemistry.

**Key words:** Density functional theory – Kohn–Sham – Electron correlation – Molecular orbital theory

## 1 Introduction

The title paper by Kohn and Sham (KS) [1] has of course to be considered in conjunction with its predecessor, the paper by Hohenberg and Kohn (HK) [2]. The HK paper established for a many-particle system with some two-particle interaction, where all particles move in a given local potential, $v(\mathbf{r})$, and with a restriction to systems that have nondegenerate ground states, that there is a one-to-one mapping between the potential, $v(\mathbf{r})$, the particle density, $\rho(\mathbf{r})$, and the ground-state wavefunction, $\Psi_0$,

$$\rho(\mathbf{r}) \leftrightarrow v(\mathbf{r}) \leftrightarrow \Psi_0 \ . \tag{1}$$

If $\Psi_0$ is a functional of the density, then so are all properties, since any property may be determined as the expectation value of the corresponding operator, $\hat{O}$ say, $O[\rho] = \langle \Psi_0[\rho] | \hat{O} | \Psi_0[\rho] \rangle$. In particular the kinetic energy

is also a functional of the density, $T[\rho]$, the electron–electron interaction energy, $W[\rho]$, and the total energy, $E[\rho]$. HK also established the existence of the total energy functional, $E_v[\rho]$, for which the ground-state energy, $E_0$, of the system with external potential $v$, is a lower bound,

$$E_v[\rho] = \langle \Psi[\rho] | \hat{T} + \hat{V} + \hat{W} | \Psi[\rho] \rangle$$
$$= T[\rho] + \int \rho v \, \mathrm{d}r + W[\rho] \geq E_0 \ . \tag{2}$$

It is easy to generalize Eqs. (1) and (2) if the ground state is degenerate [3]. A particularly elegant definition of the functional $F[\rho] = T[\rho] + W[\rho]$, which automatically covers the case of ground-state degeneracy, has been provided in Levy's constrained search formulation [4]

$$F_{\mathrm{L}}[\rho] = T[\rho] + W[\rho] = \min_{\Psi \to \rho} \langle \Psi | \hat{T} + \hat{W} | \Psi \rangle \ , \tag{3}$$

where the minimum is to be searched over all possible wavefunctions that yield the given $\rho$ as the density. There is no denying the great importance of these theorems, but it may still be argued that the step taken in the KS paper has been as important: it certainly has been for chemistry. Equation (2) holds the promise of a very efficient route to total energies of many-electron systems, from a Euler–Lagrange equation for the density, if good approximations for $T[\rho]$ and $W[\rho]$ could be found. This would effectively reduce the very high dimensional problem of the calculation of the many-particle wavefunction, $\Psi_0$, to the determination of just the simple function in 3D real space, $\rho(\mathbf{r})$; yet, Eq. (2) has found very little practical application. The reason is that it is very difficult to develop sufficiently accurate density functionals, in particular, for the kinetic energy. This is a problem that already plagued the Thomas–Fermi approach, and although the HK theorems give a much more theoretically sound basis to density functional theory (DFT), its practical importance might not have risen above that of the Thomas–Fermi model if it had not been for the KS one-electron model.

So it is the second step, made in the KS paper, that has been essential for DFT to become the widely applied method for electronic structure calculations that it is today. We will comment briefly on the historical context of the KS paper. Then a few remarks on practical (computational) aspects of the KS equations will be made and finally the conceptual implications of the KS one-electron or molecular orbital (MO) model and its importance for chemistry now and in the future will be commented on.

## 2 Historical context

In the many years that have passed since the publication of the KS paper, the way that their results are usually presented, and where the emphasis is put, has of course shifted. It is now customary to stress from the outset that the KS theory introduces a system of noninteracting electrons, moving in a local potential, $v_s(\mathbf{r})$. The ground-state wavefunction of the KS system – a single Slater determinant of the lowest $N$ orbitals – will yield precisely the same electron density as the exact interacting electron system with potential $v(\mathbf{r})$. So the KS Hamiltonian, $\hat{H}_s$, is just a sum of one-electron Hamiltonians, $\hat{h}_s$, and the wavefunction of the KS system is a simple one-determinantal wavefunction,

$$\hat{H}_s = \sum_i \hat{h}_s(i) = \sum_i \left( -\frac{1}{2}\nabla^2(i) + v_s(\mathbf{r}_i) \right)$$

$$\hat{h}_s(1)\phi_i(1) = \varepsilon_i \phi_i(1)$$

$$\Psi_s = |\phi_1(1), \phi_2(2), \ldots, \phi_N(N)| \tag{4}$$

$$\rho_s(\mathbf{r}) = \sum_{i=1}^{N} \sum_s |\phi_i(\mathbf{r}, s)|^2 = \rho^{\text{exact}}(\mathbf{r})$$

The KS paper does not even mention $\Psi_s$, but it concentrates on the orbital equations and right away specializes on systems with slowly varying density. It uses the local density approximation (LDA) from the outset. Considerable attention is given to corrections to be introduced, dependent on the gradient of the density, to take variation of the density into account. It is historically understandable, in view of the importance at the time (and for some 10–15 years after the publication date) of Slater's exchange approximation [5], that extensive comparison is made to this exchange approximation, which led to a potential proportional to $\rho(\mathbf{r})^{1/3}$. It is pointed out that the variation procedure that is inherent to the KS approach leads to an exchange potential that is two-thirds of Slater's, a point that had been shown earlier by Gáspár [6]. Slater next introduced a constant $\alpha$ in his exchange potential; this could be determined according to various criteria. There has been considerable debate over this constant, which has now long subsided, the KS treatment with its explicit

inclusion of a correlation functional and potential having been generally adopted. In their paper KS also discuss the possibility that one could use a nonlocal exchange potential, and add only a local correlation potential. Although they stress (in the title of the paper) that they introduce self-consistent equations that include correlation effects, it is, with hindsight, interesting to observe that KS only mentioned in a note added in proof that the paper has actually achieved the possible replacement of the many-electron problem with an exactly equivalent set of self-consistent one-electron equations. In this note they introduce the local exchange–correlation potential $v_{xc}(\mathbf{r}) = \delta E_{xc}[\rho]/\delta\rho(\mathbf{r})$ which features in the exact $v_s$,

$$v_s(\mathbf{r}) = v(\mathbf{r}) + v_{\text{Coul}}(\mathbf{r}) + v_{xc}(\mathbf{r}) \;, \tag{5}$$

where $v_{\text{Coul}} = \int \rho(\mathbf{r}_2)/r_{12}\, d\mathbf{r}_2$. The exchange–correlation energy, $E_{xc}$, which is a crucial quantity in DFT, is defined only within the context of the KS one-electron model. We will comment later on $E_{xc}$ more extensively, but note at this point that it is exactly this feature, of treating the complicated many-electron system in principle exactly with only the computational expense of a self-consistent-field calculation, which has been so appealing in KS theory. However enticing the promise of computational simplicity and efficiency is, the possibility to treat correlation at the one-electron level has been hard to accept for a quantum chemistry community that was steeped in the belief that correlation was by definition everything that could not be covered at the one-electron level.

## 3 Computational considerations

The KS equation shares with Slater's X$\alpha$ method the replacement of the nonlocal exchange operator of Hartree–Fock (HF) theory by a local potential. This particular feature of Slater's exchange approximation had been extensively exploited in solid-state physics, and initially the interest in quantum chemistry almost exclusively focussed on the efficiency of the scattered-wave technique borrowed from physics for solving the X$\alpha$ and LDA one-electron equations [7]. The required muffin-tin approximation of the potential, however, proved to be too severe in molecules, prohibiting reliable bond energy and structure determinations. The obvious alternative is to use basis sets [8, 9], as was common in quantum chemistry, and evaluate the matrix elements of $v_{xc}$ by numerical integration. The approximate Diophantine method was introduced by Ellis for this purpose [10]. Although this method was sufficiently precise to calculate bond energies and structures, it was not capable of high numerical precision. In the mid 1980s the problem of 3D numerical integration of molecular integrands, with their characteristic singularities at the nuclear sites, was solved simultaneously, in somewhat different ways, by Becke [11] and by Boerrigter et al. [12].

Precise 3D numerical integration affords solution of the KS equations with a basis-set-expansion method. It would, however, make a KS calculation more expensive than a HF calculation, simply adding the burden of numerical integration of the $v_{xc}$ matrix elements to the two-electron integral evaluation still required for the Coulomb potential matrix elements. The latter would cause an $n^4$ scaling of DFT calculations just as they did for HF calculations (ignoring distance cut-offs). KS DFT has often inappropriately been quoted as being relatively efficient, and in particular as scaling better than HF theory. As a matter of fact, the reputation of "local-density methods" was initially (1970s first half of 1980s) largely based on computational efficiency, the often higher accuracy than HF methods, in particular for transition-metal systems (complexes, cluster compounds) being considered a nice advantage, but not a decisive one since it was ill-understood. However, efficiency compared to the standard HF calculations of the day was achieved by the introduction of auxiliary basis sets for the expansion of the density [8]. This density-fitting, which has been employed in most DFT codes, having later been adopted by Sambe and Felton [13] and improved by Dunlap et al. [14], can also be formulated as an insertion of a resolution of the identity and is sometimes denoted as such (RI-$J$[15]). Auxiliary basis sets can just as well be used to provide better scaling (as $n^3$) of the Coulomb operator matrix elements in HF calculations. The KS equations do not, with basis-set methods, inherently scale better than the HF equations. The real virtue of the KS method is the possibility to deal with effects of electron correlation, in particular the calculation of the correlation energy, with the computational expense of a one-electron method, while post-HF methods become very expensive indeed. The success of the KS method therefore hinges on the availability of good approximations to $E_{xc}$, which have been much improved by the introduction of the generalized gradient approximations (GGA) [16, 17].

## 4 Electron correlation in a one-electron method

The great appeal of the KS scheme is, as previously mentioned, that it offers the promise of correlated energies with a one-electron method, which scales much more favourably than traditional correlation calculations. The definition of the crucial quantity, the exchange–correlation energy, $E_{xc}[\rho]$, has been somewhat confusing for quantum chemists, who have all been trained with the "standard model" where the "best" one-electron model is HF, the exchange energy is defined within that model, and the correlation energy is by definition the difference between the HF energy and the exact energy. KS introduced $E_{xc}[\rho]$, which they said is "by our definition" the exchange and correlation energy. Until rather recently [18–20], it had almost never been emphasized that KS's definition was not the standard one. The KS definition is straightforward once one puts aside the reservation that has long existed, certainly in the quantum chemistry community, against using the one-determinantal wavefunction of KS orbitals, $\Psi_s$. The general feeling has been for a long time that the KS orbitals were only mathematical constructs that one could use to build the total density, but that

these orbitals were deprived of physical meaning. They should therefore not be used to build a wavefunction like $\Psi_s$. It was considered to be against the spirit of DFT anyway to go back to a wavefunction. I believe that this conviction has considerably delayed a full appreciation of the meaning and status of the KS orbitals. They are much more useful and physically meaningful than they were held to be, and they can play a significant role in interpretation and prediction in chemistry, combining the elegance and transparency of the qualitative MO theory with the high accuracy that modern density functionals afford. We return to this point later.

KS did not stress a point which received more attention later: their basic ansatz is that for every interacting electron system with a given external local potential, $v(\mathbf{r})$, there actually exists a corresponding noninteracting system, i.e. there exists a local potential, $v_s(\mathbf{r})$, such that a noninteracting system of electrons moving in that potential will have the same density as the interacting one. In spite of occasional assertions to the contrary, it has not been proven that for every interacting system the KS potential $v_s(\mathbf{r})$ exists. What has been proven is the following. Consider an interacting electron system with potential $v(\mathbf{r})$. To this system belongs a ground-state density (nondegenerate case) or a set of pure state and ensemble representable densities in the case of degeneracy. The HK functional, $F[\rho] = T[\rho] + W[\rho]$, can be extended by Levy's constrained-search approach [4] to a functional $F_L[\rho]$ (see Eq. 3) that can be proven to have a functional derivative $\delta F_L[\rho]/\delta\rho(\mathbf{r})$, with the potential $v(\mathbf{r})$ as a tangent functional, at those densities and nowhere else [21–23]. Similarly, for noninteracting electrons the analogous constrained-search functional, $T_L[\rho]$ (only the kinetic energy remains when $\hat{W} = 0$), can be defined. Again, when the noninteracting system is characterized by a local potential $v_s(\mathbf{r})$, the differentiability of $T_L[\rho]$ has been established at its pure state or ensemble representable densities, with $v_s(\mathbf{r})$ as its functional derivative. However, these results, although sometimes quoted as such, do not constitute the desired proof, which requires the demonstration that the set of interacting densities is contained in the set of noninteracting ground-state densities. It is a sobering thought that after such a long time this proof has not been found. In practice this has not been a drawback: a breakdown of the KS ansatz has not been observed so far. Nevertheless, one should not naively expect that for every nondegenerate interacting ground-state density a nondegenerate (one-determinantal) KS system actually exists. It was pointed out by Levy [24] that there are many densities (for instance, convex combinations of densities belonging to a set of degenerate ground states) that are not pure-state representable. Given an interacting ground-state density, one therefore has to consider the possibility that this density is not representable by a pure state of a noninteracting KS system, but that it is ensemble-representable in the noninteracting case. Cases have been identified where a nondegenerate ground-state density of an interacting system was ensemble-representable by degenerate KS ground-state densities [25].

Assuming that $v_s$ exists, and therefore the KS orbitals can be calculated, the quantity $E_{xc}$ is defined as the remaining unknown part of the exact total energy, $E$, the kinetic energy, the nuclear attraction energy, and the classical electron–electron repulsion energy following straightforwardly from the knowledge of the density and orbitals

$$E = T_s[\rho] + \int v(\mathbf{r})d\mathbf{r} + \frac{1}{2}\iint \frac{\rho(\mathbf{r})\rho(\mathbf{r}')}{|\mathbf{r}-\mathbf{r}'|}d\mathbf{r}\,d\mathbf{r}' + E_{xc}[\rho] \ . \tag{6}$$

$E_{xc}$ is different from the corresponding HF quantities. Its meaning becomes very simple when we consider the energy of the single-determinantal wavefunction of KS orbitals, $\Psi_s$,

$$E^{KS} = \langle \Psi_s|\hat{H}|\Psi_s\rangle$$
$$= T_s + \int v(\mathbf{r})d\mathbf{r} + \frac{1}{2}\iint \frac{\rho(\mathbf{r})\rho(\mathbf{r}')}{|\mathbf{r}-\mathbf{r}'|}d\mathbf{r}\,d\mathbf{r}' + W_x \ , \tag{7}$$

where the exchange energy, $W_x$ has the same form as in the HF case but is to be evaluated with the KS orbitals. Defining the correlation energy, $E_c$ in DFT as the difference between the energy of the determinant with the exact KS orbitals and the exact energy, one obtains

$$E_c = E - E^{KS} = E_{xc} - W_x$$
$$E_c^{HF} = E - E^{HF} \tag{8}$$

where $E^{HF}$ is the traditional HF energy. So the DFT definitions of exchange and correlation are natural when one recognizes that KS introduced a one-electron model that may be considered an alternative to the long-established HF model, with the KS determinant taking the place of the HF determinant. The difference between the DFT and standard definitions of exchange and correlation energies was only given attention a long time after the KS paper appeared (see Ref. [26] for a review). The KS noninteracting electron system and its determinantal wavefunction have a somewhat different status than the HF model, being related to exact properties (notably the density) of the real system. A clear advantage of the HF model from an operational point of view, and therefore of the related definition of the correlation energy, is the possibility to obtain very accurate HF wavefunctions, no knowledge of the exact solution being required. Since the HF determinant minimizes the expectation value of the Hamiltonian, the DFT correlation energy will always be larger in an absolute sense, i.e. more negative, than the HF-based correlation energy,

$$\langle \Psi^{HF}|\hat{H}|\Psi^{HF}\rangle \le \langle \Psi_s|\hat{H}|\Psi_s\rangle$$
$$E_c \le E_c^{HF} \tag{9}$$

The reservation against the use of $\Psi_s$ was caused by the suspicion that the KS orbitals might not be "normal" orbitals, and should not be used in the same way as HF or extended Hückel orbitals for qualitative rationalization of experimental trends. On the other hand, the

orbitals of X$\alpha$ (i.e. exchange-only LDA) or LDA and later GGA calculations have always been used as such. It has been argued [20, 26, 27] that this is actually perfectly in order. It has emerged from detailed comparisons between HF and KS orbitals, and between components of the energy (kinetic energy, electron–nuclear attraction, classical electron–electron repulsion, exchange energy) in the two models, that the KS orbitals and determinantal wavefunction are not very different from their HF counterparts, and to the extent that they are, one might argue that they have some advantages. This may best be understood from an analysis of the KS potential $v_s(\mathbf{r})$, which may be written as the sum of a number of physically meaningful terms,

$$v_s(\mathbf{r}) = v(\mathbf{r}) + v_{Coul}(\mathbf{r}) + v_x^{hole}(\mathbf{r})r + v_c^{hole}(\mathbf{r})$$
$$+ v_{c,kin}(\mathbf{r}) + v^{resp}(\mathbf{r}) \ . \tag{10}$$

The leading terms in $v_s$ are the attractive nuclear field, $v(\mathbf{r})$, and the repulsive Coulomb potential of the electronic charge density, $v_{Coul}(\mathbf{r})$. The next important term is the exchange hole potential. It is local and therefore different from the HF exchange operator, but it also represents an exchange hole comprising one electronic charge, and taken together the first three terms make the KS potential rather similar to the HF operator. These terms determine the rough features of the spectrum of orbital energies and the shape of the MOs. KS orbitals are therefore most of the time very similar to HF orbitals. The difference is in the last three terms of $v_s$. In particular, it is interesting to note that the potential of the Coulomb hole, $v_c^{hole}$, is present in $v_s$. It has been stressed [26, 27] that in situations of strong left–right correlation, such as in dissociating bonds, this potential makes an important difference. While in that case the HF orbitals and the electron density become much too diffuse, the Coulomb hole potential causes the KS orbitals not to suffer from this deficiency. As a consequence, the correlation error in individual energy terms such as the kinetic energy or the electron–nuclear energy, is much smaller in the KS case. This illustrates in what sense and with what effect electron correlation is embodied in the one-electron KS model. It is interesting to observe that the incorporation of the potential of both the exchange and correlation holes, i.e. neglecting $v_{c,kin}$ and $v^{resp}$ in Eq. (10), gives precisely the potential advocated by Slater for use in one-electron equations just 2 years after his introduction of the exchange approximation [28]. Slater conjectured that this potential, based on the conditional probability to find other electrons around a given position of the reference electron, might give optimal orbitals for a configuration interaction. Speculating that the X$\alpha$ potential might actually incorporate part of the correlation effects led Slater to denote the X$\alpha$ potential as an exchange–correlation potential (this has been corroborated by the finding that the current exchange approximations of DFT, which are still predominantly the Slater exchange, actually incorporate the nondynamical correlation effects in chemical bonds [27]). Löwdin [29] was inspired by Slater's proposal to investigate the equations for optimal orbitals in a configuration interaction. These natural orbitals obey an equation in which the physics of the exchange–correlation hole potential, $v_{xc}^{hole}$ (sometimes referred to as the Slater–Löwdin potential), can indeed be recognized; however, the application of only $v_{xc}^{hole}$ in one-electron equations leads to too contracted orbitals and density, and the remaining potentials, $v_{c,kin}$ and $v^{resp}$ in the KS potential play an important role to build precisely the right shape of the total density. It is possible to relate special features in the potentials $v_{c,kin}$ (peak behaviour around the bond midpoint) and $v^{resp}$ (repulsive steps in atomic shells) to specific electron correlation effects [26].

It is worth noting that not just the occupied KS orbitals but also the virtual KS orbitals and orbital energies can be used for qualitative interpretation. They surely are not devoid of physical meaning, but they are directly related to excitation energies and to the electronic nature of excited states. This has become evident very recently from the development of time-dependent DFT methods for response properties, in particular excitation energies [30–32]. (see also the results in Ref. [33]).

The properties of the KS orbitals recommend them strongly for use in the qualitative MO theories of chemistry. It would be hard to find a better MO theoretical context in which to apply concepts such as "charge control" and "orbital control" than the KS one-electron model. In view of the good quality of several energy components (those directly following from the electron density are in fact exact) the KS one-electron model can be used reliably for interpretation using analysis along the lines delineated by Morokuma [34] within the HF model. Energy contributions such as classical electrostatic interaction, Pauli repulsion, and donor–acceptor interaction (orbital interaction energy) can be calculated and used for interpretation in the KS model [35–38]. We refer to Ref. [39] for a recent review with more complete discussion and examples. It is thought provoking that KS MO theory is actually related to exact energetics by way of the exchange–correlation functional, $E_{xc}$. The implication is that the concepts mentioned above, that traditionally were used only qualitatively since they were defined in a one-electron model (HF) that did not yield the exact energy, can now be used with accurate energies. Within the KS model, the MO-type analysis constitutes a complete conceptual framework. Since qualitative understanding and interpretation is the primary objective of this type of analysis, this may in practice not be such a huge advantage over MO theory based on semiempirical methods as it may seem at first sight, but it doubtless recommends the KS one-electron model as the preferred basis for MO theory.

We conclude that the 1965 paper by KS has had a singular influence in the electronic structure theory of physics in the last part of the century. Its impact in chemistry has enormously increased in the last decade. Its implications are only gradually being understood and its full depth may not yet have been fathomed. There can be no doubt that it will exert a large influence well into the next century.

## References

1. Kohn W, Sham LJ (1965) Phys Rev A 140: 1133
2. Hohenberg P, Kohn W (1964) Phys Rev B 136: 864
3. Dreizler RM, Gross EKU (1990) Density functional theory: an approach to the many-body problem. Springer, Berlin Heidelberg New York
4. Levy M (1979) Proc Natl Acad Sci USA 76: 6062
5. Slater JC (1951) Phys Rev 81: 385
6. Gáspár R (1954) Acta Phys Acad Sci Hung 3: 263
7. Johnson KH, Smith FC Jr (1972) Phys Rev B 5: 831
8. Baerends EJ, Ellis DE, Ros P (1973) Chem Phys 2: 41
9. Rosén A, Ellis DE, Adachi H, Averill FW (1976) J Chem Phys 65: 3629
10. Ellis DE (1968) Int J Quantum Chem Symp 2: 35
11. Becke AD (1988) J Chem Phys 88: 2547
12. Boerrigter PM, te Velde G, Baerends EJ (1988) Int J Quantum Chem 33: 87
13. Sambe H, Felton RH (1975) J Chem Phys 62: 1122
14. Dunlap BI, Connolly JWD, Sabin JR (1979) J Chem Phys 71: 3396
15. Eichkorn K, Treutler O, Öhm H, Häser M, Ahlrichs R (1995) Chem Phys Lett 240: 283
16. (a) Perdew JP (1986) Phys Rev B 33: 8822; (b) Perdew JP (1986) Phys Rev B 34: 7406 (erratum)
17. Becke A (1988) Phys Rev A 38: 3098
18. Sahni V, Levy M (1985) Phys Rev B 33: 3869
19. Gross EKU, Petersilka M, Grabo T (1996) In: Laird BB, Ross RB, Ziegler T (eds) Chemical applications of density functional theory. ACS Symposium Series 629. American Chemical Society, Washington, D.C., p 42
20. Baerends EJ, Gritsenko OV, van Leeuwen R (1996) In: Laird BB, Ross RB, Ziegler T (eds) Chemical applications of density functional theory. ACS Symposium Series 629. American Chemical Society, Washington, D.C., p 20
21. Lieb EH (1982) Int J Quantum Chem 24: 243
22. Englisch H, Englisch R (1983) Phys Status Solidi B 123: 711
23. Englisch H, Englisch R (1983) Phys Status Solidi B 124: 373
24. Levy M (1982) Phys Rev A 26: 1200
25. Schipper PRT, Gritsenko OV, Baerends EJ (1998) Theor Chem Acc 99: 329
26. Baerends EJ, Gritsenko OV (1997) J Phys Chem A 101: 5383
27. Gritsenko OV, Schipper PRT, Baerends EJ (1997) J Chem Phys 107: 5007
28. Slater JC (1953) Phys Rev 91: 528
29. Löwdin PO (1955) Phys Rev 97: 1474
30. Petersilka M, Gossmann UJ, Gross EKU (1996) Phys Rev Lett 76: 1212
31. Petersilka M, Gross EKU (1996) Int J Quantum Chem Symp 30: 181
32. Casida M (1995) In: Chang DP (ed) Recent advances in density functional methods, Vol 1. World Scientific, pp 155
33. Savin A, Umrigar CJ, Gonze X (1998) Chem Phys Lett 288: 391
34. Morokuma K (1977) Acc Chem Res 10: 244
35. Ziegler T, Rauk A (1979) Inorg Chem 18: 1558
36. Ziegler T, Rauk A (1979) Inorg Chem 18: 1755
37. Baerends EJ, Rozendaal A (1986) In: Veillard A (ed) Quantum chemistry: the challenge of transition metals and coordination chemistry. Reidel, Dordrecht, p 159
38. Baerends EJ (1992) In: Pacchioni G, Bagus PS (eds) Cluster models for surface and bulk phenomena. NATO ASI Series, Series B: Physics Plenum, New York, p 189
39. Bickelhaupt FM, Baerends EJ (1999) In: Boyd D, Lipkowitz K (eds) Reviews of computational chemistry, Vol. 15, Wiley, New York, to be published

Theor Chem Acc (2000) 103:270–272
DOI 10.1007/s002149900043

Theoretical
Chemistry Accounts
© Springer-Verlag 2000

*Perspective*

# Perspective on "Exchange reactions with activation energy. I. Simple barrier potential for (H, H₂)"

## Karplus M, Porter RN, Sharma RD (1965) J Chem Phys 43: 3259–3287

**George C. Schatz**

Department of Chemistry, Northwestern University, Evanston, IL 60208-3113, USA

Received: 12 January 1999 / Accepted: 14 April 1999 / Published online: 14 July 1999

**Abstract.** This paper provides an overview of the title paper by Karplus, Porter and Sharma, including a description of the history that led to this work, some details of the calculation and the results, and a discussion of subsequent developments in the field which were stimulated by this work.

**Key words:** Quasiclassical trajectory method – Reactive cross section – Thermal rate coefficient – Potential-energy surface – Quantum scattering

The paper by Karplus, Porter and Sharma (KPS) [1] is, from my perspective, the most important early (pre-1970) piece of computational work in gas-phase chemical reaction dynamics. In it, the commonly used quasiclassical trajectory (QCT) method was described for three-dimensional atom–diatom reactive collisions (i.e., A + BC → AB + C), and was applied to the H + H₂ reaction to determine cross sections and thermal rate constants. In 35 years of subsequent work on gas-phase reaction dynamics, the QCT method has remained largely the same, and it continues to be a standard tool for studying quantum state-resolved dynamical processes.

This paper also provided much of the early stimulus for developing semiclassical and quantum mechanical theories of chemical reaction dynamics, which is a research field that continues to be active. In addition, this paper provided the foundation for molecular dynamics studies of chemical reactions in condensed phases, including applications to gas–surface scattering and biomolecular simulation.

To understand the importance of this paper, let me give some background. The possibility of using first-principles theoretical methods to determine the rate constants for simple gas-phase bimolecular reactions became of interest shortly after the discovery of quantum mechanics in the 1920s. The reaction H + H₂ was a focal point of much of this work, due to the simplicity of the electronic Schrödinger equation (just three electrons), the simplicity of the nuclear motion (just three nuclei), and the fact that measurements of the thermal rate constant were available from studies of the conversion of para hydrogen into ortho hydrogen, and from studies of isotopic labeled reactions such as D + H₂ and H + D₂.

One important accomplishment of work on H + H₂ during the 1920s and 1930s was the development of a potential surface that determines the forces between the atoms during reaction. This was done for H + H₂ initially by London, Eyring and Polanyi (LEP) [2] using valence-bond methods. The resulting surface was not very accurate, but it was good enough to give useful qualitative information about the nature of nuclear motions during reactive collisions. In fact, in the mid 1930s, Hirschfelder, Eyring, and Topley [3] attempted to do the first classical molecular dynamics study of H + H₂ using this LEP surface. This required numerical computations to solve the classical equations of motion, and at the time these had to be done by hand and thus were very tedious. As a result, only a short segment of one trajectory was determined, and no results that could be compared with experiment were obtained.

Further developments in this field waited until computers were developed. In the late 1950s and early 1960s computer programs were developed to perform trajectory studies of several atom–diatom chemical reactions [4]. For the most part these calculations were simplified in important ways to make them feasible, such as by reducing the motions to one or two rather than three dimensions, or by greatly restricting the number of trajectories integrated so that only a coarse-grained description of the reaction was obtained. In addition, many of the early studies used inaccurate potential-energy surfaces, and so the results were of only qualitative use for interpreting experiments.

By the mid 1960s, computers had advanced to the point where a detailed three-dimensional classical molecular dynamics study of an atom–diatom reaction was feasible (although still requiring a substantial amount of programming and data processing). In addition the potential-energy surface for H + H$_2$ had been improved by Porter and Karplus [5] and others, such that it was now quite realistic (though still not chemically accurate). This set the stage for the work of KPS [1].

The KPS calculation was based on integrating Hamilton's equations of motion for the time evolution of the Cartesian components of the Jacobi coordinates that describe the three-atom system. A fourth-order Runge–Kutta method was used for the numerical integration, and the computations were done using an IBM 7090-4 computer at the IBM Watson Research Center and the Columbia Computing Center. The computation time per trajectory was listed as 10 s using a time step of 0.025 fs.

The QCT approach that was used by KPS involves choosing the initial H$_2$ vibrational energy to equal the quantum energy for the initial state of interest. The H$_2$ rotational angular momentum was correspondingly chosen to be an integer multiple of $\hbar$. The other variables such as vibrational and rotational phases, and the impact parameter were sampled by Monte Carlo methods, which means that they are chosen randomly from the appropriate probability distributions. The QCT procedure thus has the feature that it mimics quantum mechanics as far as the initial conditions are concerned but the collisions themselves are purely classical. No attempt was made in the KPS study to assign quantum numbers to the final conditions of the reactive trajectories, but this is now frequently done, and the term "QCT" commonly refers to calculations in which final quantum numbers are assigned based on the vibrational and/or rotational states of the products of the trajectory calculations [6].

The results presented by KPS were mostly in the form of integral cross sections as a function of collision velocity and thermal rate constants as a function of temperature. There were no experimental cross sections to compare with back then, so most of the analysis was concerned with the comparison of thermal rate constants with either experiment, or with other theories such as transition-state theory. The comparisons with experiment were actually quite good, but KPS included many "cautions" towards the end of their paper to note the many uncertainties associated with these comparisons. These uncertainties include errors in the potential surface used, uncertainties in the experimental results, and errors due to the use of classical mechanics. They conclude by saying that "no unequivocal answer [could] be given concerning ... the direct applicability of the present study to specific chemical reactions." The authors were, in retrospect, far too pessimistic about the accuracy and usefulness of their results, as I now discuss.

There have been numerous tests since the KPS study of the QCT approach, mostly based on comparisons of QCT results with those from quantum scattering theory (QST) calculations for the same potential-energy surface. These scattering calculations involve solving the Schrödinger equation for the collision dynamics, which is the quantum mechanical analog of what KPS did with classical mechanics, and thus should provide rigorously correct results.

Many of the early comparisons between QCT and QST results suggested that there were serious errors in the QCT cross sections [7], but eventually it was discovered that the errors were actually in the quantum dynamics results. Once this problem was corrected [8], the comparisons of QCT and QST results were much better, with essentially quantitative agreement in the total reactive cross sections except at very low energies where barrier tunneling is important. Even more-detailed comparisons of QCT and QST results have been made more recently [9] using more-accurate potential-energy surfaces. These studies also include extensive comparison with recently measured experimental cross sections and thermal rate constant data [10]. These comparisons show unequivocally that the QCT method is capable of describing most aspects of the reaction dynamics with sufficient accuracy to match experimental data. Alternative algorithms to the QCT method have been proposed and tested (such as schemes for constraining zero-point energy throughout the collision event [11], and schemes for imposing constraints on final state energies (so as to correct for the lack of microscopic reversibility in the QCT method) [12]); however, the QCT method developed by KPS generally gives the best agreement with quantum cross sections [13].

The KPS paper stimulated research in several new directions, and ultimately spawned new fields. Many researchers, including Karplus, got interested in the development of QST of chemical reactions, and this led to accurate quantum descriptions of the H + H$_2$ reaction [8] a decade after the KPS paper. There was also significant interest in the application of QCT methods to gas-phase reactions other than H + H$_2$, and in fact this approach is now considered to be a standard research tool for studying gas-phase reaction dynamics of relevance to laser chemistry, combustion chemistry, atmospheric chemistry, and other applications.

Another direction of research that was fostered by the KPS work was the development of semiclassical theories of chemical reactions. This development arose because the QCT method is an ad hoc procedure for mimicking quantum effects in chemical reaction dynamics wherein quantization is imposed initially and finally but not in-between. In semiclassical methods, one imposes the $\hbar \to 0$ limit of quantum mechanics in a consistent way throughout the reactive collision process. The search for a consistent semiclassical theory eventually produced classical S-matrix theory [14], which is a topic of continuing interest in gas-phase dynamics [15], and it also led to the development of Gaussian wave-packet methods for simulating chemical reactions [16].

Finally, we should note that the success of the KPS work in describing gas-phase chemical reaction dynamics has stimulated continuing research aimed at using first-principles methods to describe reaction dynamics in gas–surface scattering, in biomolecular processes, and in other processes taking place in condensed phases [17].

*Acknowledgement.* This research was supported by NSF grant CHE-9873892.

## References

1. Karplus M, Porter RN, Sharma RD (1965) J Chem Phys 43: 3259
2. (a) London F (1929) Z Elektrochem 35: 532; (b) Eyring H, Polanyi M (1931) Z Phys Chem (Munich) B 12: 279
3. Hirschfelder J, Eyring H, Topley B (1936) J Chem Phys 4: 170
4. (a) Wall FT, Hiller LA Jr, Mazur J (1958) J Chem Phys 29: 255; (b) Blais NC, Bunker DL (1962) J Chem Phys 37: 2713
5. Porter RN, Karplus M (1964) J Chem Phys 40: 1105
6. Schatz GC, ter Horst M, Takayanagi T (1998) In: Thompson DL (ed) Modern methods for multidimensional dynamics computations in chemistry World Scientific, Singapore, p 1
7. (a) Wolken G, Karplus M (1974) J Chem Phys 60: 351; (b) Choi BH, Tang KT (1974) J Chem Phys 61: 2462
8. Schatz GC, Kuppermann A (1976) J Chem Phys 65: 4668
9. (a) Aoiz FJ, Banares L, D'Mello MJ, Herrero VJ, Sáez Rábanos V, Schnieder L, Wyatt RE (1994) J Chem Phys 101: 5781; (b) Blais NC, Zhao M, Truhlar DG, Schwenke DW, Kouri DJ (1990) Chem Phys Lett 166: 11; (c) Erratum (1992) Chem Phys Lett 188: 368
10. (a) Schnieder L, Seekamp-Rahn K, Borkowski J, Wrede E, Welge KH, Aoiz FJ, Bañares L, D'Mello MJ, Herrero VJ, Sáez Rábanos V, Wyatt RE (1995) Science 269: 207; (b) Xu H, Shafer-Ray NE, Merkt F, Hughes DJ, Springer M, Tuckett RP, Zare RN (1995) J Chem Phys 103: 5157
11. (a) Guo Y, Thompson DL, Sewell TD (1966) J Chem Phys 104: 576; (b) Bowman JM, Gazdy B Sun Q (1989) J Chem Phys 91: 2859; (c) Miller WH, Darling CL (1989) J Chem Phys 91: 2863; (d) Lim KE, McCormack DA (1995) J Chem Phys 102: 1705
12. (a) Nyman G, Davidson J (1990) J Chem Phys 92: 2415; (b) Miller JA (1981) J Chem Phys 74: 5120
13. Kumar S, Sathyamurthy N, Ramaswamy R (1995) J Chem Phys 103: 6021
14. (a) Miller WH (1970) J Chem Phys 53: 1949; (b) Marcus RA (1970) Chem Phys Lett 7: 525; (c) Miller WH (1971) Acc Chem Res 4: 161; (d) Miller WH (1974) Adv Chem Phys 25: 69; (e) Miller WH (1975) Adv Chem Phys 30: 77
15. (a) Herman MF (1994) Annu Rev Phys Chem 45: 83; (b) Sepulveda MA, Grossmann F (1996) Adv Chem Phys 96: 191
16. Heller EJ (1975) J Chem Phys 62: 1544
17. Jungwirth P, Gerber RB (1999) Chem Rev (in press)

Theor Chem Acc (2000) 103:273–275
DOI 10.1007/s002149900060

# Perspective

# Perspective on "On the correlation problem in atomic and molecular systems. Calculation of wavefunction components in Ursell-type expansion using quantum-field theoretical methods"

## Čížek J (1966) J Chem Phys 45:4256

**Rodney J. Bartlett**

Quantum Theory Project, University of Florida, Gainesville, FL 32611, USA

Received: 12 March 1999 / Accepted: 4 June 1999 / Published online: 4 October 1999

**Abstract.** This is a personal perspective on the paper of Jiri Čížek that initiated the use of coupled-cluster (CC) theory in chemistry. As CC theory is now the method of choice for most highly accurate quantum chemical studies, its influence is profound.

**Key words:** Correlation problem – Coupled-cluster theory – Size-entensivity – Many-body theory – Diagrammatic methods

Looking at the modern state of the treatment of the correlation problem in the electronic structure of molecules, few papers can be said to be more instrumental to that effort than the paper of Jiri Čížek in 1966, that first detailed the equations of the coupled-cluster (CC) approach. This too-brief, 11-page paper summarizes his dissertation work of 2 years earlier [1]. As the title says his use of "quantum-field theoretical methods" was intimidating to most quantum chemists of the time, which no doubt kept many from appreciating the significance of this work in 1966.

At that time, the best-known method for electron correlation in molecules was undoubtedly configuration interaction (CI). This tool had developed in the hands of Slater and Condon, with early applications by Boys, Parr, Matsen, and their coworkers around 1950 (see Ref. [2] for an excellent review). Somewhat less known to the quantum chemistry community was the parallel development in the mid 1950s of the correlation problem in physics that originated with Brueckner [3] and Goldstone [4], termed many-body perturbation theory (MBPT) because it was applicable to many-electron systems. This feature, that we now call size-extensivity [5], was not shared by CI, but was a necessity for the physics applications to nuclear matter and the electron gas. Important questions at this time included the correlation treatment of the high- and low-density electron

gas, and in the first case it was possible to make an infinite sum of "ladder" diagrams to get a good answer, and in the other case, an infinite sum of "ring" diagrams; however, atoms and molecules have regions of both high and low electron density. Initial atomic applications of MBPT were made by Kelly [6]. Sinanoglu [7] and Nesbet [8] took some of the cluster ideas into the quantum chemistry community, though retaining many of the trappings of the more familiar CI world. Hence, the first complete solution of these diverse problems was offered by CC theory. In one convenient, conceptual, and computational framework it consolidates the infinite sum of ring and ladder diagrams and, indeed, all other kinds of diagrams, to offer a unified, size-extensive treatment of electron correlation.

The concept of an exponential wavefunction of a cluster operator

$$T = T_1 + T_2 + T_3 + T_4 + \cdots$$
$$\Psi = \exp(T)\Phi_0$$

for the correlation problem, where $\Phi_0$ represents the wavefunction for an independent particle model such as Hartree–Fock and $T_n$ the "connected" cluster operators for $n$ electrons, is implicit in the linked-diagram expansion of Brueckner and Goldstone, and was, perhaps, stated most explicitly in a paper of Hubbard [9] in 1957. Hubbard gives the formula

$$\Psi = \exp\{S'_L(0)\}\Psi_0 \ .$$

The CC wavefunction had been considered by Coester and Kummel [10] as the "exponential S" ansatz at about the same time in the nuclear physics literature; however, none of these authors took the next step to develop explicit equations for the cluster amplitudes $\{t_{ijk\ldots}^{abc\ldots}\}$, which appear in the cluster operators,

$$T_n = \frac{1}{n!} \sum_{\substack{i,j,k,\ldots \\ a,b,c,\ldots}} t_{ijk\ldots}^{abc\ldots} \{a^\dagger i b^\dagger j c^\dagger k \ldots\}$$

and are the object of a CC calculation. To do this in a very general way required more powerful tools than the familiar rules based upon the Slater–Condon rules for matrix elements of the Hamiltonian between excited determinants; hence, the use of "diagrammatic methods". Čížek took his approach from a Russian language technical note of Tolmachev on "The field form of the perturbation theory applied to many-electron problems of atoms and molecules" written in 1963. In two pages, Čížek develops all the second-quantized tools he requires, defining the operators, their anticommutation relations, their normal ordered form, and their contractions or "pairings". He emphasizes the effective use of normal ordered operators and their contractions. This, in my opinion, still offers the most economical development of even algebraic second-quantized operator derivations, and is even more effective once we employ diagrams.

In the next six pages he introduces in rapid order H, T, S, R, and M skeleton diagrams, and all but the most avid reader is likely to be lost! In addition there is a need for "weight factors", equivalent line definitions, equivalent skeleton definitions which depend upon an appreciation of "topological deformations", open lines, closed lines, sign factors, etc. This leads to statements such as "In this special case, the M skeletons consist of a set of T skeletons (the empty set being included) each of which has two open paths. The possible R skeletons, having none or two open paths, which can be obtained from these M skeletons..." Whew! This is nothing less than a course on diagrams in six pages!

Using all the above, Čížek presents the explicit, spin orbital and spin-adapted CC doubles equations (CCD) i.e., $T = T_2$ (then called coupled-pair many-electron theory) in terms of one- and two-electron integrals over an orthogonal basis set. Assisted by Joe Paldus with some computations, he also reports some CCD results for $N_2$, which though limited to only $\pi_u$ to $\pi_g$ excitations, uses ab initio integrals in an Slater-type-orbital basis. He also does the full CI calculation to assess convergence, a tool widely used in Čížek's and Paldus' work and by most of us, today. He also reports results for the minimum-basis $\pi$-electron approximation to benzene.

Though used in some semiempirical applications by Paldus and Čížek [11] and one ab initio study [12] (see later), the CCD equations were not implemented into general purpose programs until 1978 by me and Purvis [5] and Pople et al. [13]. This general implementation included allowing for the open-shell case subject to an unrestricted Hartree–Fock reference function.

In another landmark paper, Paldus, Čížek and Shavitt [12] partly considered effects of other cluster operators, $T_1$ and $T_3$, in a minimum-basis ab initio study of $BH_3$. In 1982 we presented the detailed equations and implementation of CCSD for open and closed shells [14]. Two years later, we considered the initial effects of connected triple excitations (CCSDT-1) [15] and their noniterative inclusion CCSD[T] [16]. Today full CCSDT [17] and full CCSDTQ [18] are known and applied, and their noniterative forms such as CCSDT($Q_f$) [19], and the ubiquitous CCSD(T) [20], which is a noniterative version of CCSDT-1 and a slight modification of

CCSD[T] [21]; that adds the initial perturbative effects of triples to singles along with those due to doubles that define [T].[1]

For ground states (or the lowest state of a given symmetry) single-reference CC methods occupy the critical role in the now well-known paradigm of improving quantum chemical calculations: SCF < MBPT(2) ~ MBPT(3) < CCD < CCSD ~ MBPT(4) < CCSD(T) < CCSDT < CCSDT(Q) < CCSDTQ < full CI. Combined with systematically converging basis sets such as the cc-pVNZ basis [22], this sequence enables quantum chemistry to make reliable predictions about the structure, spectra, and most transition states for molecules with an ease of application that only requires a choice of basis set, level of correlation, and multiplicity, albeit with substantial computer time for CCSD(T) and beyond. These applications depend upon further extensions in the CC treatment of analytical gradients [23] and even Hessians [24]. The former provides first-order properties as a consequence of the "relaxed" and "response" one- and two-particle density matrices [21]. To quote Dunning, who comes from the multireference CI background, "...of the methods of widespread use today, the CCSD(T) method is the only one that provides a consistently accurate description of molecular interactions for all interaction scales investigated, from more than 200 kcal/mol to 0.02 kcal/mol" [25].

Now, depending only on a choice of basis and level of correlation just as for ground states, there are generalizations for the treatment of excited, ionized and electron-attached states [21, 26] and analytical gradients for such states [27]; for the treatment of second- and higher-order properties, static [28] and frequency-dependent [29]; for relativistic problems [30]; for explicit $r_{12}$ CC theory [31]; and for various multireference generalizations of CC theory [32]. In brief, CC theory has now assumed a dominant place in the field of quantum chemistry, and it started with the pioneering efforts of Jiri Čížek and Joe Paldus.

### References

1. (a) Čížek J (1966) J Chem Phys 45: 4256; (b) Čížek J (1969) Adv Chem Phys 14: 35
2. Shavitt I (1998) Mol Phys 94: 3
3. Brueckner KA (1955) Phys Rev 97: 1353
4. Goldstone J (1957) Proc R Soc Lond A 239: 267
5. (a) Bartlett RJ, Purvis GD (1978) Int J Quantum Chem 14: 561; (b) Bartlett RJ, Purvis GD (1980) Phys Scr 21: 255
6. Kelly HP (1969) Adv Chem Phys 14: 129, and references therein
7. Sinanoglu O (1964) Adv Chem Phys 5: 35, and refererences therein
8. Nesbet RK (1969) Adv Chem Phys 14: 1 and references therein
9. Hubbard J (1957) Proc R Soc Lond A 240: 539
10. (a) Coester F (1958) Nucl Phys 1: 421; (b) Coester F, Kummel H (1960) Nucl Phys 17: 477
11. Čížek J, Paldus J, Sroubkova L (1969) Int J Quantum Chem 3: 149

---

[1] Numerically, the doubles-to-triples effect [T] is approximately 95%–105% of the result, but the additional triples-to-singles part added in (T) is often numerically significant for difficult cases, and tends to reduce an overestimate of the correlation energy

12. Paldus J, Čížek J, Shavitt I (1972) Phys Rev 5: 50
13. Pople JA, Krishnan R, Schlegel HB, Binkley JS (1978) Int J Quantum Chem 14: 561
14. Purvis GD, Bartlett RJ (1982) J Chem Phys 76: 1910
15. Lee YS, Kucharski SA, Bartlett RJ (1984) J Chem Phys 81: 5906
16. Urban M, Noga J, Cole SJ, Bartlett RJ (1985) J Chem Phys 83: 4041
17. (a) Noga J, Bartlett RJ (1987) J Chem Phys 86: 7041; (b) Noga J, Bartlett RJ (1988) J Chem Phys 89: 3401; (c) Scuseria GE, Schaefer HF III (1988) Chem Phys Lett 152: 382; (d) Watts J, Bartlett RJ (1990) J Chem Phys 93: 6104
18. Kucharski SA, Bartlett RJ (1992) J Chem Phys 97: 4282
19. Kucharski SA, Bartlett RJ (1998) J Chem Phys 108: 9221
20. Raghavachari K, Trucks GW, Pople JA, Head-Gordon M (1989) Chem Phys Letters 157: 479
21. Bartlett RJ (1995) In: Yarkony DR (ed) Modern electronic structure theory, part II. World Scientific, Singapore, pp 1047–1131 and references therein
22. (a) Dunning TH Jr (1989) J Chem Phys 90: 1007; (b) Woon DE, Dunning TH Jr, Harrison RJ (1995) J Chem Phys 103: 4572
23. (a) Adamowicz L, Laidig WD, Bartlett RJ (1984) Int J Quantum Chem Symp 18: 245; (b) Scheiner AC, Scuseria G, Rice JE, Lee T, Schaefer HF. III (1987) J Chem Phys 87: 5368; (c) Salter EA, Trucks GW, Bartlett RJ (1989) J Chem Phys 90: 1752
24. Gauss J, Stanton JF (1997) Chem Phys Lett 276: 70
25. Dunning TH, J Phys Chem (to be published)
26. (a) Mukherjee D, Pal S (1989) Adv Quantum Chem 20: 91 and references therein; (b) Kaldor U (1991) Theor Chim Acta 80: 427; (c) Bartlett RJ, Stanton JF (1994) In: Lipkowitz KB, Boyd DB (eds) Reviews of computational chemistry vol 5. VCH Weinheim, pp 65–169
27. (a) Stanton JF, Gauss J (1994) J Chem Phys 100: 4695; (b) Stanton JF, Gauss J (1994) J Chem Phys 101: 8983
28. Perera SA, Nooijen M, Bartlett RJ (1996) J Chem Phys 104: 3290
29. (a) Monkhorst HJ (1977) Int J Quantum Chem Symp 11: 421; (b) Stanton JF, Bartlett RJ (1993) J Chem Phys 99: 5178
30. (a) Kaldor U, Eliav E (1998) Adv Quantum Chem 31: 313, and references therein; (b) Lindgren I (1998) Mol Phys 94: 19
31. Noga J, Klopper W, Kutzelnigg W (1997) In: Bartlett RJ (ed) Recent advances in coupled-cluster methods. World Scientific, Singapore, pp 1–48
32. Paldus J (1992) In: Wilson S, Diercksen GHF (eds) Methods in computational molecular physics. NATO Advanced Study Institute Series B, vol 293. Plenum, New York, pp 99–194

Theor Chem Acc (2000) 103:276–277
DOI 10.1007/s002149900014

Theoretical
Chemistry Accounts
© Springer-Verlag 2000

*Perspective*

# Multireference many-body methods. Perspective on "Linked-cluster expansions for the nuclear many-body problem"

Bradow BH (1967) Rev Mod Phys 39: 771

**Uzi Kaldor**

School of Chemistry, Tel Aviv University, 69978 Tel Aviv, Israel

Received: 20 January 1999 / Accepted: 2 March 1999 / Published online: 7 June 1999

**Abstract.** The role of the title paper in the history of multireference many-body methods is reviewed. Subsequent developments are described, and unsolved outstanding problems are also discussed.

**Key words:** Many-body methods – Multireference – Coupled cluster

Many-body methods, based on the linked-cluster expansion (LCE), were first developed by Brueckner [1] and Goldstone [2] in the 1950s for nuclear physics problems. Perturbation-theory applications to atomic and molecular systems (in a numerical, one-center frame) were pioneered by Kelly [3] in the early 1960s. Basis sets were later introduced, first in second-order [4] and then in third-order [5]. The 1970s saw a proliferation of molecular applications with basis sets, under the names of many-body perturbation theory (MBPT) [6] or the Møller-Plesset method [7]. Nowadays, many-body methods offer some of the most powerful tools in the quantum chemistry arsenal, in particular the coupled-cluster (CC) method, and are available in many widely used quantum chemistry program packages.

All the early applications, as well as the vast majority of many-body calculations performed to this day, are single-reference in character. They start from an appropriate single determinant, usually (but not necessarily) Hartree–Fock, and include correlation by finite-order perturbation or infinite-order summation of certain perturbation terms (the CC approach). The starting determinant may be closed-shell or open-shell: the latter leads to contaminated spin states and occasionally to broken spatial symmetry [8], but acceptable results are obtained in most cases. States involving degeneracy or quasidegeneracy, where a single determinant

does not provide an adequate starting point, are not amenable to this treatment.

The early 1960s saw several attempts to extend the linked-cluster expansion to multireference cases, to make possible the treatment of general open-shell systems. These attempts met with partial success, and it was Brandow [9] who first proved the LCE for multireference methods in his 1967 paper, opening the way to new types of applications. The first (short) part of the article is an elegant time-independent proof of the linked-cluster theorem for single-reference cases using diagrammatic methods (previous proofs used time-dependent evolution operators and infinite time integrations). Most of the paper is devoted to a detailed discussion of the degenerate case. A novel type of diagram is described; the so-called "folded" diagrams solved the difficulties which had stymied earlier attempts. It was now possible for the first time to have a linked-cluster, energy-independent expansion for degenerate systems (it is interesting to note that Sandars [10] came up with a similar solution 2 years later, apparently unaware of Brandow's paper: he used the term "backward" diagrams). A detailed proof of the LCE is given, rules for application are described, and subjects such as calculation of properties other than energy and handling of quasidegeneracy (as distinct from exact degeneracy) are also discussed in this comprehensive landmark paper.

The route to applications involving states of multireference character was now open. The first application of Brandow's method in molecular physics treated the excited states of $H_2$ at several internuclear separations [11]. Several other molecules followed. In the coming years, emphasis in the high-accuracy small-molecules end of quantum chemistry shifted from finite-order MBPT to the all-order CC method, thanks largely to the work of Bartlett, Pople and their coworkers [12]. Here the problem of degeneracy or quasidegeneracy and the need for multireference methods appeared again. Brandow's folded diagrams were used by Lindgren [13] to derive a linked (and therefore size-extensive) multireference CC method. Shavitt [14] proposed in a 1983 workshop that "a multireference CC formalism is probably the single most promising approach" to molecular computations.

What is the current situation in multireference many-body methods? On the low-order side, the second-order CASPT2 method [15] based on a large active space is highly successful. At the high-order or CC end, several methods exist, each with its advantages and shortcomings (for a review see Ref. [16]). The Fock-space method, which starts from a closed-shell state and reaches the state(s) of interest by adding and/or removing electrons from the reference, is the most widely used [17]. Its strong point is preservation of the full symmetry (both spin and spatial) of the system. The method has been particularly successful in calculating spectra of heavy atoms in the relativistic regime [18]. The main drawback is the existence of so-called "spectator" lines, with the same (valence) orbital going into and coming out of the diagram, creating terms which are formally $n$-excited but physically describe lower excitation numbers. The Hilbert-space approach [19] does not suffer from this problem, but it breaks symmetry and may generate substantially different energies for states which should be degenerate (e.g., $M_s = 1$ and 0 states of a triplet) [20]. A widely used alternative for open-shell systems which may be described by a single determinant is to start from the UHF or ROHF function. This may be problematic, leading in some cases to breaking geometrical symmetry of open-shell molecules [8]. The QRHF-CC method [21] starts from an uncorrelated function with correct spin and spatial symmetry, but breaks symmetry in the correlation part. It should be noted that different approaches yield, in general, different energies and other properties for the same system.

To conclude, much progress has been achieved in applying multireference many-body methods in quantum chemistry since Brandow made it possible by proving the linked-cluster expansion, but the problem cannot be considered fully solved. Further work is needed to realize the hope expressed by Shavitt [14] 15 years ago.

## References

1. Brueckner KA (1955) Phys Rev 97: 1353
2. Goldstone J (1957) Proc Roy Soc Lond Ser A 239: 267
3. Kelly HP (1969) Adv Chem Phys 14: 129, and references therein
4. Schulman JM, Kaufman DN (1970) J Chem Phys 53: 477
5. Kaldor U (1973) Phys Rev A 7: 427
6. Bartlett RJ, Silver DM (1974) Phys Rev A 10: 1927
7. Pople JA, Binkley JS, Seeger R (1976) Int J Quantum Chem Symp 10: 1
8. Kaldor U (1991) Chem Phys Lett 166: 599; (b) Kaldor U (1991) Chem Phys Lett 185: 131
9. Brandow BH (1967) Rev Mod Phys 39: 771
10. Sandars PGH (1969) Adv Chem Phys 14: 365
11. (a) Kaldor U (1973) Phys Rev Lett 31: 1338; (b) Kaldor U (1975) Phys Rev A 63: 2199
12. Bartlett RJ (1981) Ann Rev Phys Chem 32: 359, and references therein
13. Lindgren I, Morrison J (1982) Atomic many-body theory. Springer, Berlin Heidelberg New York
14. Shavitt I (1984) In: Dykstra CE (ed) Advanced theories and computational approaches to the electronic structure of molecules. Reidel, Dordrecht p 192
15. Andersson K, Roos BO (1975) In: Yarkoni DR (ed) Modern electronic structure theory, vol 1. World Scientific, Singapore, p 55
16. Mukherjee D, Pal S (1989) Adv Quantum Chem 20: 292
17. Kaldor U (1991) Theor Chim Acta 80: 427, and references therein
18. Kaldor U, Eliav E (1998) Adv Quantum Chem 31: 313, and references therein
19. Balková A, Kucharski SA, Meissner L, Bartlett RJ (1991) Theor Chim Acta 80: 321
20. (a) Berkovic S, Kaldor U (1992) Chem Phys Lett 199: 42; (b) Berkovic S, Kaldor U (1993) J Chem Phys 98: 3090
21. Rittby M, Bartlett RJ (1988) J Phys Chem 92: 3033

Theor Chem Acc (2000) 103:278–280
DOI 10.1007/s002149900039

Theoretical
Chemistry Accounts
© Springer-Verlag 2000

*Perspective*

# Finding the way through intermolecular forces. Perspective on "Permanent and induced molecular moments and long-range intermolecular forces"

## Buckingham AD (1967) Adv Chem Phys 12: 107–142

Clifford E. Dykstra

Department of Chemistry, Indiana University-Purdue University Indianapolis, 402 N. Blackford Street, Indianapolis, IN 46202, USA

Received: 15 February 1999 / Accepted: 16 March 1999 / Published online: 9 September 1999

**Abstract.** This is an overview of the title paper by A.D. Buckingham and a description of its impact and of many of the developments that it fostered.

**Key words:** Intermolecular forces – Permanent moment interactions – Polarization – Weak and long-range interaction

We almost all start learning chemistry from the standpoint of atoms being the stuff of matter, and molecules being the stable combinations of atoms we find and synthesize. Hence, molecular structure and bonding define chemistry. However, at another level, there is structure and bonding – weak bonding – of molecules aggregating, and thereby forming clusters, droplets, and condensed phases. Molecules (and atoms) can and do stick to each other without forming "chemical" bonds. In many ways, this is its own chemistry, with rules of structure, energetics, and dynamics that are different – with a different physical basis – from the rules we invoke for individual molecules and their reactions. The two are intertwined. Chemical bonding somehow affects properties that play a role in weak interaction, and the weak interaction of surrounding molecules can have a profound impact on chemical reactions. The path to our understanding of the chemistry associated with intermolecular interaction has been complicated and challenging. A key report in that path is one written by A.D. Buckingham for *Advances in Chemical Physics* [1], a report that appeared about two-thirds of the way through the century that is closing and a report that is still relevant for embarking on the study of intermolecular interaction phenomena.

The basic parts of chemical bonding, such as orbital and configuration mixing, are not the primary features of weak and long-range interaction. Some may disagree with that, but even they are likely to see weak interaction as a juxtaposition of several elements. Thus, anyone seeking fundamental understanding of weak interaction and intermolecular forces at some point has to examine and weigh competing and offsetting contributions. Buckingham's paper is an essential source for finding the way through this juxtaposition.

To establish the setting for David Buckingham's guiding review, one might go back a number of decades to the first ideas of nonbonding interactions between atoms and molecules; however, a fine place to start is just 10 years before the title paper when Coulson looked with great interest at the water–water interaction. A short article by Coulson [2] on this work is one of the best of those I know in recognizing very early why we would someday need to analyze in detail the contributions to hydrogen bonding, or more generally to weak interaction. He started by saying, "The hydrogen bond plays a very conspicuous role in human life, for it is responsible for the adherence of dirt to our skin, the structure of proteins, the action of glues and adhesives, the rigidity of many synthetic polymers, such as the polyamides, and a good many other biological phenomena." He concluded by foreshadowing, perhaps, current-day biomolecular simulations when he said that water "is so small that it can fit into quite small interstices between the chains in any multiple helix" and it can "lie between two helices, and be quite tightly bound to them." The weak and long-range interactions between molecules could be profoundly important, though the coarse level of analysis of the time needed improvement.

From among the things Coulson discussed, there was plenty to examine in depth. Buckingham clearly saw what deserved a high level of attention in the physics of weak interaction; it was electrical effects. One might say that Coulson provided the "why" and Buckingham defined the "what" for the story that was unfolding. Another major step was taken by Morokuma [3], and by others, though with Morokuma's ways taking the strongest hold and being developed the most extensively.

This step was the "how" – how to extract the pieces that make up weak interaction, with the source being the then new things known as ab initio electronic structure calculations. In the 1980s, work of Applequist [4] yielded the "machinery" for weak interaction through his brilliant structuring of electrical interaction analysis suited to extensive calculations and simulations. Hence, in roughly a quarter of a century, the time spanned by Refs. [1–4] insight and inspiration, fundamental understanding, and mathematical/computational technology had all come together to find a way through the tangle of competing effects that comprise intermolecular forces.

Buckingham's work is crucial in the saga. He started his 1967 report [1] with the statement "There is now general agreement that the significant forces between atoms and molecules have an electric origin." He was building a solid case for a consistently important role of electrical interaction, which he said consisted of the interactions of permanent moments and the interaction energy associated with distortion of the electronic structure due to a neighboring molecule's permanent moments. He referred to the first as the electrostatic energy and the second as the induction energy. He included a third attractive force, dispersion, pointing out its connection with polarizabilities. These terms are widely used in just that way today.[1] His overall thesis was that "detailed knowledge of molecular charge distributions and polarizabilities is essential for an understanding of intermolecular forces." The power and scope of this statement could only emerge over a few more decades as theoretical and computational techniques evolved that could really provide the detailed knowledge.

Buckingham's paper [1] gives a collection of formulas and symmetry analyses for evaluating electrical interaction energies and electromagnetic scattering, including treatment of periodically varying fields and special pairs of interacting systems. At the least, a lasting impact of this enumeration is in showing such analysis to be straightforward and readily adaptable to diverse molecular systems; it was doable and workable. This also made clear that chemists needed to think about electrical interaction more completely than that of two interacting molecular dipoles. That report connects several of the other major, contemporaneous developments from Buckingham [5–11] that together pushed spectroscopic study toward determination of precise quadrupole moment values, the study of many other electrical response properties with their anisotropies and frequency dependence, and optical activity generally.

Another key feature of the review [1] was the detailed connection of intermolecular interaction with the quantum mechanics of electronic structure. Buckingham offered a perturbative approach, taking the interaction as the perturbing Hamiltonian, and thereby associating the elements that make up the interaction with specific terms and specific orders of perturbation. This type of analysis of weak interaction has expanded significantly, reaching a high and fruitful level of sophistication via ab initio methodology [12].

In the last decade or so, ab initio methodology has made possible high quality determinations of electrical response properties, i.e., multipole moments, multipole polarizabilities, and multipole hyperpolarizabilities. This has opened the door to utilizing Buckingham's electrical interaction ideas to the fullest. One product was the Buckingham–Fowler model [13] for weakly bound complexes. This combined hard-sphere repulsion with permanent charge field interaction energies for the essentially attractive part of the potential. The model has yielded very nice determinations overall for shapes and structures of many complexes [13–16]. This is where my interest enters the story, since one of our developments is a model called molecular mechanics for clusters. It incorporated electrical interaction in a way that was open-ended with respect to the order of hyperpolarization and multipole [17]. Simple dispersion and exchange repulsion energies are added to the electrical energies to yield the complete potential. We had concluded from using our implementation of Morokuma's partitioning scheme that Buckingham's 1967 review had exactly the right direction. Electrical interaction was a crucial element for many types of clusters and it was something that could be well determined from properties intrinsic to the interacting species. We used Applequist's structure [4] to include polarization response, and we have found many times that, indeed, shapes and structures of clusters can be nicely determined, and also vibrational frequencies, certain transition moments, and property changes (dipole moments, nuclear quadrupole coupling constants, etc.).

The role of induction or polarization that Buckingham addressed [1] over 30 years ago is turning out to be every bit as important as anticipated. Drawing on work we have done for examples, we find electrical polarization effects account for an unexpected correlation among four experimentally measured quantities for a series of carbon monoxyheme proteins, the $^{13}C$ and $^{17}O$ chemical shifts of the CO, the CO group's stretching frequency, and the $^{17}O$ quadrupole coupling constant [18]. I think this fits what Coulson foresaw. We can achieve insight, in this case insight that aids structural analysis for these types of proteins, through simulation based on fundamental understanding of contributing effects in weak and long-range interaction, even intramolecular electrical interaction. In a recent analysis, very sizable medium effects in a representative segment of solid-state polyacetylene were found to be due to mutual polarization [19]. Clearly, the polarization or induction part of electrical interaction is proving to be a key aspect of nonbonding interaction, and this is why Buckingham's 1967 review [1] has had a strong impact.

Other successful models for weak interaction exist and more continue to appear, often with different application objectives and different ways of assembling and weighting components of intermolecular interaction. More and more, a foundational element for modeling

---

[1] Sometimes polarization energy is used for induction energy to avoid confusion with induction of current by a magnetic source. Also, permanent charge field interaction energy is sometimes used instead of electrostatic energy, reserving electrostatics for use in a collective sense that covers everything not varying in time, including polarization. Normally, usage is clear from the context.

efforts is electrical interaction. As well, simpler models such as those for molecular mechanics and molecular dynamics simulations are often being improved through more complete electrical analysis, such as incorporation of polarization effects.

Where are we heading as we enter the next millenium? Ten to fifteen years ago, we could study intermolecular interaction quite accurately and in detail with ab initio methods, but usually for systems with only two molecules – two small molecules. With the tremendous advances in computer technology and ab initio methods since then, what could be done for two small molecules can be done for tens of molecules now. Hence, models that were essential for, say, trimers and tetramers are being supplanted, and one might conclude that all this physical insight developed for weak interaction will have diminishing practical value. I do not share that conclusion. Instead, I see the expansion of the problems tackled by ab initio methods as being matched by an expansion in the scale of problems where detailed interaction modeling is invoked. For instance, where continuum modeling of solvents might have been the only manageable calculational route for condensed-phase simulations of reactions, explicit representation of individual solvent molecules is now tractable by many different means. In this way, I expect that Buckingham's ideas about the significant role of electrical interaction will broaden into the problems of proteins and water, of polymers, and maybe even as Coulson also thought, of glues and adhesives. Greater detail and more complete analysis have become possible through our more complete picture of intermolecular forces, and that offers more significant applications than we have yet seen.

*Acknowledgements.* Writing this report brings to mind the infectious enthusiasm for small complexes coming from W.H. Flygare and his group during my first years on the faculty at the University of Illinois. Bill was wonderfully supportive and inspiring for me, offering good guidance that included the important suggestion to look into a number of Buckingham's papers.

## References

1. Buckingham AD (1967) Adv Chem Phys 12: 107
2. Coulson CA (1957) Research (Lond) 10: 149
3. Morokuma K (1971) J Chem Phys 55: 1236
4. (a) Applequist J (1983) J Math Phys 24: 736; (b) Applequist J (1984) Chem Phys 85: 279
5. Buckingham AD, Pople JA (1955) Proc Phys Soc Lond Sect A 68: 905
6. Buckingham AD (1959) J Chem Phys 30: 1580
7. Buckingham AD (1959) Q Rev (Lond) 13: 189
8. Buckingham AD, Stephens PJ (1964) Mol Phys 7: 481
9. Buckingham AD, Longuet-Higgins HC (1968) Mol Phys 14: 63
10. Buckingham AD, Orr BJ (1969) Trans Faraday Soc 65: 673
11. Barron LD, Buckingham AD (1971) Mol Phys 20: 1111
12. Jeziorski B, Moszynski R, Szalewicz K (1994) Chem Rev 94: 1887
13. Buckingham AD, Fowler PW (1985) Can J Chem 63: 2018
14. Buckingham AD, Fowler PW, Stone AJ (1986) Int Rev Phys Chem 5: 107
15. Hurst GJB, Fowler PW, Stone AJ, Buckingham AD (1986) Int J Quantum Chem 29: 1223
16. Buckingham AD, Fowler PW, Hutson JM (1988) Chem Rev 88: 963
17. (a) Dykstra CE (1989) J Am Chem Soc 111: 6168; (b) Dykstra CE (1988) J Comput Chem 9: 476
18. Augspurger JD, Dykstra CE, Oldfield E. (1991) J Am Chem Soc 113: 2447
19. Kirtman B, Dykstra CE, Champagne B Chem Phys Lett (1999) 305: 132

Theor Chem Acc (2000) 103:281–285
DOI 10.1007/s002149900070

Theoretical
Chemistry Accounts
© Springer-Verlag 2000

*Perspective*

# Perspective on "Molecular collisions. VIII"

## Curtiss CF (1968) J Chem Phys 49:1952–1957

**Donald J. Kouri**[1], **David K. Hoffman**[2]

[1] Department of Chemistry and Department of Physics, University of Houston, Houston, TX 77204-5641, USA
[2] Department of Chemistry and Ames Laboratory, Iowa State University, Ames, IA 50011, USA

Received: 19 April 1999 / Accepted: 2 July 1999 / Published online: 4 October 1999

**Abstract.** We present an overview of the influence of
C.F. Curtiss on the theory of molecular collisions, as
exemplified by the title paper. Both authors were
graduate students of Curtiss and, as such, were strongly
influenced by his ideas and approaches to theoretical
chemistry. This resulted in a subsequent collaboration
that provided the rigorous basis for understanding the
success of the so-called centrifugal sudden and energy
sudden approximations (the two combined being the
"infinite order sudden" approximation).

**Key words:** Centrifugal sudden approximation – Infinite
order sudden approximation – Sudden approximations –
Inelastic collisions – Rotationally inelastic collisions

## 1 Background

A fundamental approach, now routinely employed in
essentially all quantal treatments of molecular collisions
(both inelastic and reactive), is to use rotating coordinate systems which are generalizations of those
commonly used to describe rigid-body dynamics. Additionally, the most accurate and widely used quantal
approximations for rotationally inelastic molecular
collisions are those based on the so-called "sudden
assumption" (essentially a time-scale criterion in which
an internal degree of freedom is assumed to be slow
compared to the time scale or suddenness of the
collision). We give a brief summary of these ideas,
focussing on Curtiss' role both in his research and as a
mentor. We conclude with a summary of subsequent
developments which show the success of Curtiss'
research and mentoring.

The paper "Molecular collisions. VIII" is a landmark
in the theory of inelastic scattering of molecules because
it was the first to bring together two of the most useful
and powerful techniques in the field: the use of body-
fixed or rotating frames and the sudden approximation
as applied to internal and/or orbital angular momentum
operators. Unfortunately, for reasons we discuss later,
Curtiss' work has not received the acclaim that it deserves. This is in large measure due to the rather formal,
mathematical style that typifies all of Curtiss' research
papers, and the fact that computational demonstrations
of the power and accuracy of the methods were first
given by others. (In fact, these individuals all have
connections to Curtiss and to the Theoretical Chemistry
Institute at the University of Wisconsin!) The recognition of the accuracy of these approximate methods was
also delayed because of the unavailability of numerically
converged quantal results to serve as "gold standards".

To appreciate the beauty of this paper (and the
monumental series of papers in which it appeared), we
begin by noting that the idea of using rotating or body-
fixed frames arose very early in the history of quantum
mechanics. The earliest work of which we are aware
is that of Eckart [1, 2], but his main focus was on
describing the bound rotational–vibrational states of
polyatomic molecules. As far as we are aware, the first
ones to consider the possibility of rotating frames with
the ultimate goal of treating collisions (which involve
large separations of particles not relevant for the bound
molecular states considered by Eckart) were Hirschfelder and Wigner [3]. It is no coincidence that Curtiss
acquired an interest in body-frame approaches to molecular collisions, since his Ph. D. thesis advisor was
Hirschfelder. Indeed, the first papers published by Curtiss on this subject were joint ones with Hirschfelder [4,
5]; however, subsequently Curtiss went on to make the
body-frame approach his own specialty with the initial
papers of his long running Molecular collisions series [6].
The present authors were graduate students under
Curtiss in the early 1960s, with Kouri doing research in
atom–atom collisions using a rotating-frame description,
and in numerical approaches to solving the Schrödinger
equation for scattering. Hoffman's research with Curtiss
was in classical and quantal nonequilibrium statistical
mechanics (in addition, he gained further exposure to the

*Correspondence to*: D.J. Kouri

kinetic theory of nonspherical molecular fluids and gases while a postdoctoral fellow with John Dahler, another student of Hirschfelder's).[1]

Because of the necessity of evaluating collision integrals as a part of a kinetic theory of gases and liquids, Curtiss maintained research efforts in both fields. It is particularly significant that much of the more analytical work in kinetic theory involved the use of simple molecular models (e.g., rigid ellipsoids of revolution) interacting through a hard core potential. Except for the possible "chattering" collisions, such classical scattering dynamics satisfies, to the best degree possible, the condition that the collision is sudden compared to internal and/or relative orbital rotation. Much of Curtiss' computational effort in scattering focussed on classical and semiclassical treatments of the dynamics, and he did very little on numerically converged quantum approaches, with the exception of Kouri's thesis research [7]. However, the preceding discussion indicates why Curtiss was destined to have a particularly important impact on inelastic scattering. Of course, there were other researchers exploring the use of rotating frames for collision processes [8–11], but it was the "Wisconsin school" led by Curtiss and Hirschfelder that ultimately provided the decisive research that remains at the core of our understanding of inelastic molecular collisions, and which has also been of great importance in the treatment of quantum reactive scattering.

The first accurate quantum scattering results for molecular collisions were obtained by Allison and Dalgarno, and independently by Lester and Bernstein (at Wisconsin!) using an approach due to Arthurs and Dalgarno [12–14]. Both treatments used a total angular momentum representation but quantized along a space-fixed, center-of-mass $z$-axis. It is to some degree ironic that results obtained by this "space-fixed" formalism ultimately provided the standard used to establish the body-frame approximate results as highly accurate for appropriate interactions, and in the process showed that the body-frame approaches were more efficient even for converged full quantal calculations. We now turn to a description of Curtiss' landmark paper.

## 2 Curtiss' body-frame/sudden approximation ideas

In this section, we outline the basic ingredients of Curtiss' paper. Rather than go through detailed derivations, we concentrate on the key ideas and simply refer to the specific equations in "Molecular collisions. VIII" that resulted. First, we note that Curtiss virtually always preferred to treat diatom–diatom scattering as his "lowest level" system, pointing out that the atom–diatom and atom–atom scattering cases resulted simply by assigning appropriate limiting values to the various internal angular momentum quantum numbers. As a result, the first major difficulty one has in reading the "Molecular collisions" series of papers is notational. In the case of "Molecular collisions. VIII", this is further compounded by his use of the diagrammatic treatment of angular momentum coupling [15], which was (and still is, for the most part) unfamiliar to theoretical chemists. The second stumbling block facing the reader attempting to understand this paper is the introduction of numerous new functions (defined for mathematical convenience, but seldom, if ever, accompanied by motivation). The starting equation is Eq. (VIII.5), which is written as an integral equation using Green's functions.[2] This further contributed to a certain opacity of Curtiss' papers on scattering because he was one of the few theoretical chemists to use this mathematical tool. The equation cited is one of a set of coupled equations in which a very compact notation is used for representing the various quantum numbers needed to characterize the system. Note that barred quantum numbers are precollision and unbarred ones are postcollision. The asymptotic form of the exact scattering solution is given in Eq. (VIII.6). The second section of the paper deals with the diagonalization of the potential matrix: the eigenvalues are found to be simply the potential in the body-frame coordinate representation. This is followed by a section that introduces auxilliary sets of scattering-type infinite order sudden (IOS) functions that are used formally to solve the full scattering problem. This is done by applying the diagonalizing transformation for the potential to the exact equation, written in differential-equation form. The IOS Hamiltonian is essentially used to define a reference Green's function having a well-defined orbital rotational energy. In particular, Eq. (VIII.30) defines a function $\psi(\bar{l}S_aS_b)$, which is identical to the so-called initial-$l$-, initial-$k$-labeled IOS approximation wavefunction satisfying standing-wave boundary conditions. Note that the use of the eigenvalue form of the centrifugal energy operator corresponds to a "centrifugal sudden" or "coupled states" (CS) approximation. The use of an effective radial kinetic energy eigenvalue corresponds to the "energy sudden" approximation. The exact differential equation describing the scattering is Eq. (VIII.29), and Curtiss expresses its exact solution in terms of the IOS functions, obtaining Eq. (VIII.37). Here, he has made use of the eigenstates of the potential matrix and IOS-type scattering functions to construct a Green's function by using the linear independence of the causal and anticausal IOS states. This procedure for constructing Green's functions may be found in the "bible" of mathematical physics written by Morse and Feshbach [17]. A crucial feature of this equation is the appearance of an infinite ranged perturbation, $V(\bar{l}l,r)$, given explicitly in Eq. (VIII.36). Thus, analysis of higher-order

---

[1] It is perhaps worth our pointing out that graduate students of Curtiss all had the same major hurdle to surmount in order to do a Ph.D. thesis under his direction. It was not passing the usual courses or Candidacy Exams, but rather a consequence of the fact that Curtiss' research program was very strongly integrated so that each of his research projects built in some way on his previous work. Thus, the major difficulty was to read and understand one of the previous papers in Curtiss' publication list. The project you ended up working on depended on which paper it was that you finally understood

[2] We follow the convention that equations quoted from Molecular collisions. VIII are denoted by their equation number in that paper, but preceeded by the Roman numeral VIII

contributions to the scattering requires the use of singular or boundary perturbation theory, but this can be carried out.[3] The rest of this section deals with how one obtains asymptotically correct behavior for the exact solutions, and is not needed for the usual IOS level of treatment. Finally, in the last section of the paper, Curtiss presents a careful asymptotic analysis, with the main result being obtained for the situation where the change in the internal rotational energies is small. The most readily recognized equations are the expression for the IOS S matrix, Eq. (VIII.53), the expression for the IOS **T** matrix, Eq. (VIII.56), and the IOS approximation for the moments of the cross sections, Eq. (VIII.58). The essential feature of the **S**- or **T**-matrix expressions is that they are given as averages over the rotational states of the **S**-matrix, which depends on orientation-dependent scattering phase shifts. It is especially interesting to note that Curtiss' approach automatically ensured that the basic nature of the quantum numbers is preserved, and so orbital angular momentum quantum numbers are not approximated by the total angular momentum quantum number. Thus, Curtiss' results do not suffer from basic difficulties resulting from the use of "$J$-labeling" as opposed to "$l$-labeling" [18, 23]. In Sect. 3 we describe what we believe to be the best way of understanding physically why the approximations are, in fact, so useful. This will necessitate a brief summary of work done by not only the present authors, but by other researchers. The remarkable aspect we wish to stress is that these other studies are dominated by students of either Curtiss or Hirschfelder!

## 3 Exact quantum scattering and the propensity to conserve the $z$ component of internal angular momentum

The advances of the next few years after the appearance of "Molecular collisions. VIII" are summarized in a chapter in a book edited by Bernstein (also at Wisconsin and a collaborator with Curtiss) [23]. The next major impetus to this approach to molecular collisions came in the period 1973–74 with the publication of two papers [18, 19] dealing especially with the CS approximation. One paper is by a former student of Hirschfelder and the other involves one of the present authors. The relationship between these two papers can best be seen by a brief consideration of how each treated the orbital rotational kinetic energy operator. Both papers used rotating-frame ideas and the Hamiltonian written either in the form

$$H = T_R + T_r + \frac{\mathbf{j}^2}{2\mu r^2} + \frac{(\mathbf{J} - \mathbf{j})^2}{2MR^2} \; , \qquad (1)$$

or in the form

$$H = T_R + T_r + \frac{\mathbf{j}^2}{2\mu r^2} + \frac{\mathbf{l}^2}{2MR^2} \; . \qquad (2)$$

---

[3]Such an analysis has been carried out by V. Khare and D.J. Kouri but is unpublished

Equation (1) is the form used in Pack's approach while McGuire and Kouri (MK) actually refer to both forms. As noted in both papers, the evaluation of the cross term, $\mathbf{J} \cdot \mathbf{j}$, by raising and lowering operators leads to coupling in the magnetic states of the diatom rotation, which physically are the Coriolis effects. Pack suggested approximating the centrifugal energy operator by the eigenvalue of the total angular momentum, $J(J + 1)\hbar^2/2MR^2$. MK instead replaced it by the centrifugal energy eigenvalue form $l(l + 1)\hbar^2/2MR^2$. It is clear that the latter is in the exact same spirit as Curtiss' treatment, but both are sudden approximations. MK [18] also included detailed calculations of complete degeneracy-averaged differential and integral state-to-state cross sections for the scattering of He off $H_2$, and the accuracy of the results was unprecedented! Of course, the converged close-coupling results used to establish the accuracy of the CS approximation were not available when Curtiss' paper was published. Once the accuracy of the approach was proven, there was great activity in the field focussed on its implementation and further testing. MK also discussed conditions under which the method could be expected to perform well. In particular, they pointed out that strongly backscattered systems should be well described, which translates to scattering dominated by a repulsive core potential; however, they also pointed out that for systems satisfying an approximate total cross section conservation rule, the optical theorem implies that the elastic forward scattering ought also to be treated accurately.

The next major advance in understanding came with a paper by Secrest [24], another student of Hirschfelder. This work follows even more closely the approach used by Curtiss, in that it essentially involves the approximate diagonalization of the rotational part of the Hamiltonian using the transformation that diagonalizes the potential matrix. However, unlike Curtiss, there was no attempt to write the exact solution in terms of the approximate ones, so there was no systematic way to compute corrections to the sudden approximations. However, Secrest's paper was much more accessible to the general audience and thus increased interest and activity in the methods.

It should be pointed out that the CS approximation was soon shown not to be a general solution to inelastic rotational scattering in another paper by Kouri and McGuire [25]. The system studied was the scattering of $Li^+ + H_2$. In this system, one has a very long ranged, attractive anisotropy and the scattering is no longer dominated by the short-range repulsive interaction. This, of course, was in agreement with the ideas proposed by MK. At about this same time, numerous other groups tried to extend the CS and IOS approximations to treat other kinds of cross sections than the degeneracy-averaged differential and integral cross sections. The results were confusing because extremely large errors were found in calculations done for line-broadening and other phase-sensitive cross sections [23]. The first clarification came in a study by Goldflam, Kouri and co-workers [26, 27]. They showed that if such generalized relaxation cross sections were computed using the MK expression for the differential scattering amplitude (which clearly depended strongly on subtle phase relationships), then the same high accuracy was also obtained for these phenomena. (This was true for almost all the generalized cross sections except for those that are associated with the effects of external magnetic field gradients on transport coefficients. These turn out to depend on the very small differences between large opacity-like sums, and the CS and IOS approximations failed miserably in these cases!) The way to understand this was arrived at independently, and at about the same time by Shimoni and Kouri [20] (see also Ref [28]) and by Parker and Pack [21]. The result essentially was that the most consistent treatment of the basic CS and IOS $T$-matrix requires that the approximate solutions be superposed so that the correct incident plane wave and angular momentum coupling of the rotor and orbital angular momenta is satisfied. When this is done, one finds that the CS and IOS amplitudes must be multiplied by Clebsch–Gordan coefficients, and summed over the effective orbital angular momentum quantum number. When this is done, it is seen that although the equations determining the CS and IOS wavefunctions are diagonal in the rotor magnetic quantum number, magnetic transitions are predicted in all other quantization frames. The degeneracy-averaged cross sections were shown to be identical whether one used the "$l$-initial" or "$l$-final" choice of approximate orbital angular momentum quantum number. Pfeffer and Secrest [29] suggested using the arithmetic average of the initial and final orbital angular momentum quantum numbers for determining the centrifugal potential in the CS and/or IOS calculations, but they did not analyze the implications of this choice. In fact, the fundamental question became "is there any quantization frame for which there is a true strong propensity to conserve $j_z$?". Closely related to this issue was another raised by Stolte and Reuss [30] and by Dickinson and Richards [31]. They were able to prove that either an initial-$l$- or final-$l$-labeled CS approximation predicts qualitatively incorrect behavior in the near-forward elastic ($\Delta j = 0$) scattering, particularly in the diffraction–oscillation region. In a series of papers by Khare and Kouri and coworkers [32], it was shown that such difficulties could be removed by use of the $l_{av} = \frac{1}{2}(l_{initial} + l_{final})$ choice of CS parameter. This brings us to the final, ultimate resolution of the physical basis of the sudden approximations. During a trip by Hoffman to Houston, he, Kouri and Khare discussed the possibility of collaborating on the explication of the physical basis of the angular sudden approximations. Drawing on his experience with the classical kinetic theory of gases, Hoffman suggested that the way to attack the problem was in terms of quantization along the apse vector. In scattering by a spherical potential, the apse vector is directed along the difference between the initial and final relative linear momenta. When there is inelasticity (as in a nonspherical interaction between an atom and a diatom), the apse direction is dependent on the initial and final internal states of the diatom. Consequently, in collisions of nonspherical molecules, there is a different apse direction for each $\Delta j$ transition; however, in the case of the IOS approximation all rotor states are considered degenerate and one again has a

single apse vector for all transitions. First Khare, Kouri and Hoffman (KKH) [33] were able to show explicitly that this "geometric-apse" quantization was the natural result of using the $l_{av}$ CS parameter. Thus, when one parameterizes the CS in terms of $l_{av}$, $j_z$ is rigorously conserved along the geometric apse! However, of even greater interest was the fact that Hoffman showed that if the true, state-dependent "kinematic-apse" is used as the quantization axis, $j_z$ is exactly conserved for the classical collision of a rigid, non-spherical molecule with an atom! The proof is exceptionally simple and is reproduced here. The total classical angular momentum for such a collision is given exactly by

$$\mathbf{J} = \mathbf{r}_a \times \mathbf{p}_a + \mathbf{r}_m \times \mathbf{p}_m + \mathbf{j} \; , \tag{3}$$

where the subscripts **a** and **m** denote the atom and the molecule, respectively. Then the first two terms in Eq. (3) give the orbital angular momentum of the atom and molecule, and the last term is the intrinsic angular momentum of the molecule. However, the total angular momentum is conserved in the absense of external torques, so one requires that

$$\Delta \mathbf{J} \equiv 0 \tag{4}$$

for the overall collision. Therefore,

$$0 = \Delta(\mathbf{R} \times \mathbf{P}) + \Delta \mathbf{j} \; , \tag{5}$$

where **R** is the usual scattering vector, **P** is the linear momentum for the relative motion of the atom and diatom, and $\Delta \mathbf{j}$ is the change in the diatom's angular momentum. Now for an exactly impulsive (hard core) collision, the collision time is zero (if chattering collisions are ignored) and there is no change in the collision geometry. This requires that $\Delta \mathbf{R}$ vanishes identically. Then Eq. (5) implies that

$$\Delta \mathbf{j} = -(\mathbf{R} \times \Delta \mathbf{P}) \; . \tag{6}$$

This result shows that $\Delta \mathbf{j}$ must be orthogonal to the apse vector, since the apse is defined to be along the vector $\Delta \mathbf{P}$! The final step in establishing that this is indeed the correct physical basis of the sudden approximations is to carry out numerically converged fully quantal calculations of polarization transition cross sections for a number of systems and see if there is such a propensity to conserve the component of **j** along the apse. In fact, such studies were carried out by KKH [34], in which they tested converged close-coupling cross sections for polarization transitions using four different choices of quantization: namely, the standard space-fixed axes, the helicity axes (where the initial states are quantized relative to the incident relative linear momentum and the final states are quantized along the final relative linear momentum), the geometric apse (GA), and the kinematic apse (KA). The first results [34] were for the polarization-state-resolved integral cross sections for He − CO, Ne − HD and He − H$_2$. It was only for the GA and KA quantization schemes that a completely systematic propensity to preserve $j_z$ was found. Finally, KKH [35] reported a detailed study of numerically exact quantum differential state-to-state scattering cross sections for the He + CO and He + HCl systems. Again, they showed that only for the KA and GA choices of quantization was there a systematic propensity to conserve $j_z$. Furthermore, it was found in both the integral and differential cross sections that the KA was the best in terms of the extent of the $j_z$-preserving propensity. A comparison of these two choices of quantization showed that they agreed well except in the extreme forward scattering direction (angles smaller than about 15°). Since the integral cross section involves a Jacobian weight that vanishes in the exact forward direction, it is clear that the GA and KA will agree very well. This is borne out in Ref. [34]. Another very interesting result obtained in Ref. [35] is the fact that one can obtain reasonably good results for the differential cross section quantized in the space-fixed scheme by transforming only the $\Delta j_z = 0$ KA amplitude and neglecting all other KA amplitudes for a given $j_i \rightarrow j_f$ rotational transition. This completes the story regarding the physical basis of the angular-sudden approximations. There is much more to tell in the area of reactive scattering.[4] There still remains a need for research into the best way to implement such approximations for this problem. Certainly the issue has been studied but that would lead us into another mass of literature. Suffice it to say that even though Curtiss never considered the problem, his and Hirschfelder's mentoring influence has been strong.

*Acknowledgements.* Supported by NSF Grants and in part, by R.A. Welch Foundation grant E-608. The Ames Laboratory is operated for the Department of Energy by Iowa State University, under contract no. 2-7405-ENG82.

# References

1. Eckart C (1934) Phys Rev 46: 383
2. Eckart C (1935) Phys Rev 47: 552
3. Hirschfelder JO, Wigner EP (1935) Proc Natl Acad Sci USA 21: 113
4. Curtiss CF, Hirschfelder JO, Adler FT (1950) J Chem Phys 18: 1638
5. (a) Curtiss CF, Hirschfelder JO (1952) Proc Natl Acad Sci USA 38: 235; (b) Curtiss CF, Adler FT (1952) J Chem Phys 20: 249
6. (a) Gioumousis G, Curtiss CF (1958) J Chem Phys 29: 996; (b) Gioumousis G, Curtiss CF (1961) J Math Phys 2: 96; (c) Gioumousis G, Curtiss CF (1961) J Math Phys 2: 723; (d) Gioumousis G, Curtiss CF (1962) J Math Phys 3: 1059; (e) Curtiss CF, Hardisson A (1967) J Chem Phys 46: 2618; (f) Curtiss CF (1968) 48: 1725; (g) Biolsi L, Curtiss CF (1968) J Chem Phys 48: 4508. These papers are I to VII of Curtiss' "Molecular collisions" series
7. Kouri DJ (1965) PhD thesis. University of Wisconsin
8. (a) Bates DR (ed) (1961) Quantum theory. I. Academic, New York (b) Bates DR (ed) (1962) Atomic and molecular processes, 2nd edn. Academic, New York
9. Thorson WR (1965) J Chem Phys 42: 3878
10. Jacob M, Wick GC (1959) Ann Phys 7: 404
11. Lawley KP, Ross J (1965) J Chem Phys 43: 2930
12. Arthurs AM, Dalgarno A (1960) Proc R Soc Lond Ser A 256: 540
13. Allison AC, Dalgarno A (1967) Proc R Soc 90: 609
14. (a) Lester WA, Bernstein RB (1968) J Chem Phys 48: 4896; (b) Lester WA, Bernstein RB (1967) Chem Phys Lett 1: 207
15. Percival IC, Seaton MJ (1957) Proc Camb Philos Soc 53: 654
16. (a) Yutsis AP, Levinson IB, Vanagas VV (1962) Mathematical apparatus of the theory of angular momentum. Israel Program for Scientific Translations, Jerusalem; (b) Massot JN, El-Baz E, Lafoucriere J (1967) Rev Mod Phys 39: 288
17. Morse PM, Feshbach H (1953) Methods of theoretical physics, vols I and II. McGraw-Hill, New York
18. McGuire P, Kouri DJ (1974) J Chem Phys 60: 2488
19. Pack RT (1974) J Chem Phys 60: 633
20. Shimoni Y, Kouri DJ (1977) J Chem Phys 66: 2841
21. Parker GA, Pack RT (1977) J Chem Phys 66: 2850
22. Khare V, Kouri DJ (1978) J Chem Phys 69: 4916
23. Kouri DJ, (1979) In: Bernstein RB (ed) Atom–molecule collision theory. Plenum, New York, pp 301–358
24. Secrest D (1975) J Chem Phys 62: 710
25. Kouri DJ, McGuire P (1974) Chem Phys Lett 29: 414
26. (a) Goldflam R, Kouri DJ (1977) J Chem Phys 66: 542; (b) Goldflam R, Kouri DJ (1977) J Chem Phys 66: 2452
27. (a) Goldflam R, Green S, Kouri DJ (1977) J Chem Phys 67: 225; (b) Goldflam R, Green S, Kouri DJ (1977) J Chem Phys 67: 4149; (c) Goldflam R, Kouri DJ, Green S (1977) J Chem Phys 67: 5661; (d) Goldflam R, Kouri DJ (1979) J Chem Phys 70: 5076
28. Khare V (1978) J Chem Phys 67: 3897
29. Pfeffer G, Secrest D (1977) J Chem Phys 67: 1394
30. Stolte S, Reuss J (1979) In: Bernstein RB (ed) Atom-molecule collision theory. Plenum, New York, pp 201–237
31. (a) Dickinson AS, Richards D (1980) J Phys B 13: 1293; (b) Dickinson AS, Richards D (1980) J Phys B 13: 3189
32. (a) Khare V, Kouri DJ (1980) J Chem Phys 72: 2007; (b) Khare V, Kouri DJ (1980) J Chem Phys 72: 2017; (c) Khare V, Kouri DJ, Pack RT (1978) J Chem Phys 69: 4419; (d) Khare V, Fitz DE, Kouri DJ (1980) J Chem Phys 73: 2802; (e) Fitz DE, Khare V, Kouri DJ (1980) J Chem Phys 73: 3147; (f) Fitz DE, Khare V, Kouri DJ (1981) Chem Phys 56: 267; (g) Khare V, Fitz DE, Kouri DJ, Evans D, Hoffman DK (1981) In: Truhlar DG (ed) Potential energy surfaces and dynamics calculations. Plenum, New York, pp 717–736; (h) Fitz DE, Kouri DJ, Evans D, Hoffman DK, Liu WK, McCourt FR (1982) J Phys Chem 86: 1087; (i) Kouri DJ, Fitz DE (1982) J Chem Phys 86: 2224
33. Khare V, Kouri DJ, Hoffman DK (1981) J Chem Phys 74: 2275
34. Khare V, Kouri DJ, Hoffman DK (1981) J Chem Phys 74: 2656
35. Khare V, Kouri DJ, Hoffman DK (1982) J Chem Phys 76: 4493
36. (a) Zhang JZH, Miller WH (1989) J Chem Phys 91: 1528; (b) D'Mello M, Manolopoulos DE, Wyatt RE (1991) J Chem Phys 94: 5985; (c) Neuhauser D, Judson RS, Kouri DJ, Adelman DE, Shafer NE, Kliner DAV, Zare RN (1992) Science 257: 519
37. (a) Launay JM, LeDourneuf M (1990) Chem Phys Lett 169: 473; (b) Neuhauser D, Judson RS, Jaffe RL, Baer M, Kouri DJ (1991) Chem Phys Lett 176: 546; (c) Mielke SL, Lynch GC, Truhlar DG, Schwenke DW (1993) Chem Phy Lett 213: 10; (d) Manolopoulos DE, Stark K, Werner H-J, Arnold DW, Bradforth SE, Neumark DM (1993) Science 262: 1852; (e) Baer M, Faubel M, Martinez-Haya B, Rusin L, Tappe U, Toennies JP (1999) J Chem Phys 110: 10231

---

[4] A number of reactive scattering cross section calculations have been done since the classic works of Schatz and Kuppermann, and of Elkowitz and Wyatt in the middle 1970s. General methods capable of treating nonhydrogenic atoms and energies above the threshold for vibrational excitation of reactants and/or products became available only after about 1985. References [36, 37] contain a few examples of such calculations that have been carried out using body-frame formulations; we emphasize that the list is far from complete and is given simply to provide the reader with an indication of how things have progressed. All the studies are based on the centrifugal potential operator being written in the form of Eq. (1). For some hydrogenic reaction studies, see Ref. [36] and for some studies of $F + H_2$ reaction cross sections see Ref. [37]

Theor Chem Acc (2000) 103:286–288
DOI 10.1007/s002149900041

Theoretical
Chemistry Accounts
© Springer-Verlag 2000

## *Perspective*

# Perspective on "Benzynes, dehydroconjugated molecules, and the interaction of orbitals separated by a number of intervening $\sigma$ bonds"

## Hoffmann R, Imamura A, Hehre WJ (1968) J Am Chem Soc 90: 1499–1509

### K.D. Jordan

Department of Chemistry, University of Pittsburgh, Pittsburgh, PA 15260, USA

Received: 8 March 1999 / Accepted: 29 March 1999 / Published online: 28 June 1999

**Abstract.** This paper provides an overview of the title paper by Hoffmann, Imamura, and Hehre, and the impact this paper has had on the area of long-range intramolecular interactions. The author has made extensive use of the through-bond/through-space decomposition of Hoffmann et al. in his work on long-range interaction in bichromopheric molecules. In particular he has applied these ideas in analyzing electron transmission spectra of such molecules.

**Key words:** Through-bond interactions – Orbital ordering

Chemists have long been fascinated with interactions between remote functional groups. One of the most intriguing aspects of such interactions is that they can proceed directly via through-space (TS) coupling between the functional groups, or indirectly via orbitals of the intervening bridge. Through-bond (TB) coupling (although not specifically referred to as such) is the basis of the McConnell model [1] introduced in 1961, and which has proven to play a central role in electron-transfer theories. A systematic procedure for dissecting net interactions into their TB and TS components and for analyzing how the TB coupling depends on the number and relative orientation of the bonds of the intervening bridges, was first laid out in the pioneering paper of Hoffmann, Imamura, and Hehre (HIH) [2]. Although the HIH paper focused on the orbital interactions in benzynes and dehydroconjugated molecules, the strategies introduced therein have a much wider range of applicability [3–13].

In reviewing the main contributions of the HIH paper, it will suffice to focus on one of the molecules, *p*-benzyne, considered in that study. *p*-Benzyne has two half-occupied localized lone-pair orbitals, $\phi_1$ and $\phi_2$,

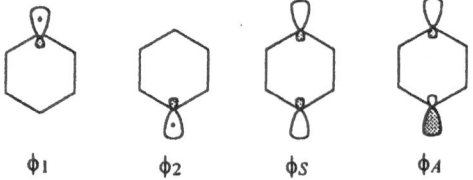

$$\phi_1 \qquad \phi_2 \qquad \phi_S \qquad \phi_A$$

depicted in Scheme 1. These give rise to the symmetry-adapted molecular orbitals (MO)

$$\phi_S = \phi_1 + \phi_2$$

$$\phi_A = \phi_1 - \phi_2 ,$$

where S and A denote, respectively, the symmetric and antisymmetric combinations of the localized lone-pair orbitals, and $\phi_1$ and $\phi_2$ have been assumed to be orthogonal. Of primary interest is the splitting energy $\Delta E = \varepsilon_S - \varepsilon_A$, as this provides a measure of the coupling between the localized orbitals.

The major conceptual advance provided by HIH was the dissection of $\Delta E$ into TS and TB components. As defined by HIH, TS refers to the direct interaction between the localized $\phi_1$ and $\phi_2$ orbitals and TB to the indirect interaction via the intervening $\sigma$ bonds. The TS interaction and a subset of the TB pathways for coupling the lone-pair orbitals of *p*-benzyne are shown in Scheme 2.

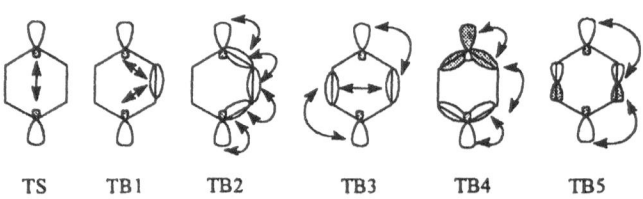

$$\text{TS} \qquad \text{TB1} \qquad \text{TB2} \qquad \text{TB3} \qquad \text{TB4} \qquad \text{TB5}$$

Note, in particular, that with the HIH definition, pathways TB1, TB3, TB4, and TB5, which involve interactions via $\sigma$ orbitals not directly coupled to $\phi_1$ or $\phi_2$ or which skip over bonds, are classified as TB.

MO calculations on *p*-benzyne place the A orbital energetically below the S orbital, counter to one's expectations [2]. The TB/TS decomposition provides an explanation for this surprising result, namely that the TB interactions with the $\sigma$ and $\sigma^*$ orbitals of the bridge destabilize $\phi_S$ relative to $\phi_A$ and, moreover, that the TB interactions are sufficiently large so as to reverse the "natural", TS ordering of the A and S orbitals. HIH showed how this can be understood in terms of a perturbative treatment, describing the interactions of the symmetry-adapted $\phi_S$ and $\phi_A$ orbitals with the symmetry-adapted orbitals of the benzene "bridge".

To simplify the analysis we consider the two symmetry-adapted bridge orbitals $\sigma_S$ and $\sigma_A^*$ shown in Scheme 3.

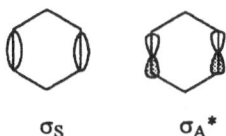

$\sigma_S$        $\sigma_A^*$

These are singled out because they interact especially strongly with $\phi_S$ and $\phi_A$ (due to the interactions with the back lobes of the lone-pair orbitals. Specifically $\sigma_S$ mixes with $\phi_S$ destabilizing the latter, while $\sigma_A^*$ mixes with $\phi_A$ stabilizing the latter. This results in an inverted ordering of the "mixed" lone-pair orbitals as shown in Fig. 1. Of course, a quantitative prediction of the $\phi_S/\phi_A$ splitting requires inclusion of the other possible TB coupling pathways. The relevency of the orbital splittings for the singlet/triplet gaps in $p$-benzyne and other aryl biradicals is discussed in a recent paper of Squires and Cramer [14].

The HIH paper also made important contributions to our understanding of how the electronic coupling depends on the length and conformation of the bridge. Specifically, it was shown why the relative energy of the S and A orbitals depends on whether there is an odd or even number of carbon–carbon linkages in a bridge, and why the coupling is generally greater through bridges with all-trans orientation of the carbon–carbon $\sigma$ bonds. The TB coupling model is also able to account for the inverted ordering of the $\pi$ MOs ($\pi_A < \pi_S$) in 1,4-cyclohexadiene and a wide range of other hydrocarbons [3, 15]. In the case of 1,4-cyclohexadiene, the TB coupling proceeds via the bridging methylene groups.

The ideas laid down in the HIH paper have also proven pivotal for analyzing the interactions responsible for long-range electronic coupling in bichromophoric systems such as 1–5 shown in Scheme 4 [8–13].

For example, both MO calculations and experiment (photoelectron spectra) reveal that the splitting between the $\pi$ orbitals of 4 is much smaller than that in 1 (0.44

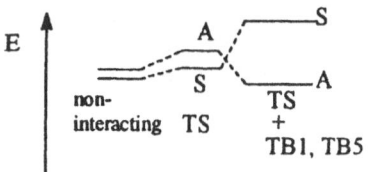

**Fig. 1.** Energies of the $\phi_S$, and $\phi_A$, orbitals of $p$-benzyne, in the absence of interactions, with through-space (*TS*) coupling only, and with both TS and through-bond (*TB*) TB1 + TB5 coupling

versus 0.87 eV) [16, 17]. This is understandable in terms of the greater effectiveness of all-trans bridges at relaying the coupling as discussed by HIH (sometimes referred to as the "all-trans" effect [2, 3, 4, 7]. Also, the HIH analysis explains the "inverted" ($\pi_A < \pi_S$) ordering of the $\pi$ orbitals of 5.

There has been immense interest in understanding how electronic couplings fall off with increasing bridge length, for example, along the sequence of molecules 1, 2, and 3. MO calculations on this series of molecules and other bichromophoric species show that both the $\pi_S$, $\pi_A$ and $\pi_S^*$, $\pi_A^*$ splittings fall off nearly exponentially with the length of the bridge [9–11, 19, 20]. Although, this is consistent with the predictions of the McConnell model [1], which includes only nearest-neighbor interactions, the strong coupling through the bridges in fact derives from pathways that skip over bonds [9–13, 19, 20]. This conclusion is based on an analysis in which the net TB coupling is dissected into contributions due to individual pathways, very much in the spirit of HIH.

Scheme 5 shows the nearest-neighbor pathway **a** and a non-nearest-neighbor pathway **b** for a six-bond bridge. Although the contribution of **b** to the $\pi_S$, $\pi_A$ splitting is indeed smaller than that of **a**, this is compensated by the fact that there are multiple pathways that skip over bonds. Moreover, the number of pathways that involve "bond-skipping" grows rapidly with increasing chain length [9, 11].

(a)             (b)

Recent theoretical studies have also elucidated the role of constructive/destructive interference between various TB coupling pathways [12, 21]. For example, it has been shown that the smaller $\pi_S$, $\pi_A$ splitting in 1 compared to 6 is due to the destructive interference between pathways that jump between two parallel bridges in the former [12, 21]. The theoretical analysis leading to this conclusion is a natural extension of the perturbative approach presented by HIH.

Although the examples discussed previously focused on systems in which the chromophores are covalently bonded to the bridge, it has been demonstrated that TB coupling can be sizable through noncovalently linked bridge units, for example, the model system comprising two ethylene molecules separated by $n$ methane molecules in van der Waals contact [9, 20]. This implies that solvent molecules can be very effective at relaying electronic coupling. Not surprisingly, several research groups have been designing bichromophoric systems which allow testing of this idea [22, 23].

In summary, the pioneering paper of HIH not only provided a conceptual and computational framework for dissecting electronic coupling between remote functional groups into TB and TS components, but also a qualitative understanding of how the TB coupling depends on the number and orientation of the $\sigma$-bond linkages. Moreover, the perturbative approach used by HIH forms the basis of modern theoretical studies of the contributions of various TB pathways to the net coupling in D–B–A and other bichromophoric systems.

*Acknowledgements.* Our research on long-range interactions was supported by the National Science Foundation. I also wish to acknowledge my long-term collaboration in this area with Michael Paddon-Row.

## References

1. McConnell HM (1961) J Chem Phys 35: 508
2. Hoffmann R, Imamura A, Hehre WJ (1968) J Am Chem Soc 90: 1499
3. Hoffmann R (1971) Acc Chem Res 4: 1
4. Martin HD, Meyer B (1983) Angew Chem Int Ed Engl 22: 283
5. Heilbronner E, Schmelzer A (1975) Helv Chim Acta 58: 936
6. Dougherty DA (1991) Acc Chem Res 24: 88
7. Paddon-Row MN (1982) Acc Chem Res 15: 245
8. Balaji V, Ng L, Patney HK, Jordan KD, Paddon-Row MN (1987) J Am Chem Soc 109: 6957
9. Jordan KD, Paddon-Row MN (1992) Chem Rev 92: 395
10. Paddon-Row MN, Shephard MJ, Jordan KD (1993) J Phys Chem 97: 1743
11. Shephard MJ, Paddon-Row MN, Jordan KD (1993) Chem Phys 176: 289
12. Shephard MJ, Paddon-Row MN, Jordan KD (1994) J Am Chem Soc 116: 5328
13. Jordan KD, Paddon-Row MN (1998) In: Schleyer PvR (ed) Enclopedia of computational chemistry, Vol 2. Wiley, New York, p 826
14. Squires RR, Cramer CJ (1998) J Phys Chem 102: 9072
15. Hoffmann R, Heilbronner E, Gleiter R (1960) J Am Chem Soc 82: 5450
16. Paddon-Row MN, Patnew HK, Brown RS, Houk KN (1982) J Am Chem Soc 103: 5575
17. Martin HD, Schwesinger R (1974) Chem Ber 107: 3143
18. Kurnikov IV, Beratan DN (1996) J Chem Phys 105: 9561
19. (a) Naleway CA, Curtis LA, Miller JR (1991) J Phys Chem 95: 8434 (b) Naleway CA, Curtis LA, Miller JR (1993) J Phys Chem 97: 4050
20. (a) Liang C, Newton MD (1992) J Phys Chem 96: 2855 (b) Liang C, Newton MD (1993) J Phys Chem 97: 3199
21. Paddon-Row MN, Shephard MJ (1997) J Am Chem Soc 119: 5355
22. Kumar K, Lin Z, Waldeck DH, Zimmt MB (1996) J Am Chem Soc 118: 243
23. Roest MR, Verhoeven JW, Schuddeboom W, Warman JM, Lawson JM, Paddon-Row MN (1996) 118: 1762

Theor Chem Acc (2000) 103:289–291
DOI 10.1007/s002149900054

**Theoretical
Chemistry Accounts**
© Springer-Verlag 2000

## *Perspective*

# Perspective on "Intermolecular orbital theory of the interactions between conjugated systems." I General theory; II Thermal and photochemical cycloadditions

## Salem L (1968) J Am Chem Soc 90: 543, 553

**Odile Eisenstein**

Laboratoire de Structure et Dynamique des Systèmes Moléculaires et Solides, UMR 5636, Case Courrier 14,
Université de Montpellier 2, 34095 Montpellier Cedex 5, France

Received: 12 April 1999 / Accepted: 5 May 1999 / Published online: 9 September 1999

**Abstract.** At the time where a reaction path between molecules could not be calculated easily, the interaction between conjugated molecules was calculated using perturbation theory. The theory of Lionel Salem gave a very interesting approximation to a reaction path between rather large conjugated systems.

**Key words:** Molecular orbitals – Perturbation theory – Intermolecular interactions

## 1 Personal historical perspective

It was the fall of 1968. Universities opened late this term in France. Paris-Jussieu (called Halles aux Vins because the university was built on the site of an old wine whole sale market) was still closed and I registered at the University of Orsay, 25 km southwest of the city. Nice trees, plenty of grass, a stream going through the campus, it was one of the most beautiful campuses in France. In the program of chemical physics, I sat through classes of quantum mechanics, spectroscopy, thermodynamics, kinetics, and structure and bonding of molecules. The professor for "structure and bonding of molecules" was young, tall and thin. He looked at us with piercing eyes under dark bushy eyebrows. He came to class with heavy piles of yellow scratch papers covered with handwritten notes but hardly looked through them except for writing something in the middle of his presentation. His name was Lionel Salem. As we went through the material during the semester, he often repeated "Molecular orbitals, they don't exist but that should not stop us from drawing and using them". In the following semester, during the class for organic chemistry, I discovered the full meaning of this sentence. Very luckily for us, the professor, Nguyen Trong Anh, decided on a very unusual program. In place of the

expected order stereochemistry, organic functions and "if time permits modern approaches", he started with the last section. This is why we were introduced to organic chemistry through the Woodward–Hoffmann rules. Nothing could have been better and we learned what nonexisting molecular orbitals could do for chemistry. At the end of this year I had my BA. I joined Nguyen Trong Anh's research group under the condition that I was going to work on a project related to Woodward–Hoffmann rules, and I was going to "play" with these nonexisting molecular orbitals which could still be drawn.

In my project, I had to understand the regiochemistry of the addition of unsymmetrically substituted dienes and dienophiles. The major product (shown in Scheme 1

$R_1 = NEt_2$, Ph, $CO_2H$, $^tBu$

$R_2 = CO_2R$, Ph, CN

for one type of reactants) could not be explained by classical electrostatic arguments since new bonds were made between carbon centers with identical charges. Steric effects were also in contradiction with the nature of the major isomer [1, 2].

The selection rules of the pericyclic reactions were established for idealized nonsubstituted molecules [3]. In the particular case of cycloaddition, the two bonds were thus supposed to form in a synchronous manner. There was certainly no reason for the two bonds to be made synchronously for unsymmetrically substituted molecules although the reaction was concerted, i.e. there was no detectable intermediate. There was in fact proof that the two bonds were not formed synchronously [4, 5]. The use of perturbation theory and frontier orbitals within the hypothesis of synchronous formation of the two bonds gave disastrous results for "predicting" the major isomer. I had to assume asynchronicity . How to determine the amount of asynchronicity? This is how I came to the paper of Lionel Salem.

In the1960s, reaction mechanisms were not studied through localization of transition states as has been done since the 1980s. Two methods were possible at this time. The extended Hückel method [6] could be as easily used for a single molecule as for two molecules in interaction considered as a composite system, sometimes also called a super-system [3, 7, 8]. However, to quote Lionel, "the numerical resolution of a large secular equation does not provide any information on the important atom–atom or orbital–orbital interactions..." and "It is difficult, without repeating the entire diagonalisation at many different interatomic distances to establish whether a given bond closure is more favorable than another one and whether a given reaction must proceed symmetrically or not". It is a mark of the time that Lionel also wrote, "Furthermore for very large molecules, the secular determinant procedure itself becomes unwieldy. The interaction of two pyrene molecules requires the resolution of a $132 \times 132$ determinant for each dimer configuration when there is no plane of symmetry allowing for its factorization".

## 2 General theory

Lionel Salem developed "a theory which provides explicit expressions for the interaction energy of two conjugated molecules as a function of the various atomic orbital overlaps (a reaction surface of sorts). Such a theory should provide insight into the important orbital interactions and should allow ready calculations of reaction paths". The essential features of the proposed theory were as follows.

1. The wave functions were built out of intermolecular orbitals covering the entire system of interacting molecules. The Hamiltonian was an effective one-electron Hamiltonian.
2. The molecular orbitals and experimental energies of the separate molecules were chosen as starting points. This theory was going to seek only the small changes brought about by the interaction.
3. Perturbation theory was used to determine the changes in the molecular orbitals and in the energies of states.
4. Expansion in powers of the overlap allows for explicit analytical energy expressions which were numerically tractable.

The theory was thus developed for conjugated systems with well-separated $\sigma$ and $\pi$ orbitals. While this approximation is not strictly true for small olefinic systems such as ethylene, it is valid for larger conjugated molecules.

Coulombic repulsion terms were neglected thus prohibiting the use of the theory for highly polar systems. All overlaps between interacting atomic orbitals $\phi_r$ and $\phi_{r'}$ were considered to be small with respect to unity ($S_{rr'} < 0.2$). This is especially valid for conjugated systems when the two molecules approach each other in roughly parallel planes (Scheme 2).

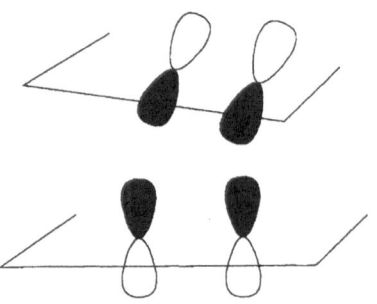

$$E_{\text{int}} = -\sum_{rr'}(q_r + q_{r'})\eta_{rr'}S_{rr'}$$
$$-2\sum_{j}^{\text{occ}}\sum_{k'}^{\text{unocc}}\frac{\left(\sum_{rr'}c_{jr}c_{k'r'}\eta_{rr'}\right)^2}{E_{k'} - E_j}$$
$$-2\sum_{j'}^{\text{occ}}\sum_{k}^{\text{unocc}}\frac{\left(\sum_{rr'}c_{kr}c_{j'r'}\eta_{rr'}\right)^2}{E_k - E_{j'}} \tag{1}$$

The expression (Eq. 1) is made of two terms with clear physical meaning:

1. A repulsive term proportional to the square of the overlap and to the sum of the $\pi$ charge densities on the interacting atoms.[1] This repulsion arises because each conjugated molecule has a closed-shell structure in which all the bonding orbitals are full. The larger the $\pi$ charge density on an atom, the larger the size of the exclusion shell into which other electrons are forbidden from penetrating.
2. An attractive term due essentially to the mixing of the occupied orbitals of one molecule with the unoccupied orbitals on the other. This term depends on the inverse of the energy difference between interacting molecular orbitals and is thus important for occupied and empty orbitals that are close in energy. This attraction takes place because conjugated molecules although being closed shells from the strict point of view of occupancy of bonding orbitals are also open shells since they possess a half-filled band of $\pi$ levels.

This equation illustrates the dual nature of conjugated molecules: a closed shell but yet available low-lying antibonding orbitals. The great reactivity arises from the existence of half-filled bands of $\pi$ electrons even though the valency shell of each carbon atom is full.

---

[1] In this equation, $\eta_{rr'}$ is the resonance integral which is proportional to the overlap $S_{rr'}$, $q_r$ is the charge density at atom $r$, $E_j$ is the energy of the molecular orbital $\psi_j$

Lionel Salem also established the equation for the interaction energy when one molecule is excited. The interaction contains a term that is linear in the overlap in addition to terms that depend on $S^2$ and higher orders. For identical molecules the linear term dominates. "The stabilizing interaction depends essentially on the magnitude of the interaction between the molecular orbital $\psi_j$ which loses the electron and its degenerate partner $\psi_{j'}$, and between the molecular orbital $\psi_k$ which gains the electron and its partner $\psi_{k'}$. For different molecules the linear term disappears and the terms of higher order in $S$ take precedent.

These equations were used subsequently to account for the thermal and photochemical dimerization of butadiene, the Diels–Alder addition between two acrolein molecules, where the asynchronous nature of the cycloaddition was demonstrated, and the endo mechanism. These equations provided a demonstration of the symmetry rules for the cycloaddition reactions, and could account for the photodimerization of tropone.

While this method allows a number of different reactions occurring between conjugated systems to be treated successfully, Lionel Salem pointed out some strengths and weaknesses of this theory. A major flaw, inherent to all molecular orbital theories, is the lack of correlation between the order of the orbitals and the order of the states. Other weaknesses are

1. The neglect of explicit intermolecular Coulomb interactions resulting in erroneous predictions of reaction paths, particularly when large net charges force a pathway different from that favored purely from the overlap viewpoint.
2. The neglect of explicit interaction with the $\sigma$ electrons.
3. The use of a parameter $k$ which links the resonance integral and the overlap (as is done in extended Hückel theory).

This theory was in the "air" at this time. Fukui [9, 10] had presented a perturbation treatment of the interaction between conjugated molecules but which neglected overlap explicitly. Closely related papers were published by Klopman [11, 12]. In these papers a perturbation treatment explicitly including the ionic terms was designed to treat hard and soft acids and bases (HSAB theory).

This type of approach had a major impact in popularizing the theoretical approach amongst a large group of people, especially experimentalists. The shape and energy of $\pi$ orbitals could be obtained from EHT programs and even through simple Hückel programs. The physical meaning of the terms had a close relationship with concepts used by experimental chemists. While the Woodward–Hoffmann rules gave a Yes/No answer to what was allowed or forbidden, this approach gave a

numerical estimate of the interaction between reagents and a rough idea of the reactions paths. It was certainly a great tool. In the case of Diels–Alder reactions, the asynchronicity was easy to demonstrate. In fact assuming full asynchronicity gave the major isomer for a large number of systems [13]. A generalization of the perturbational approach which explained nicely how any substituent could influence the diene and dienophine molecular orbital energies and coefficients was presented shortly after by Houk [14].

## 3 The present time

There is no longer explicit use of this theory to probe reaction paths because it is now possible to calculate reaction paths for rather large systems with quantitative methods. For instance, the transition states for a large number of cycloadditions were calculated with ab initio methods [15, 16]; however, it is not realistic and not useful to calculate paths for every reaction and perturbation theory still remains the best tool for explaining a large number of experimental results. The power of this tool is clearly illustrated in the books of Fleming [17] for organic chemistry, of Albright, Burdett and Whangbo [18] for a general approach to chemistry of Hoffmann [19] and of Iung and Canadell [20] for solid state chemistry.

## References

1. Sauer J (1967) Angew Chem Int Ed Engl 6: 23
2. Feuer J, Herndon WC, Hall LH (1968) Tetrahedron 24: 2575
3. Hoffmann R, Woodward RB (1965) J Am Chem Soc 87: 2046
4. Woodward RB, Katz T (1959) Tetrahedron 5: 70
5. Berson JA, Remanick A (1961) J Am Chem Soc 83: 4947
6. Hoffmann R (1963) J Chem Phys 39: 1397
7. Pople JA, Santry DP (1963) Mol Phys 7: 269
8. Herdon WC, Hall LH (1967) Theor Chim Acta 7: 4
9. Fukui K (1966) Bull Chem Soc Jpn 39: 498
10. Fukui K, Fujimoto H (1966) Bull Chem Soc Jpn 39: 2166
11. Klopman G (1968) J Am Chem Soc 90: 223
12. Klopman G, Hudson RF (1967) Theor Chim Acta 8: 165
13. Eisenstein O, Lefour JM, Anh NT (1971) Chem Commun 969
14. Houk KN (1973) J Am Chem Soc 95: 4092
15. Townshend RE, Ramunni G, Segal G, Hehre WJ, Salem L (1976) J Am Chem Soc 98: 2190
16. Houk KN, Li Y, Evanseck JD (1992) Angew Chem Int Ed Engl 31: 682
17. Fleming I (1978) Frontier orbitals and organic chemical reactions. Wiley
18. Albright TA, Burdett JK, Whangbo MH (1975) Orbital interactions in chemistry. Wiley, New York
19. Hoffmann R (1988) Solids and surfaces: A chemist's view of bonding in extended structures VCH New York
20. Iung C, Canadell E (1997) Description orbitalaire de la structure electronique des solides eduscience Paris

Theor Chem Acc (2000) 103:292–293
DOI 10.1007/s002149900096

Theoretical
Chemistry Accounts
© Springer-Verlag 2000

*Perspective*

# Two landmarks in polymer physics: the Edwards model and de Gennes' observation

**D. Thirumalai**

Institute for Physical Science and Technology, University of Maryland, College Park, MD 20742, USA

Received: 4 July 1999 / Accepted: 7 July 1999 / Published online: 17 January 2000
© Springer-Verlag 2000

**Abstract.** The impact of two landmark papers by Edwards and de Gennes on the field of polymer physics is highlighted.

**Key words:** Flory theory – Edwards model – n-vector model

A major advance in the physics of polymers occurred in 1949 when Flory [1] provided a simple but profound argument for the swelling (compared to the ideal chain size) of flexible polymer chains due to excluded-volume interactions. In essence, the Flory result for the dependence of the radius of gyration, $R_g$, on the degree of polymerization, $N$, is obtained by minimizing the elastic energy (due to chain connectivity) and the repulsive energy arising from the volume excluded by a given monomer for all other monomers. The resulting prediction for the exponent, $v$, defined by $R_g \approx aN^v$ is remarkably accurate in all space dimensions, $d$. For all practical purposes the Flory result for $v = \frac{3}{d+2}$ may be considered exact [2]. Similarly, the Flory argument is also found to be nearly exact for describing sizes of $D$-dimensional objects embedded in $d$ spatial dimensions, such as tethered membranes with $D = 2$ [3]. For polymers $D = 1$.

A fundamental understanding of the reasons for the success of the theory due to Flory is still lacking. In an attempt to derive the Flory exponent Edwards proposed a model for polymers that bears his name in 1965 [4]. This paper brought to bear, for the first time, methods of functional integrals and many-body theory on problems in polymer physics. Edwards proposed a very simple form for the short-range repulsive potential describing the interactions between the monomers. He suggested replacing the actual potential by $v\delta[r(s) - r(s')]$, where $v$ is the strength of the excluded-volume interactions, $r(s)$ is a path of the polymer chain and $s$ and $s'$ are the positions of two monomers along the positions of the chain. The use of the $\delta$ function pseudopotential should

not (see later) affect the long-wavelength properties of the polymer chain. With this replacement Edwards formally showed that polymer statistics boils down to summing over all possible paths weighted by the Hamiltonian given by the sum of the "kinetic energy" (representing chain connectivity) and the pseudopotential. The resulting path integral is non-Markovian, which is a reflection of the nature of the excluded-volume interactions. The formal analogy to the path integral allowed the use of many approximations devised in the context of quantum mechanics to problems in polymers.

Several studies utilizing the Edwards model for polymers followed [5]. In addition, using enumerations of self-avoiding walks using lattice models [6] and through the ingenious use of exact relations for Ising models [7] many new results for polymer statistics were obtained; however, an understanding of the varied universal behavior of polymer solutions was lacking. This state of affairs in polymer physics was to change dramatically after the profound discovery by de Gennes, who showed a connection between polymer statistics and phase transitions in 1972 [8]. This short and lucid paper followed right at the heels of the discovery of the renormalization group in the context of second-order phase transitions. de Gennes showed that the $n$-vector magnetic spin problem with $n = 0$ is equivalent to the excluded-volume problem considered by Flory [1] and formalized in terms of path integral methods by Edwards [4]. The connection between the excluded-volume problem and phase transitions also clarified the reasons for the independence of the values of $v$ on the details of the interaction potentials as long as they are short ranged. This, in retrospect, justified the Edwards choice of delta function interaction between two monomer segments. With this profound observation the entire machinery developed for understanding critical phenomena could be imported to obtain a vast number of new results. Thus, the concept of scaling was born in polymer physics and it continues to dominate the thinking of many scientists in this area.

The marriage of the Edwards model and de Gennes' observation brought an onslaught of several field theoretical methods to derive various scaling laws describing the static properties of dilute and semidilute polymers solutions. The Edwards model was also generalized to poor solvent conditions so that polymer collapse could be described. These developments are summarized in a beautiful monograph by des Cloizeaux and Jannink [9]. It is fair to say that these two landmarks in polymer physics have enabled us to understand many structural aspects of polymers in solution.

There still are challenges which have come about in extending the Edwards model to tethered membranes ($D = 2$) [3]. The demonstration of the renormalizability of the resulting model is a topic of current research [10]. In this context there does not appear to be an equivalent spin model which describes self-avoidance in such objects. Further extension of these models to membranes and charged species is expected to be an important problem in the general area of soft-condensed-matter physics. A perusal of the literature on these topics is sufficient to appreciate the deep influence of the two landmark papers [4, 8] on polymer physics.

*Acknowledgements*: I am grateful to M. E. Fisher and H. Orland for useful discussions.

## References

1. Flory PJ (1949) J Chem Phys 17: 303
2. de Gennes PG (1985) In: Scaling concepts in polymer physics. Cornell University Press, Ithaca
3. Kantor Y, Kardar M, Nelson DR (1987) Phys Rev A 35: 3056
4. Edwards SF (1965) Proc Phys Soc 85: 613
5. Freed KF (1972) Adv Chem Phys 22: 1
6. Domb C (1963) J Chem Phys 38: 2957
7. Fisher ME (1966) J Chem Phys 44: 616
8. de Gennes PG (1972) Phys Lett A 38: 229
9. des Cloizeaux J, Jannink G In: Polymers in solution. Oxford University Press, Oxford
10. David F, Wiese KJ (1998) Nuc Phys B 535: 555

Theor Chem Acc (2000) 103:294–296
DOI 10.1007/s002149900024

Theoretical
Chemistry Accounts
© Springer-Verlag 2000

*Perspective*

# Perspective on "Ab initio calculation of force constants and equilibrium geometries in polyatomic molecules. I. Theory"

## Pulay P (1969) Mol Phys 17: 197–204

H. Bernhard Schlegel

Department of Chemistry, Wayne State University, Detroit, MI 48202, USA

Received: 26 February 1999 / Accepted: 5 April 1999 / Published online: 21 June 1999

**Abstract.** This article provides an outline of the title paper by Peter Pulay and discusses some of the methodology that grew from it, and the impact that it has had on the development of computational chemistry.

**Key words:** Gradients – Force constants – Hessians – Energy derivatives – Geometry optimization

## 1 The paper

The title indicates that this paper is about the calculation of vibrational force constants and the geometry optimization of polyatomic molecules; however, its primary impact on computational chemistry comes from the methodology for calculating analytic first derivatives with respect to molecular coordinates at the Hartree–Fock (HF) level of theory. Applications of first and higher derivatives of the energies obtained by molecular orbital (MO) calculations have revolutionized computational chemistry, allowing molecular structures and properties to be computed efficiently and reliably [1–5]. Almost all electronic structure codes compute analytic first derivatives of the energy, and Pulay's paper was the first to describe a practical calculational approach.

In the linear combination of atomic orbitals to form MOs (LCAO-MO) approach, the HF energy can be expressed in terms of the one-electron density matrix $\mathbf{D}$ and integrals over the basis functions:

$$
\begin{aligned}
E = \langle \psi | H | \psi \rangle &= 2\Sigma_j \big( \phi_j | H_1 | \phi_j \big) + \Sigma_{jk} \big( \phi_j^2 | 1/r_{jk} | \phi_k^2 \big) \\
&= 2\Sigma_{mn} (m | H_1 | n) D_{mn} \\
&\quad + \Sigma_{mnrs} (mn | rs)(2 D_{mn} D_{rs} - D_{mr} D_{ns}) \ ,
\end{aligned}
\tag{1}
$$

where $H_1$ is the one-electron part of the Hamiltonian, $\phi$ are the MOs and $m$, $n$, $r$, and $s$ refer to the basis functions (shown for a spin-restricted, closed-shell system with real orbitals). The first derivative of the energy is

$$
dE/dq_i = \langle \psi | dH/dq_i | \psi \rangle + 2\langle d\psi/dq_i | H | \psi \rangle \ .
\tag{2}
$$

The first term is the Hellmann–Feynman contribution and is readily calculated as an expectation value of a one-electron operator. The second part can be called the wavefunction derivative term (it has also come to be known as the Pulay term). Typical MO wavefunctions are constructed from basis functions that are centered on the atoms and follow them rigidly. Such wavefunctions do not obey the Hellmann–Feynman theorem and the wavefunction derivative term is not zero. Many molecular properties can also be written as energy derivatives [1–4]. For some properties, it is advantageous to make the wavefunction depend on the electric or magnetic field; for these cases the wavefunction derivative term must also be calculated.

The wavefunction derivative term depends on the derivatives of the one- and two-electron integrals over the basis functions, and on the derivative of the density matrix. The self-consistent-field (SCF) process finds the density matrix that minimizes the energy under the constraint of orthonormality of the MOs; hence, only the changes required to maintain orthonormality are needed for the density matrix derivative contributions to the wavefunction derivative term. The constraint that the molecular orbitals are orthonormal is equivalent to the requirement that the density matrix be idempotent:

$$
\mathbf{D} \, \mathbf{S} \, \mathbf{D} = \mathbf{D} \ ,
\tag{3}
$$

where $\mathbf{S}$ is the overlap matrix. Equation (3) can be differentiated to obtain the change in $\mathbf{D}$ that maintains idempotency

$$
d\mathbf{D}/dq_i \, \mathbf{S} \, \mathbf{D} + \mathbf{D} \, d\mathbf{S}/dq_i \, \mathbf{D} + \mathbf{D} \, \mathbf{S} \, d\mathbf{D}/dq_i = d\mathbf{D}/dq_i \ .
\tag{4}
$$

A simple solution that satisfies Eq. (4) is

$$
d\mathbf{D}/dq_i = -\mathbf{D} \, d\mathbf{S}/dq_i \, \mathbf{D} \ .
\tag{5}
$$

The derivative of the energy, including both the Hellmann–Feynman and wavefunction terms, is

$$\begin{aligned}
dE/dq_i = {} & 2\Sigma_{mn}(m|dH_1/dq_i|n)D_{mn} \\
& + 2\Sigma_{mn}[(dm/dq_i|H_1|n) + (m|H_1|dn/dq_i)]D_{mn} \\
& + \Sigma_{mnrs}[(dm/dq_i n|rs) + (m\,dn/dq_i|rs) \\
& + (mn|dr/dq_i s) + (mn|r\,ds/dq)] \\
& \times (2D_{mn}D_{rs} - D_{mr}D_{ns}) \\
& + 2\Sigma_{mn}(m|H_1|n)dD_{mn}/dq_i + \Sigma_{mnrs}(mn|rs) \\
& \times (2dD_{mn}/dq_i\,D_{rs} + 2D_{mn}dD_{rs}/dq_i \\
& - dD_{mr}/dq_i\,D_{ns} - D_{mr}dD_{ns}/dq_i)\;.
\end{aligned} \qquad (6)$$

Insertion of Eq. (5) and recognition that the terms multiplying the density derivative constitute the Fock matrix, $\mathbf{F}$, yields the following compact expression for the analytic first derivative of the HF energy

$$\begin{aligned}
dE/dq_i = {} & 2\Sigma_{mn}\,d(m|H_1|n)/dq_i\,D_{mn} \\
& + \Sigma_{mnrs}\,d(mn|rs)/dq_i\,(2D_{mn}D_{rs} - D_{mr}D_{ns}) \\
& - 2\Sigma_{mn}\,dS_{mn}/dq_i\,W_{mn}\;,
\end{aligned} \qquad (7)$$

where $\mathbf{W} = \mathbf{D}\,\mathbf{F}\,\mathbf{D}$ (borrowing the notation of later papers on energy derivatives).

Pulay's paper goes on to calculate force constants (second derivatives of the energy with respect to geometrical parameters) by numerically differentiating the analytical first derivatives. At the time this was by far the best compromise between efficiency and accuracy in computing the force constants. The second use of geometric derivatives described in Pulay's paper is for geometry optimization. Optimization is best carried out in internal coordinates, but the calculation of analytic derivatives is most practical in Cartesian coordinates. The forces are the negative of the first derivatives of the energy. The relation between Cartesian and internal displacements ($\mathbf{dx}$ and $\mathbf{dq}$) and forces ($\mathbf{f}$ and $\phi$) can be written in terms of the Wilson $\mathbf{B}$ matrix [6]

$$\mathbf{dq} = \mathbf{B}\,\mathbf{dx}, \quad \mathbf{f} = \mathbf{B^t}\phi \qquad (8)$$

The $\mathbf{B}$ matrix is rectangular; however, a suitable left inverse can be constructed to transform Cartesian forces into internal coordinates.

$$\phi = (\mathbf{B}\,\mathbf{m}\,\mathbf{B^t})^{-1}\,\mathbf{B}\,\mathbf{m}\,\mathbf{f} \qquad (9)$$

The internal forces and a suitable approximation to the force constants, $\mathbf{F}_0$, can be used to relax the molecule to its equilibrium geometry.

$$\mathbf{q}_{\text{new}} = \mathbf{q}_{\text{old}} + \mathbf{F}_0^{-1}\phi_{\text{old}} \qquad (10)$$

## 2 Perspective

Pulay's paper is an early landmark in the explosive growth in computational chemistry that we have seen in the past quarter century. The method for calculating first derivatives as outlined in the article forms the basis for the subsequent development of first, second and higher energy derivatives for many different theoretical methods (for reviews, see Refs. [1–5]). The advances brought about by energy derivative methods have enabled theoretical calculations to become practical and efficient

methods for determining molecular structures, exploring potential-energy surfaces and computing molecular properties.

Pulay demonstrated that analytic first derivatives with respect to geometric parameters can be calculated easily and efficiently for HF energies. Derivatives of correlated methods followed a number of years after SCF derivatives [4, 5]. Extensions of the SCF derivatives to density functional theory methods were straightforward. In the three decades since Pulay's article, hundreds of papers on energy derivatives have been published, and all can trace their roots back to his paper. Energy derivatives have become so useful for calculating molecular structures and properties that, almost universally, first derivatives are formulated and coded soon after a new theoretical method is developed for the energy.

Pulay's paper opened the way for analytic second and higher derivatives of the SCF energy. Earlier papers had suggested that this might be prohibitively expensive [7], but the development of an efficient method to solve the couple perturbed HF (CPHF) equations, made the calculation of SCF second derivatives practical [8]. As a consequence, vibrational force constants and frequencies could be calculated routinely and efficiently. Third and fourth geometric derivatives of the SCF energy followed after a few years [9–12]. The solution of the CPHF equations (in their full or reduced Z-vector form [13]) also made post-SCF first derivatives practical and cost-effective.

Analytic first derivatives with respect to geometrical parameters have proven to be extremely useful for exploring potential-energy surfaces [14]. Almost all studies using electronic structure methods involve geometry optimization at some level of theory. With the possible exception of diatomics, energy derivative methods are much less expensive than energy-only methods for obtaining equilibrium geometries. Calculations on molecules with hundreds and thousands of atoms are now possible, and energy derivatives provide an enormous amount of information about the potenital-energy surface at very little additional cost. Pulay's paper also outlined the transformation of first derivatives from Cartesian to internal coordinates. Optimization in internal coordinates is significantly more efficient than in Cartesian coordinates, and it is now the standard approach. Energy derivatives are indispensable for exploring other aspects of potential-energy surfaces. For example, they are essential for finding transition states and following reaction paths [14]. Recently it has become practical to compute classical trajectories using energy derivatives directly from the electronic structure calculations without first fitting a global potential-energy surface (for a review, see Ref. [15]).

Many molecular properties, such as vibrational frequencies, IR and Raman intensities, NMR shielding constants, etc., can be formulated in terms of second and higher derivatives with respect to geometry and applied fields [1–4]. Such calculations are now practical and routine using analytic derivatives at the SCF level and a few correlated methods. For some levels of theory, analytic second and higher derivatives are not yet available or are too complicated to code. In these cases, the 'force method' described by Pulay's paper is still the method of choice, i.e. differentiating once analytically and the remaining times numerically.

## References

1. Pulay P (1995) In: Yarkony DR (ed) Modern electronic structure theory. World Scientific, Singapore, p 1191, and references therein
2. Helgaker T, Jørgensen P (1988) Adv Quantum Chem 19: 183
3. Dykstra CE (1988) Ab initio calculation of the structures and properties of molecules. Elsevier, Amsterdam
4. Jørgensen P, Simons J (eds) (1986) Geometrical derivatives of energy surfaces and molecular properties. Reidel, Dordrecht
5. Shepard R (1995) In: Yarkony DR (ed) Modern electronic structure theory. World Scientific, Singapore, p 345, and references therein
6. Wilson EB, Decius JC, Cross PC (1955) Molecular vibrations. McGraw-Hill, New York
7. (a) Thompsen K, Swanstron P (1973) Mol Phys 26: 735; Thompsen K, Swanstron P (1973) Mol Phys 26: 751
8. Pople JA, Krishnan R, Schlegel HB, Binkley JS (1979) Int J Quantum Chem Symp 13: 225
9. Pulay P (1983) J Chem Phys 78: 5043
10. Gaw JF, Yamaguchi Y, Schaefer HF III, Handy NC (1986) J Chem Phys 85: 5132
11. Augspurger JD, Dykstra CE (1991) J Phys Chem 95: 9230
12. Colwell, SM, Jayatilaka D, Maslen, PE, Amos RD, Handy NC (1991) Int J Quantum Chem 40: 179
13. Handy NC, Schaefer HF III (1984) J Chem Phys 81: 5031
14. Schlegel HB (1995) In: Yarkony DR (ed) Modern electronic structure theory. World Scientific, Singapore, p 459, and references therein
15. Bolton K, Hase WL, Peshlherbe GH (1998) In: Thompson DL (ed) Modern methods for multidimensional dynamics computation in chemistry. World Scientific, Singapore, pp 143

Theor Chem Acc (2000) 103:297–299
DOI 10.1007/s002149900026

**Theoretical**
**Chemistry Accounts**
© Springer-Verlag 2000

*Perspective*

# Perspective on "Semiclassical theory of atom–diatom collisions: path integrals and the classical $S$ matrix"

## Miller WH (1970) J Chem Phys 53: 1949–1959

**Reinhard Schinke**

Max-Planck-Institut für Strömungsforschung, Bunsenstrasse 10, D-37073 Göttingen, Germany

Received: 26 February 1999 / Accepted: 5 April 1999 / Published online: 21 June 1999

**Abstract.** Miller's papers on classical $S$-matrix theory had a profound influence on the understanding of inelastic atom–molecule collisions. This perspective discusses the historical background, the content, some applications, and new developments.

**Key words:** Semiclassical dynamics – Rainbows

When I joined the department for Molecular Interactions at the Max-Planck-Institut für Strömungsforschung as an undergraduate student in 1972, the head of that department, Professor J.P. Toennies, gave me a couple of papers [1, 2] to read, which had great influence on many researchers in the field of molecular dynamics; they certainly shaped my own scientific thinking for many years to come.

Around 1970 there was a flourishing interest in state-resolved scattering cross sections and at several locations crossed molecular beams apparatus were under construction; our institute was one of them. Although it was clear from the beginning that quantum mechanics is the ultimate tool for describing collisions between atoms and molecules, exact quantum mechanical calculations for realistic systems, especially for chemical reactions, were essentially impossible in those times. On the other hand, purely classical mechanics, i.e., trajectory calculations a la Karplus, Porter, and Sharma [3] (see the perspective by Schatz in this issue) were believed to be not fully adequate, since important quantum effects such as tunneling, zero-point energy, and interferences are – by definition – not incorporated. Thus, a theory was sought which amalgamates the simplicity of classical calculations with the essential concepts of quantum mechanics.

In the first of the papers mentioned above Miller established a general semiclassical theory for inelastic molecular scattering. This approach was the natural extension of the celebrated semiclassical theory for elastic atom-atom scattering [4] which had been developed about a decade before Miller's papers (see the perspective given by Miller in this issue). The basic concept of the semiclassical theory is the incorporation of quantities solely derived from the solution of the classical (Newton's or Hamilton's) equations of motion into the quantum mechanical principle of superposition of probability amplitudes (as opposed to the summation of probabilities). Building the theory on the superposition principle guarantees that the quantum mechanical effects neglected in classical mechanics are at least qualitatively included. Solving the classical equations of motion for a system with several degrees of freedom, in order to extract the necessary ingredients such as action integrals or generalized deflection functions, even three decades ago was not a problem and thus the classical $S$-matrix theory, as the semiclassical theory was termed, was considered to be a promising alternative to exact quantum mechanical methods for calculating state-resolved inelastic integral and differential cross sections for atom–molecule collisions. One must remember that around that time the first exact (close coupling) calculations had just appeared in the literature [5]. For a more comprehensive historical survey of semiclassical methods for bound states and scattering problems see the review articles by Miller [6, 7] or the monograph of Child [8].

While the first of the two Miller papers was a bit formal, especially for a young undergraduate student making his first steps into science, in the second paper it was demonstrated by means of a simple collinear $A + BC(n_1) \rightarrow A + BC(n_2)$ scattering system how the general theory must be applied [2]. Within the simplest version of classical $S$-matrix theory the probability for making a vibrational transition from an initial state $n_1$ to a final state $n_2$ is given by

$$P_{n_2,n_1} = p_1 + p_2 + 2(p_1 p_2)^{1/2} \sin(\Delta\phi) , \qquad (1)$$

298

where the

$$p_i = \left| 2\pi \frac{dn_2(\bar{q}_1)}{d\bar{q}_1} \right|_{\bar{q}_1 = \bar{q}_1^i}^{-1} \qquad (2)$$

are purely classical probabilities and $\Delta\phi$ is the difference between the action integrals along the two different classical trajectories, specified by the initial phase angles $\bar{q}_1^1$ and $\bar{q}_1^2$, which correspond to the particular $n_1 \rightarrow n_2$ transition. The particular form of Eq. (1) is only valid if there are two trajectories that – in the classical sense – contribute. The interference term $\sin(\Delta\phi)$ is the manifestation of the quantum mechanical superposition principle (addition of probability amplitudes rather than probabilities).

The connection between the classical trajectories and the quantum transition is established by the classical excitation function $n_2(\bar{q}_1)$, an example of which is depicted in Fig. 1a. $n_2(\bar{q}_1)$ is the final classical vibrational quantum number (not necessarily an integer) as a function of the initial phase of the oscillator. It is calculated by running – for a particular collision energy – trajectories with different initial phase angles $\bar{q}_1$. The particular trajectories which contribute to the probabilities in Eq. (1) are found using the equation

$$n_2 = n_2(\bar{q}_1) , \qquad (3)$$

where $n_2 = 0, 1, \dots$. In the particular case shown in Fig. 1 exactly two trajectories contribute when $n$ is smaller than the maximum of the excitation function (classically allowed case).

The interference of these different roots to Eq. (3) leads to the pronounced oscillations of the quantum mechanical probability shown in Fig. 1b. In contrast to the quantum mechanical curve the classical probability is a smooth function of $n_2$. If $n_2$ is larger than the maximum of the excitation function there are no real-valued trajectories and the transition is classically forbidden.

An extension of the semiclassical theory into the non-classical regime is possible by analytical continuation and complex-valued trajectories [9–11]. If $n_2$ is close to $n_2^{max}(\bar{q}_1)$ the classical probabilities become singular and the "primitive" semiclassical theory breaks down; however, this failure can be corrected on a more sophisticated (uniform) level of the semiclassical theory. Thus, the probability has three different regimes: the classically allowed region, in which the probability shows quantum interference oscillations; the classically forbidden region, in which the probability decays exponentially to zero; between these two regimes the probability shows a pronounced rainbowlike maximum.

About a decade later, the general predictions concerning the shape of transition probabilities, that naturally emerge from the semiclassical theory, were beautifully confirmed in rotational-state-resolved differential-scattering cross sections. These cross sections exhibit a classically forbidden region, a dominant rainbow maximum (rotational rainbow), and interference oscillations in the classically allowed region [12]. An experimental example, highlighting the supernumerary rotational rainbows, is shown in Fig. 2 for the He + Na$_2$ collision system together with the results of theoretical calculations employing an accurate potential-energy surface [13]. The rotational rainbow occurs at very small angles in this particular example and is not observable. However, the quite regular supernumerary oscillations, which are so nicely predicted by Eq. (1), are clearly observable; these oscillations are distinctively different from the rainbow oscillations observed in elastic collisions [4]. The semiclassical picture has also been found extremely helpful in understanding final product state distributions following the fragmentation

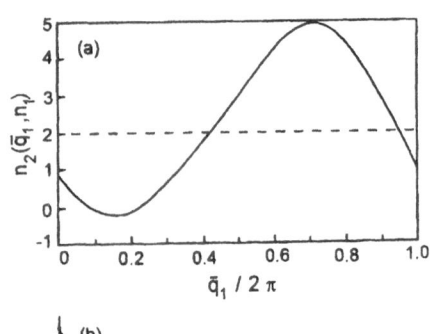

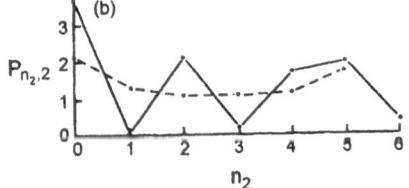

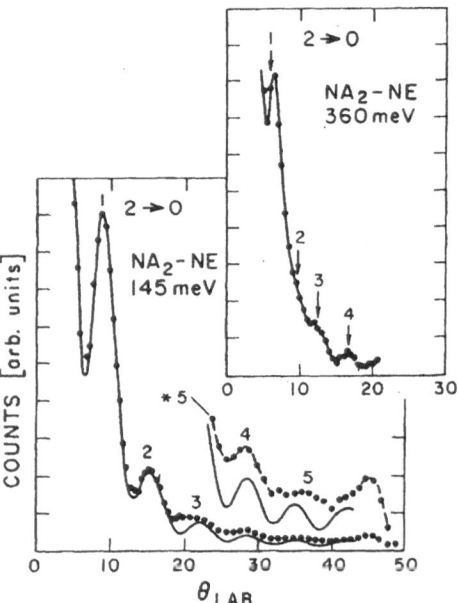

**Fig. 1. a** Vibrational excitation function $n_2(\bar{q}_1)$ as a function of the initial phase angle of the BC oscillator $\bar{q}_1$. Redrawn from Fig. 2 of Ref. [2]. **b** Comparison between the quantum mechanical (*solid line*) and the classical (*dashed line*) transition probabilities. Redrawn from Fig. 1 of Ref. [2]

**Fig. 2.** Supernumerary rotational rainbow oscillations for the $j = 0 \rightarrow 2$ transition in He + Na$_2$ collisions. The *dots* are the experimental data and the *solid lines* are the results of quantum mechanical scattering calculations. Reproduced, with permission of the American Institute of Physics, from Ref. [13]

of polyatomic molecules, provided the dissociation is fast and direct [14]. Another, very fine application of semiclassical $S$-matrix theory has been provided by Rost in the context of electron–atom scattering, particularly the ionization of atomic hydrogen [15]. This application is especially fascinating because electrons are not considered to be well described by classical mechanics.

In the early 1970s, the semiclassical approach to inelastic scattering was mainly applied to the collinear $A + BC$ model system defined by Secrest and Johnson [5], which included only two degrees of freedom. The emphasis at that time was mainly in testing the various implementations of the classical $S$-matrix theory. Although the agreement with the exact quantum mechanical results was excellent, applications of the semiclassical theory to more realistic and therefore necessarily more complex collision systems was sparse. The reasons why the classical $S$-matrix theory did not become a major tool for numerical calculations for realistic systems are – in my opinion – twofold. First, if more than two degrees of freedom are involved, finding the correct trajectories which lead to the desired quantum transition is "thorny" (root search problem). Second, beginning in the mid-1970s better and better numerical algorithms for solving the Schrödinger equation (the time-independent or the time-dependent one) were developed, so that today it is possible to treat any triatomic molecular system in an essentially exact quantum mechanical manner. Of course, the unforseen advances in computer technology also had an enormous impact on exact scattering calculations.

Applications of exact quantum mechanics to systems with four atoms involved is still a major numerical burden, not to mention application to even bigger systems, and so approximate methods based on the solution of the classical equations of motion are nevertheless highly desired. In this context the semiclassical initial value representation (IVR), a precursor of which had already been derived in Miller's 1970 paper [2], may turn out to become a significant tool in molecular dynamics (see Ref. [16] for a comprehensive list of recent references). In the IVR approach the quantum mechanical $S$-matrix elements are approximated by a (multi-dimensional) integral over the initial classical phase space and the integrand contains only ingredients from clas-

sical trajectories. In this way, the awkward root searching procedure is avoided. The price to be paid is that instead of a few trajectories many have to be calculated. Recent applications are very promising and show that quantum mechanical effects such as interferences, for example, are reproduced perfectly [16].

In summary, Miller's 1970 papers [1, 2] on classical $S$-matrix theory had a profound influence on the theory of molecular collisions and related topics such as photodissociation. Following earlier work on elastic scattering, they demonstrated how the results of classical mechanics can be built into a quantum mechanical framework of inelastic collisions. In my view the greatest asset of the classical $S$-matrix theory is its interpretative power. The general shape of transition probabilities or collisional cross sections can be easily understood in terms of classical trajectories and their quantum mechanical interference. Exact quantum mechanical programs are like "black boxes" and the results are often difficult to understand without the help of classical mechanics or semiclassical analyses. The new developments such as the IVR are likely to become major tools for systems consisting of many atoms.

# References

1. Miller WH (1970) J Chem Phys 53: 1949
2. Miller WH (1970) J Chem Phys 53: 3578
3. Karplus M, Porter RN, Sharma RD (1965) J Chem Phys 43: 3259
4. Ford KW, Wheeler JA (1959) Ann Phys (NY) 7: 259
5. Secrest D, Johnson BR (1966) J Chem Phys 45: 4556
6. Miller WH (1974) Adv Chem Phys 25: 69
7. Miller WH (1975) Adv Chem Phys 30: 77
8. Child MS (1991) Semiclassical mechanics with molecular applications. Clarendon, Oxford
9. Miller WH (1970) Chem Phys Lett 7: 431
10. Miller WH, George TF (1972) J Chem Phys 56: 5668
11. Stine JR, Marcus RA (1972) Chem Phys Lett 15: 536
12. Korsch J, Schinke R (1980) J Chem Phys 73: 1222
13. Gottwald E, Bergmann K, Schinke R (1987) J Chem Phys 86: 2685
14. Schinke R (1993) Photodissociation Dynamics. Cambridge University Press, Cambridge
15. Rost J-M (1998) Phys Rep 297: 271
16. Skinner DE, Miller WH (1999) Chem Phys Lett 300: 20

Theor Chem Acc (2000) 103:300–305
DOI 10.1007/s002149900017

Theoretical
Chemistry Accounts
© Springer-Verlag 2000

*Perspective*

# Quantum wavepacket approach to chemical reaction dynamics. Perspective on "Dynamics of the collinear H + H₂ reaction. I. Probability density and flux"

## McCullough EA Jr, Wyatt RE (1971) J Chem Phys 54: 3578

**John Z.H. Zhang[1], D.H. Zhang[2]**

[1] Department of Chemistry, New York University, New York, NY 10003, USA
[2] Department of Computational Science, National University of Singapore, Singapore 119260, Singapore

Received: 26 February 1999 / Accepted: 16 March 1999 / Published online: 21 June 1999

**Abstract.** This paper presents an overview of the time-dependent quantum wavepacket approach to chemical reaction dynamics. After a brief review of some early works, the paper gives an up-to-date account of the recent development of computational methodologies in time-dependent quantum dynamics. The presentation of the paper focuses on the development of accurate or numerically exact time-dependent methods and their specific applications to tetraatomic reactions. After summarizing the current state-of-the-art time-dependent wavepacket approach, a perspective on future development is provided.

## 1 Background

In 1971, McCullough and Wyatt published a paper [1] in which they described an application of time-dependent (TD) quantum mechanics to the collinear H + H₂ reaction by solving the TD Schrödinger equation. In their approach, McCullough and Wyatt used the finite-difference method to approximate the spatial derivatives of the wavefunction and carried out the time propagation of the wavefunction by the Crank–Nicholson method [2]. Through explicit numerical calculations, the authors showed [1] how one can directly obtain real-time dynamics information of the reaction process by examining the scattering wavefunction and the reactive flux as functions of both time and space. This work demonstrated the numerical feasibility of solving reaction dynamics problems by using the TD quantum wavepacket approach.

However, at the time of their publication the exact TD quantum wavepacket methods were not viewed as

efficient numerical techniques for solving quantum scattering problems in chemical dynamics. This is because the coupled-channel scattering problems one could solve numerically at the time were mainly collinear atom-diatom scattering and had only a limited number of coupled channels, which were more efficiently solved by time-independent scattering methods. As a result, the emphasis of the paper by McCullough and Wyatt was mainly to demonstrate the usefulness of the TD approach as an analytical tool to understand the reaction dynamics but not as an efficient computational tool for numerical calculations. Consequently, only a few applications of TD quantum mechanics to chemical reactions were reported afterwards [3–5].

Around the middle of the 1980s, more efficient and accurate computational techniques were introduced into TD wavepacket calculations. These include the split-operator method [6], the fast Fourier transform method [6, 7], the Chebychev polynomial expansion method [8], the short iterative Lanczos method [9], etc. The utilization of these numerical methods made TD quantum wavepacket calculations more accurate and efficient. More importantly as computer speed significantly increased, the computational advantage of the TD approach became better appreciated for solving large-scale coupled-channel problems. Mowrey and Kouri [10] argued that since the TD wavepacket approach solves an initial-value problem and the computational time scales quadratically with the number of coupled channels, it should be computationally more advantageous than the traditional boundary-value time-independent scattering approach for solving large-scale coupled-channel scattering problems. Since then, more applications of the TD wavepacket approach to atom–diatom reactive scattering have appeared [11–13] including reactions of H + O₂ [14, 15] and O(¹D) + H₂ [16], both with deep potential wells. In addition, the TD wavepacket approach has also been widely applied to other dynamics problems [17], including photodissociation of tritomic molecules [18–23], vibra-

*Correspondence to*: J.Z.H. Zhang

tional predissociation [24, 25], and gas–surface reactions [26–31]. The introduction of the absorbing potentials in the numerical computation [12] has further widened the scope of TD applications to practical dynamics problems.

## 2 TD approach to AB + CD reactions

### 2.1 Introduction

Almost without exception, the previously mentioned TD applications to gas-phase problems are all for triatomic systems. By the beginning of 1990, the accurate quantum dynamics calculation for gas-phase triatomic systems of the type A + BC essentially became a solved problem. The new challenge to computational dynamics is to carry out accurate dynamics studies for polyatomic systems with more than three atoms. In particular, the rigorous quantum dynamics treatment for tetraatomic systems of the type AB + CD became the new benchmark for dynamics studies. Computationally, the quantum dynamics of the AB + CD system is not a trivial extension of the triatomic system of A + BC. The number of internal degrees of freedom suddenly increases from 3 for the A + BC system to 6 for the AB + CD system which translates into a drastic increase in the number of coupled channels and therefore huge increases in computational cost. Also, technically, the Hamiltonian for the AB + CD system is more complicated to handle than for triatomic systems. For example, one has to deal with the coupling of three angular momenta for the AB + CD system instead of two for the triatomic A + BC system. Thus, the difficulties in both the theoretical treatment and the numerical computation make the exact dynamics calculation for the AB + CD system a new challenge in quantum dynamics. Consequently, the time-independent scattering methods, which have been very successful in treating the A + BC reaction, were proven to be difficult to apply to the tetraatomic AB + CD reaction due to their computational limitations. For example, in the algebraic variational approach, one is required to invert the Hamiltonian matrix to solve linear algebraic equations. Even for a simple tetraatomic reaction such as $H_2$ + OH, the size of the Hamiltonian matrix is prohibitively large to be inverted directly using today's computers. It therefore requires alternative approaches such as iterative methods to solve linear algebraic equations due to their lower computational scaling than direct matrix inversion.

Thus, the critical measure of the applicability of a method to polyatomic reaction dynamics is the scaling of its computational cost with respect to the number of basis functions or degrees of freedom. Since the standard time-independent scattering methods solve boundary-value problems, they scale as $N^3$ with the number of basis functions $N$, and are thus difficult to extend to large systems. Until a few years ago, the reduced-dimensionality approach (RDA) [32, 33] provided the only means for treating the four-atom reactive scattering problem in which a four-atom reaction system is reduced to an effective atom–diatom system through the elimi-

nation of three internal coordinates, either by applying an adiabatic approximation for three internal angular variables [32] or by restricting the system to certain geometric configurations [33]. The results of the RDA calculations are mixed, however [36–40].

The TD wavepacket approach has very attractive features for large scale numerical calculations. Since it solves an initial-value problem and calculates the wavefunction for one initial state at a time, the computational cost is proportional to $N^\alpha$ ($1 < \alpha < 2$), where $N$ is the number of basis functions (which can be a very large number for polyatomic systems). This reduction in computational scaling is crucial for large-scale quantum dynamics calculations. In addition, a single wavepacket calculation can give dynamical quantities such as $S$ matrix elements or reaction probabilities over a wide range of energies contained in the initial wavepacket. Thus it is not surprising to see that the TD wavepacket approach has recently made significant advance in treating dynamics problems for tetraatomic reactions [41].

The accurate TD approach to tetraatomic systems was first applied to the photofragmentation dynamics of $H_2HF \rightarrow H_2$ + HF [42, 43] and HOOH → OH + OH [44], in which the two diatomic vibrations were frozen but all the other four internal degrees of freedom were treated exactly. These two theoretical studies established the numerical feasibilities for the accurate TD wavepacket treatment for tetraatomic dynamics problems. Within quick succession, accurate quantum dynamics calculations for the tetraatomic reaction AB + CD → A + BCD were reported for the benchmark reaction $H_2$ + OH → H + $H_2O$ [45–47] and its isotopic reactions DH + OH → D + $H_2O$, H + DOH [48] and $D_2$ + OH → D + DOD [49]. For the $H_2$ + OH reaction, cumulative reaction probabilities have also been computed by calculating the flux directly without summing over individual reaction probabilities [50–52]. Additionally, accurate dynamics calculations for reactions of HO + CO → H + $CO_2$ [53], $H_2$ + CN → H + HCN [54, 56], and $D_2$ + CN → D + DCN [55] have been reported. With the exception of Refs. [50, 56] which use the time-independent iterative approach to calculate cumulative reaction probabilities, the other works all applied the TD wavepacket approach. In addition, the TD calculation for the reverse reaction of H + $H_2O$ → $H_2$ + OH has been reported [57]. More recently, state-to-state TD calculations have been reported for the $H_2$ + OH reaction [58, 59] and its reverse reaction H + $H_2O$ [60].

### 2.2 Mathematical formulation

In the TD wavepacket approach, one solves the TD Schrödinger equation

$$i\hbar \frac{\partial}{\partial t} \Psi(t) = H\Psi(t) \tag{1}$$

starting from a given $L^2$ integrable initial wavefunction $\Psi(0)$. In the following, we specialize our discussions specifically for tetraatomic systems of the type AB + CD. Most of the discussions, however, should be applicable to A + BCD systems as well with some modifications.

For AB + CD systems the most straightforward choice of coordinates to describe the dynamical system is the Jacobi coordinates corresponding to the diatom–diatom arrangement. As shown in Fig. 1, the three vectors $(\mathbf{R}, \mathbf{r}_1, \mathbf{r}_2)$ denote, respectively, the vector $\mathbf{R}$ from the center of mass (CM) of diatom AB to that of CD, the AB diatomic vector $\mathbf{r}_1$, and the CD diatomic vector $\mathbf{r}_2$. The full Hamiltonian expressed in this set of coordinates is written as

$$H = -\frac{\hbar^2}{2\mu}\frac{\partial^2}{\partial R^2} + \frac{(\vec{J} - \vec{j}_{12})^2}{2\mu R^2} + h_1(r_1) + h_2(r_2)$$
$$+ \frac{\vec{j}_1^2}{2\mu_1 r_1^2} + \frac{\vec{j}_2^2}{2\mu_2 r_2^2} + V(\vec{r}_1, \vec{r}_2, \vec{R}) \ , \quad (2)$$

where $\mu$ is the reduced mass between the CM of AB and CD, $\mu_1$ is the reduced mass of AB, and $\mu_2$ is the reduced mass of CD. The vector $\vec{J}$ is the total angular momentum operator and $\vec{j}_1$ and $\vec{j}_2$ are, the rotational angular momentum operators of AB and CD respectively. The latter two are coupled to form the total internal angular momentum via $\vec{j}_{12} = \vec{j}_1 + \vec{j}_2$. The reference diatomic Hamiltonian $h_i(r_i)$ $(i = 1, 2)$ in Eq. (2) is defined as

$$h_i(r_i) = -\frac{\hbar^2}{2\mu_i}\frac{\partial^2}{\partial r_i^2} + V_i(r_i) \ , \quad (3)$$

where $V_i$ is a chosen reference diatomic vibrational potential. The eigenfunctions and eigenvalues of the reference Hamiltonian $h_i(r_i)$ are denoted by $\phi_{v_i}$ and $\varepsilon_{v_i}$, respectively.

The TD wavefunction can be expanded in the body-fixed rovibrational basis functions as [46]

$$\Psi_{v_0 j_0 K_0}^{JM\epsilon}(\vec{R}, \vec{r}_1, \vec{r}_2, t) = \sum_{n, v, j, K} F_{nvjK, v_0 j_0 K_0}^{JM\epsilon}(t) u_n^{v_1}$$
$$\times \phi_{v_1}(r_1)\phi_{v_2}(r_2)$$
$$\times Y_{jK}^{JM\epsilon}(\hat{R}, \hat{r}_1, \hat{r}_2) \ , \quad (4)$$

where $n$ is the translational basis label, $v$ denotes $(v_1, v_2)$, $j$ denotes $(j_1, j_2, j_{12})$, $(v_0, j_0)$ is the initial rovibrational state, and $\epsilon$ is the parity of the system. The determination of the TD coefficient $F_{nvjK, v_0 j_0 K_0}^{JM\epsilon}(t)$ gives the solution of the TD Schrödinger equation.

The coupled total angular momentum eigenfunctions $Y_{jK}^{JM\epsilon}$ in Eq. (4) can be written as [44, 46, 61],

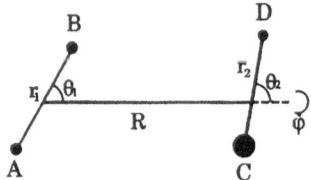

**Fig. 1.** Jacobi coordinates for the reaction AB + CD → A + BCD. The angle $\phi$ is the out-of-plane torsional angle

$$Y_{jK}^{JM\epsilon} = (1 + \delta_{K0})^{-1/2}\sqrt{\frac{2J+1}{8\pi}}$$
$$\times \left[ D_{K,M}^{J*} Y_{j_1 j_2}^{j_{12}K} + \epsilon(-1)^{j_1 + j_2 + j_{12} + J} D_{-K,M}^{J*} Y_{j_1 j_2}^{j_{12}-K} \right] \ , \quad (5)$$

where $D_{K,M}^{J}(\Theta\Phi\Psi)$ is the Wigner rotation matrix [62] with three Euler angles $(\Theta, \Phi, \Psi)$, and $Y_{j_1 j_2}^{j_{12}K}$ is the angular momentum eigenfunction of $j_{12}$ [44, 46],

$$Y_{j_1 j_2}^{j_{12}K}(\theta_1, \theta_2, \phi) = \sum_{m_1} \langle j_1 m_1 j_2 K - m_1 | j_{12} K \rangle$$
$$\times y_{j_1 m_1}(\theta_1, 0) y_{j_2 K - m_1}(\theta_2, \phi) \ , \quad (6)$$

where $y_{jm}$ are spherical harmonics. Note in Eq. (5) the restriction $\epsilon(-1)^{j_1 + j_2 + j_{12} + J} = 1$ for $K = 0$.

The split-operator propagator [6] is used to carry out the time propagation of the wavepacket,

$$\Psi^{JM\epsilon}(\vec{R}, \vec{r}_1, \vec{r}_2, t + \Delta) = \exp\left(\frac{-iH_0\Delta}{2}\right)\exp(-iU\Delta)$$
$$\times \exp\left(\frac{-iH_0\Delta}{2}\right)$$
$$\times \Psi^{JM\epsilon}(\vec{R}, \vec{r}_1, \vec{r}_2, t) \ , \quad (7)$$

where the reference Hamiltonian $H_0$ is defined as

$$H_0 = -\frac{\hbar^2}{2\mu}\frac{\partial^2}{\partial R^2} + h_1(r_1) + h_2(r_2) \quad (8)$$

and the effective potential operator $U$ in Eq. (7) is defined as

$$U = \frac{(\vec{J} - \vec{j}_{12})^2}{2\mu R^2} + \frac{\vec{j}_1^2}{2\mu_1 r_1^2} + \frac{\vec{j}_1^2}{2\mu_2 r_2^2} + V(\vec{r}_1, \vec{r}_2, \vec{R})$$
$$= V_{\text{rot}} + V(\vec{r}_1, \vec{r}_2, \vec{R}) \ . \quad (9)$$

The matrix version of Eq. (7) for the expansion coefficient vector $\mathbf{F}$ is then given by

$$\mathbf{F}(t + \Delta) = \exp(-i\mathbf{H_0}\Delta/2)\exp(-i\mathbf{U}\Delta)$$
$$\times \exp(-i\mathbf{H_0}\Delta/2)\mathbf{F}(t) \ , \quad (10)$$

where $\mathbf{H_0}$ is the diagonal matrix defined in Ref. [46].

From the propagation of an initial wavepacket $|\chi_i(0)\rangle$, the time-independent wavefunction $\psi_i^+(E)$ can be obtained by Fourier transforming the TD wavefunction [46]

$$\psi_i^+(E) = \frac{1}{2\pi a_i(E)} \int_{-\infty}^{\infty} \exp\left[\frac{i}{\hbar}(E - H)t\right]\chi_i(0)dt \ , \quad (11)$$

and similarly for the derivative of the wavefunction $\psi_i^{+'}(E)$. The coefficient $a_i(E)$ is easily evaluated from the free-energy-normalized asymptotic function $\phi_l(E)$ as $a_i(E) = \langle\phi_i(E)|\chi_i(0)\rangle$ [46]. The total reaction probability from a given initial state $i$ can be calculated by using the flux formula [46]

$$P_i^R(E) = \sum_f |S_{fi}^R|^2 = \langle\psi_i^+(E)|\hat{F}|\psi_i^+(E)\rangle \ . \quad (12)$$

The initial wavepacket $|\chi_i(0)\rangle$ is usually chosen to be a Gaussian function with an average momentum $k_0$ traveling toward the interaction region

$$\phi_{k_0}(R) = \left(\frac{1}{\pi\delta^2}\right)^{1/4} \exp\left[-(R-R_0)^2/2\delta^2\right] \exp(-ik_0R) \quad (13)$$

multiplied by the selected initial rovibrational eigenfunction. In actual propagation, the TD wavefunction is absorbed at the edges of the grid to avoid boundary reflections.

## 3 Results of the $H_2$ + OH reaction

The TD wavepacket method described earlier is applied to the benchmark $H_2$ + OH reaction and its isotopic reactions. Figure 2 shows computed total (final state-summed) reaction probabilities of the three isotopic reactions on the Walch–Dunning–Schatz–Elgersma potential energy surface (PES) [63] as a function of the incident kinetic energy from the initial ground state of the reagents with zero total angular momentum ($J = 0$). The reaction probability is of the order $P(H_2) > P(HD) > P(D_2)$ at fixed kinetic energies; however, considering the vibrational energy difference among $H_2$, HD, and $D_2$, the reaction probability is about the same magnitude as a function of total energy (kinetic energy + zero-point energy).

The reaction probability for the $H_2$ + OH shows a strong steric effect as reflected in its sensitive dependence on the initial rotational states of the reagents, especially the reactive diatomic $H_2$. As shown in Fig. 3, the reaction probability initially increases quite significantly as the reagent rotation increases. In particular the maximum of the reaction probability always shows up for the $j = 1$ state of $H(D)_2$, which is believed to be a general phenomenon for collinearly dominated reactions at zero

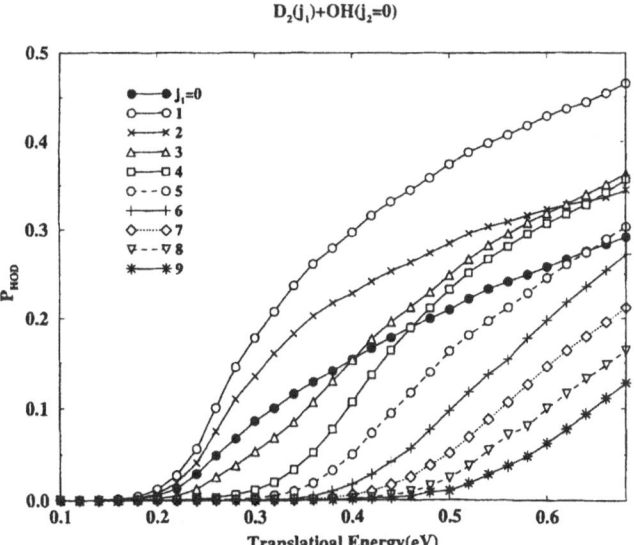

**Fig. 2.** Reaction probabilities of the three isotopic reactions of $H_2$ + OH from the initial ground state of the reagents with zero total angular momentum ($J = 0$)

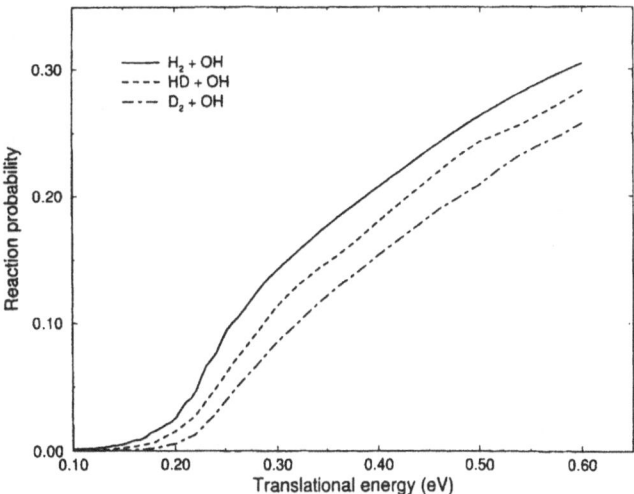

**Fig. 3.** Reaction probabilities of the reaction $D_2$ + OH from differential initial rotational states of the $D_2$ reagent with zero total angular momentum ($J = 0$)

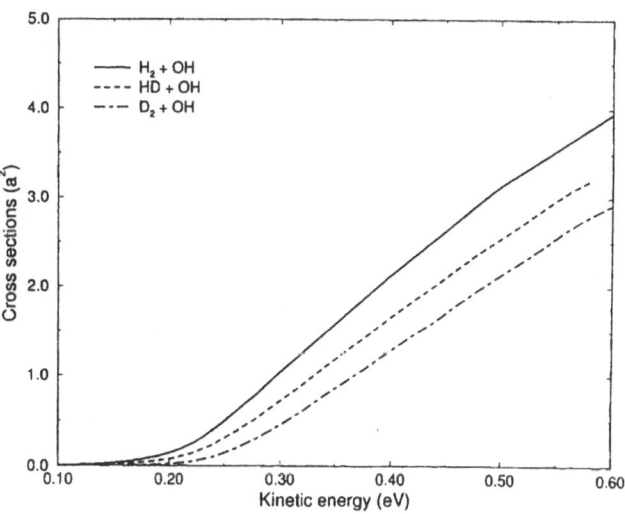

**Fig. 4.** Integral cross sections of the three isotopic reactions of $H_2$ + OH from the initial ground state of the reagents

total angular momentum as explained in Ref. [48]. In addition, the vibrational excitation of the reactive $H_2$ is found to significantly enhance the reaction probability while vibrational excitation of the nonreactive OH(D) bond has little effect on this reaction [46].

The total reaction cross sections and rate constants for the three reactions of $H_2$ + OH show a clear isotopic effect as shown in Figs. 4 and 5.

## 4 Future prospects

We believe that the TD approach currently provides the most viable means for carrying out accurate quantum dynamics studies of polyatomic reaction dynamics due to its relatively low scaling of computational cost with the number of basis function $N$ [computing time $\propto N^\alpha (1 < \alpha < 2)$]; however, we need to further develop more efficient numerical treatments to make accurate

304

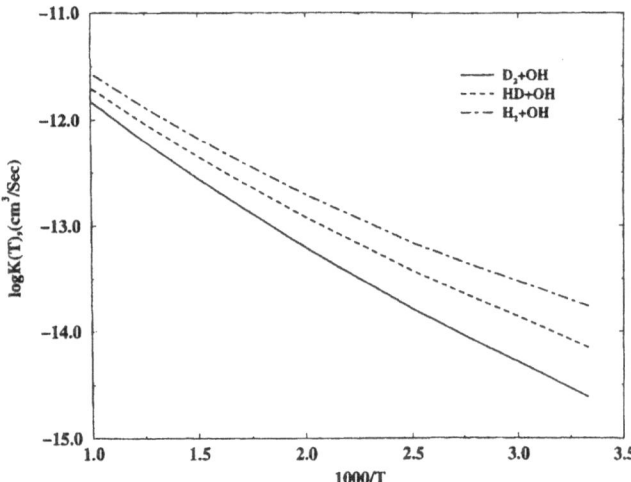

**Fig. 5.** Rate constants of the three isotopic reactions of $H_2$ + OH from the initial ground state of the reagents

dynamics calculations possible for relatively large poly-atomic systems. An obvious, but reliable, approach is to first eliminate all the nonreactive (spectator) vibrational modes of molecules from the dynamics calculation by treating them adiabatically as shown in Ref. [46] for treating the nonreactive OH bond in the $H_2$ + OH reaction. This will instantaneously reduce significantly the total number of active degrees of freedom in the dynamics calculation with little loss of accuracy.

Another important task is to develop more accurate and efficient methods to generate PESs for quantum dynamics calculations. Currently, there is a severe shortage of reliable PESs for dynamics studies. As ab initio quantum chemistry computations become more reliable and affordable, the fitting of discrete ab initio energy points into high-dimension global potential energy functions becomes a bottleneck in dynamics studies due to numerical difficulties. It is desirable to be able to carry out quantum dynamics calculations through automatic numerical fitting of the ab initio points directly to give potential energies at desired dynamical grid points (quadrature points). Recently, a number of methods have been proposed to accomplish this task. The Shepard interpolation method proposed by Ischtwan and Collins [64] and modified by others [65, 66] uses both potential energies and their derivatives from ab initio calculations to interpolate the potential energy, and employs classical trajectory calculations extensively to improve the PES. The SOFA method of Wang et al. [67] uses a sequential one-dimensional fitting approach to automate the task of multi-dimensional fitting of PESs within a precomputed set of ab initio data. The SOFA method is better suited for quantum dynamics studies because it only requires the ab initio energy, and but not their derivatives. Although in their early stages of development, both approaches seem to be quite promising for future ab initio quantum dynamics calculations.

*Acknowledgements.* We acknowledge financial support from the National Science Foundation through a presidential Faculty Fellows Award and The Donors of Petroleum Research Fund. D.H.Z. acknowledges academic research grant no. RP970632 of The National University of Singapore for partial support of this work.

## References

1. McCullough EA Jr, Wyatt RE (1971) J Chem Phys 54: 3578
2. Varga RS (1962) Matrix iterative analysis. Prentice-Hall, Englewood Cliffs, N.J., pp 262–266
3. Askar A, Cakmak S (1978) J Chem Phys 8: 2794
4. Kulander KC (1978) J Chem Phys 69: 5064
5. (a) Heller EJ (1978) J Chem Phys 68: 389; (b) Kulander KC, Heller EJ (1978) J Chem Phys 69: 2439
6. Fleck JA Jr, Morris JR, Feit MD (1976) Appl Phys 10: 129
7. Kosloff R, Kosloff D (1983) J Chem Phys 79: 1823
8. Tal-Ezer H, Kosloff D (1984) J Chem Phys 81: 3967
9. Park TJ, Light JC (1986) J Chem Phys 85: 5870
10. Mowrey RC, Kouri DJ (1986) J Chem Phys 84: 6466
11. Zhang JZH, Kouri DJ (1986) Phys Rev A 34: 2687
12. Neuhauser D, Baer M (1989) J Chem Phys 91: 4651
13. (a) Neuhauser D, Baer M, Judson RS, Kouri DJ (1989) J Chem Phys 90: 5882; (b) Neuhauser D, Judson RS, Kouri DJ, Adelman DE, Shafer NE, Kliner DAV, Zare RN (1992) Science 257: 519
14. Zhang DH, Zhang JZH (1994) J Chem Phys 101: 3671
15. Goldfield M, Gray SK (1996) Comput Phys Commun 98: 1
16. Peng T, Zhang DH, Zhang JZH, Schinke R (1996) Chem Phys Lett 248: 37
17. Zhang DH, Zhang JZH (1996) In: Wyatt RE, Zhang JZH (ed) Dynamics of molecules and chemical reactions. Dekker, New York, p 231
18. Engel V, Schinke R, Staemmler V (1988) J Chem Phys 88: 129
19. Jiang XP, Heather R, Metiu H (1989) J Chem Phys 90: 2555
20. Zhang JZ, Imre DG (1989) J Chem Phys 90: 1666
21. Le Quere F, Leforestier C (1990) J Chem Phys 92: 247
22. Guo H, Schatz GC (1990) J Chem Phys 93: 393
23. Huang ZH, Guo H (1992) J Chem Phys 97: 2110
24. (a) Gray SK, Wozny CE (1989) J Chem Phys 91: 7671; (b) Eray SK, Wozny CE (1991) J Chem Phys 94: 2817
25. (a) Zhang DH, Zhang JZH (1991) J Chem Phys 95: 6449; (b) Zhang DH, Zhang JZH (1992) J Phys Chem 96: 1575
26. Jackson B, Metiu H (1987) J Chem Phys 86: 1026
27. (a) Chiang CM, Jackson B (1987) J Chem Phys 87: 5497; (b) Cruz A, Jackson B (1991) J Chem Phys 84: 5715
28. (a) Hand MR, Holloway S (1989) J Chem Phys 91: 7209; (b) Darling GR, Holloway S (1990) J Chem Phys 93: 9145
29. (a) Mowrey RC (1991) J Chem Phys 94: 7098; (b) Mowrey RC (1993) J Chem Phys 99: 7049
30. (a) Sheng J, Zhang JZH (1992) J Chem Phys 96: 3866; (b) Dai J, Sheng J, Zhang JZH (1994) J Chem Phys 101: 1555; (c) Dai J, Zhang JZH (1994) Surface Science 319: 193
31. Ge JY, Dai J, Zhang JZH (1996) J Phys Chem 100: 11432
32. (a) Sun Q, Bowman JM (1990) J Chem Phys 92: 5201; (b) Sun Q, Yan DL, Wang NS, Bowman JM, Lin MC (1990) J Chem Phys 93: 4730
33. (a) Brook AN, Clary DC (1990) J Chem Phys 92: 4178; (b) Clary DC (1991) J Chem Phys 95: 7298
34. (a) Bowman JM, Wang D (1992) J Chem Phys 96: 7852; (b) Wang D, Bowman JM (1992) J Chem Phys 96: 8906; (c) Wang D, Bowman J (1993) 98: 6235
35. (a) Clary DC (1992) J Chem Phys 96: 3656; (b) Clary DC (1992) Chem Phys Lett 192: 34; (c) Nyman G, Clary DC (1993) J Chem Phys 99: 7774
36. (a) Szichman H, Last I, Baram A, Baer M (1993) J Phys Chem 97: 6436; (b) Szichman H, Baer M (1994) J Chem Phys 101: 2081; (c) Szichmen H, Baer M (1995) Chem Phys Lett 285: 242
37. Balakrishnan N, Billing GD (1994) J Chem Phys 101: 2785
38. Echave J, Clary DC (1994) J Chem Phys 100: 402

39. Thompson WH, Miller WH (1994) J Chem Phys 101: 8620
40. Goldfield EM, Gray SK, Schatz GC (1995) J Chem Phys 102: 8807
41. Zhang JZH, Dai J, Zhu W (1997) J Phys Chem 101: 2746, and references therein
42. (a) Zhang DH, Zhang JZH, Bacic Z (1992) J Chem Phys 97: 927; (b) Zhang DH, Zhang JZH, Bacic Z (1992) Chem Phys Lett 194: 313
43. Zhang DH, Zhang JZH, Bacic Z (1992) J Chem Phys 97: 3149
44. (a) Zhang DH, Zhang JZH (1993) J Chem Phys 98: 6276; (b) Cai ZT, Zhang DH, Zhang JZH (1994) J Chem Phys 100: 5631
45. (a) Zhang DH, Zhang JZH (1993) J Chem Phys 99: 5615; (b) Zhang DH, Zhang JZH (1994) J Chem Phys 100: 2697
46. Zhang DH, Zhang JZH (1994) J Chem Phys 101: 1146
47. Neuhauser D (1994) J Chem Phys 100: 9272
48. (a) Zhang DH, Zhang JZH (1995) Chem Phys Lett 232: 370; (b) Zhang DH, Zhang JZH, Zhang YC, Wang D, Zhang Q (1995) J Chem Phys 102: 7400
49. Zhang Y, Zhang D, Li W, Zhang Q, Wang D, Zhang DH, Zhang JZH (1995) J Phys Chem 99: 16824
50. (a) Manthe U, Seideman T, Miller WH (1993) J Chem Phys 99: 10078; (b) Manthe U, Seideman T, Miller WH (1994) J Chem Phys 101: 4759
51. Zhang DH, Light JC (1997) J Chem Phys 106: 551
52. Matzkies F, Manthe U (1998) J Chem Phys 108: 4828
53. Zhang DH, Zhang (1995) J Chem Phys 103: 6512
54. Zhu W, Zhang JZH, Zhang YC, Zhang YB, Zhan LX, Zhang SL (1998) J Chem Phys 108: 3509
55. Zhu W, Zhang JZH, Zhang DH (1998) Chem Phys Lett 292: 46
56. Manthe U, Matzkies F (1998) J Chem Phys 282: 442
57. Zhang DH, Light JC (1996) J Chem Phys 104: 4544
58. Zhu W, Dai J, Zhang JZH (1996) J Chem Phys 105: 4881
59. Dai J, Zhu W, Zhang JZH (1996) J Phys Chem 100: 13901
60. Zhang DH, Light JC (1996) J Chem Phys 105: 1291
61. Zhang JZH (1989) Theory and application of quantum molecular dynamics. World Scientific, Singapore, pp 333
62. Rose ME (1957) Elementary theory of angular momentum. Wiley, New York,
63. (a) Walch SP, Dunning TH (1980) J Chem Phys 72: 1303; (b) Schatz GC, Elgersma H (1980) Chem Phys Lett 73: 21; (c) Schatz GC J (1981) Chem Phys 74: 1133
64. Ischtwan J, Collins MA (1994) J Chem Phys 100: 8080
65. Nguyen KA, Ross I, Truhlar DG (1995) J Chem Phys 103: 5522
66. Ishida T, Schatz GC (1997) J Chem Phys 107: 3558
67. Wang DY, Peng T, Zhang JZH (1999) Phys Chem Chem Phys 1: 1067

Theor Chem Acc (2000) 103:306–307
DOI 10.1007/s002149900052

Theoretical
Chemistry Accounts
© Springer-Verlag 2000

*Perspective*

# Perspective on "Molecular dynamics study of liquid water"

## Rahman A, Stillinger FH (1971) J Chem Phys 55: 3336–3359

### Peter Kollman

Department of Pharmaceutical Chemistry, University of California at San Francisco, San Francisco, CA 94143-0446, USA

Received: 29 March 1999 / Accepted: 14 April 1999 / Published online: 9 September 1999

**Abstract.** The first molecular dynamics simulation of liquid water by Rahman and Stillinger was the cornerstone of much subsequent research, including molecular dynamics of proteins and other biological systems

**Key words:** Molecular dynamics – Liquid water

The paper by Rahman and Stillinger was the first molecular dynamics simulation of a polar molecule: the single most important one for sustaining life, water. It built on the technique for classical molecular dynamics introduced for hard spheres by Alder and Wainwright [1]. A Monte Carlo study of the structural properties of the liquid appeared around the same time [2]. This paper signaled the beginning of classical molecular dynamics studies of molecules of biological interest and also provided great insight into the structure and dynamics of liquid water. Rahman was subsequently a key member in the CECAM workshop on molecular dynamics in France which facilitated the application of molecular dynamics methods to proteins and led to the first molecular dynamics study on a protein, which was reported by McCammon et al. [3].

Rahman and Stillinger used a four-point-charge model and Lennard-Jones parameters, noting that their model gave a reasonable representation of water dimer energy and structure, which were being first studied by ab initio quantum mechanical calculations around the same time. They used a 0.4-fs time step and carried out 2-ps of production dynamics at an average temperature of 34°C. By integrating the equations of motion in dimensionless form they were able to suggest a scale factor of 1.06 for the potential energy that would improve the agreement with experiment for the average energy and diffusion coefficient and that would correspond to a temperature of 52°C. They also carried out a more limited simulation at −8°C. The simulations were carried out at a constant density of 1 gr/cm³ and the dynamical equations used rigid water monomers and a fifth-order

integration algorithm for translation and a fourth-order one for rotation.

The results were analyzed for a large number of structural and dynamical properties of the liquid and quite good agreement with available experimental results was achieved for all properties. These included average liquid energy, diffusion coefficients, radial distribution functions and their relation to experimental X-ray scattering intensities, the Kirkwood $g$ factor, hydrogen-bond energy distribution, dipole-direction relaxation, dielectric relaxation, and proton motion and neutron scattering.

Most interestingly, this article contributed to the appreciation of the usefulness of computer graphics visualization of molecular structures and dynamics for complex molecular systems [4]. Rahman and Stillinger saved ten intermediate structures and then viewed them in stereo. Equally important as the quantitative analysis noted above were the qualitative insights into the water structure they found this way, as noted on pages 3348–3349:

(1) There is a very clear tendency for neighboring molecules to be oriented into rough approximations to tetrahedral hydrogen bonds, but the average degree of bending away from bond linearity and ideal approach directions is considerable.
(2) Except on the smallest scale, the random molecular configurations are rather homogenous in density. No large "clusters" of anomalous density seem to occur.
(3) No recognizable patterns characteristic of the known ices or clathrates appear, beyond occasional polygons of hydrogen bonds. Such polygons occur with 4, 5, 6, 7 (and perhaps more) sides, but they tend to be distorted out of their most natural conformations.
(4) Dangling OH bonds exit, which are not included in hydrogen bonds. These entities persists far longer than water molecule vibrational periods, and hence may hold the key to the structurally sensitive band shapes that arise in infrared and Raman spectroscopy of water and its solutions.
(5) No obvious separation of molecules into "network" vs "interstitial" types suggests itself. This fact is consistent with the single peak character of the hydrogen-bond coordination number distributions exhibited in Fig. 19. It also seems to diminish the validity of the interstitial models that have been proposed to explain liquid water.
(6) In the case of moderately well-formed (i.e., undistorted) hydrogen bonds, all angles of rotation of the molecules about the bond axis seem to be frequently represented. This behavior may have direct relevance in study of nonelectrolyte solvation,

where the geometric requirements attendant upon formation of a hydrogen-bonded solvation cage forces the rotation angles into "eclipsed" configurations only, thus lowering configurational entropy.

(7) No significant examples of network interpenetration were found, analogous to the interpenetration known to obtain in ice VII and ice VIII.

These insights enabled one to discriminate among the many phenomenological models of liquid water that had been proposed and to argue both against the presence of "ice-bergs" and the validity of "mixture" models. Their conclusions were further supported by subsequent simulations of liquid water as well.

In their discussion, Rahman and Stillinger anticipated further studies on molecular dynamics of water solutions under various conditions, including studies of the hydration of macromolecules and membranes.

Thus, in summary, the article by Rahman and Stillinger provided the first accurate description of liquid water and that description holds today; their study laid the ground-work for molecular dynamics studies of macromolecules in aqueous solution and thus is the cornerstone of a large body of subsequent research, which is becoming ever more powerful and useful, and their use of computer graphics visualization anticipated the exciting insights such methods could give in understanding structures and dynamics of liquids and complex biological molecules.

## References

1. Alder BJ, Wainwright TE (1959) J Chem Phys 31: 459
2. Barker JA, Watts RO (1969) Chem Phys Lett 3: 144
3. McCammon JA, Gelin BR, Karplus M (1977) Nature 267: 585
4. Levinthal C (1966) Sci Am 214: 42

Theor Chem Acc (2000) 103:308–310
DOI 10.1007/s002149900068

Theoretical
Chemistry Accounts
© Springer-Verlag 2000

*Perspective*

# Sugar anomerism – a short and sweet digression
# Perspective on "The application of ab initio molecular orbital theory to the anomeric effect"

## Jeffrey GA, Pople JA, Radom L (1972) Carbohydr Res 25:117

**Christopher J. Cramer**

Department of Chemistry and Supercomputer Institute, University of Minnesota, 207 Pleasant St. SE, Minneapolis, MN 55455-0431, USA

Received: 5 February 1999 / Accepted: 20 May 1999 / Published online: 9 September 1999

**Abstract.** The subject paper presented the first conformational analysis of the anomeric effect within the context of molecular orbital theory, discussed the utility of Fourier decomposition of a torsional coordinate as a method for analyzing disparate electronic influences on that coordinate, and helped settle debate on the nature of anomeric stabilization.

**Key words:** Carbohydrate – Conformational analysis – Stereoelectronic effects – Hyperconjugation

Sugars are the stepchildren of the biomonomer family. Unlike their more popular siblings peptides and nucleic acids, the study of which occupies a swarm of researchers, sugars appeal to a coterie of devotees.

Yet sugars are wonderfully appealing to those attracted to puzzles. Capable of exhibiting multiple tautomeric forms at equilibrium, festooned with functional groups generating potentially thousands of rotational isomers, and profoundly influenced by aqueous solvation, they are a theorists dream (or nightmare) vis-á-vis challenge!

I choose to focus here on the contributions of Jeffrey, Pople, and Radom (JP&R) [1] to fostering our early understanding of one particularly fascinating feature of sugar energetics, namely the anomeric effect [2–7]. This stereoelectronic effect was first noted in sugars, where the axial anomer of mannopyranose, for instance, is observed [8] to predominate over the equatorial anomer (this situation being contrary to expectations based on standard precepts of conformational analysis as applied to substituted cyclohexanes). However, the generality of the anomeric effect, or "negative hyperconjugation" as it is sometimes called, has come to be widely recognized, and it is a key component in the conformational analysis

of a diverse array of organic and inorganic molecules [9, 10]. Its physical basis was poorly understood in 1972.

At the time two hypotheses (that were regarded for the most part as being mutually exclusive) had been put forward to account for the anomalous preference of electron-accepting substituents to orient themselves axially at the 2 positions in pyranose rings (i.e., the position next to the ring oxygen). In one analysis, the preference was ascribed to dipole–dipole and/or other Coulombic effects between the substituent group and the ring oxygen atom with its associated bonds [11, 12]. However, these models were of limited utility for predicting either experimental equilibria or geometrical differences between various pyranoside anomers; the latter were becoming increasingly available from X-ray crystallographic studies [13].

Indeed, crystal structure observations of lengthened axial bonds in 2-halopyranosides with concomitant shortening of the ring $C(2)$–$O$ bonds led to an alternative proposal rationalizing the anomeric effect by invoking double bond–no bond resonance structures [14], i.e., negative hyperconjugation. This model, however, failed to account for another crystallographic observation, namely the shortening of axial $C(2)$–$O_{exo}$ bonds in 2-oxygenated pyranosides [15] – this phenomenon acquired the sobriquet "exo-anomeric effect".

It was at roughly this stage that theory first entered the game. In 1971, two papers appeared that were germane to this topic. The first, by Wolfe et al. [16], showed that the hydroxyl rotational coordinate of fluoromethanol ($FCH_2OH$) had a deep minimum for the gauche FCOH dihedral angle ($60°$), but no minimum for the trans dihedral angle ($180°$). That same year, similar results were obtained independently by Radom et al. [17] not only for fluoromethanol, but also for dihydroxymethane, which is the smallest molecule incorporating the acetal linkage that characterizes a pyranoside.

A limitation of these 1971 studies was that they did not attempt to optimize the molecular geometries (analytic derivatives not yet being available, such optimiza-

**Fig. 1.** Anomeric equilibrium in a model pyranose

tions were not routine) but rather employed ideal bond lengths and valence angles, adjusting only the torsion angles to specific, also nonoptimized values. As such, analysis hinted at explanations for the anomeric effect, but could not be regarded as definitive.

Key advances offered by the 1972 paper of JP&R included

1. Providing a complete 2D potential-energy surface for both C—O rotations in dihydroxymethane.
2. Optimization of the C—O bond lengths for various torsion angles so as to compare to analogous linkages in experimentally known crystal structures.
3. Decomposition of the C—O rotational coordinates in a Fourier sense to assess the relative contributions of dipole–dipole interactions, hyperconjugative stabilization, and steric interactions.

Another critical strength of this paper was that it used the electronic calculations in conjunction with a simple steric model to successfully rationalize 18 different pyranose crystal structures. While this latter aspect cannot be discussed here in any detail owing to space limitations, this careful comparison of the model to experiment had a tremendous impact on experimental carbohydrate chemists, who were impressed to see theory being used to make predictions about exactly the same crystal structures they were concerned with from their own perspectives. Indeed, JP&R set a remarkably high standard to be emulated in future theoretical papers, and contributed to the generally good reputation theory has enjoyed within the carbohydrate community. The calculations of JP&R, described below, were carried out at the HF/4-31G level [18].

With respect to the 2D analysis, JP&R clearly described for the first time the simultaneous hyperconjugative effects deriving from delocalization of lone-pair density on each oxygen atom into the $\sigma^*_{CO}$ antibonding orbitals associated with their counterpart oxygens. While they did not refer to this by name, in essence this was the first demonstration of simultaneous anomeric and exo-anomeric effects.

The various changes in C—O bond lengths associated with the torsional coordinate were substantial. Thus, JP&R predicted variations in the C—O bond length of up to 0.032 Å depending on whether the oxygen atoms were acting as lone-pair donors or as part of acceptor antibonding orbitals. Results of this magnitude were in good agreement with the range observed in crystal structure data, and provided a critical validation of the theoretical approach.

JP&R properly associated these bond length deviations with hyperconjugation (which had been suggested before based on intuition and analogy [13]) but did so by a more convincing analysis than had hitherto been advanced. That analysis involved the decomposition of

individual oxygen rotational coordinates in a Fourier sense. As described by Radom et al. [19] that same year, such an analysis can be applied in simple instances to qualitatively assess the relative dependence of the torsional coordinate on dipole–dipole interactions (which should show a onefold symmetry over the full rotation angle of $2\pi$ rad), hyperconjugation (which should show a twofold periodicity since it is locally maximized for eclipsed and antiplanar arrangements by virtue of most favorable overlap between donor and acceptor orbitals), and $sp^3$-$sp^3$ steric interactions (which are threefold periodic by the tetrahedral nature of the atoms involved). JP&R showed for dihydroxymethane that the twofold periodic term was nontrivial, and discussed its general importance for sugar conformational analysis. (To be fair, their analysis did underestimate the magnitude of the importance of hyperconjugation because, inter alia, they used ideal bond angles – the anomeric effect causes an increase in the valence angle at the central carbon, and without accounting for this the gauche structures were somewhat higher in energy than they should have been – such approximations are difficult to criticize given the computational resources available at the time.)

It is important to note that the results of JP&R significantly influenced the thinking of subsequent researchers. In the interests of brevity I note here only the key papers of Tvaroska and Bleha [20, 21], which further demonstrated the utility of Fourier analysis, the particularly elegant paper in 1979 by Wolfe et al. [22], which refined many of the ideas laid out in JP&R's prior work and extended its range of applicability, and the efforts of Thøgersen et al. [23], who built upon the work of JP&R in the construction of an empirical potential function for oligosaccharide conformational analysis.

To conclude on a personal note, I first encountered the JP&R paper several years ago when I became interested in the characterization of solvation effects on various anomeric equilibria in pyranosides [24, 25]. It goes without saying that phenomenal advances in computational power now permit us to address considerably larger model systems than those accessible to JP&R and moreover to do so at much more refined levels of electronic structure theory [26–28]. Nevertheless, subsequent work has been guided by JP&R's seminal 1972 contribution, which convincingly elucidated the key physical underpinnings of the anomeric effect, successfully predicted the most important qualitative consequences, and set the foundation stone for future quantum chemical conformational analysis of sugars.

(In the spirit of the paper's title, I feel I must digress at the end point out one other important contribution of John Pople to the general area of sugar conformational analysis. He, together with Dieter Cremer, developed the systematic classification of ring geometries that continues to be used today to describe the variety of shapes furanose and pyranose rings can adopt [29]. This scheme, which expresses these shapes in terms of the Cremer–Pople ring puckering coordinates, is in my opinion an extraordinary example of the kind of impact a theoretician can have on a particular field when he or she focuses upon it as a new challenge.)

310

## References

1. Jeffrey GA, Pople JA, Radom L (1972) Carbohydr Res 25: 117
2. Jungius CL (1905) Z Phys Chem 52: 97
3. Lemieux RU (1964) In: de Mayo P (ed) Molecular rearrangements. Interscience, New York, p 709
4. Pearson CG, Rumquist O (1968) J Org Chem 33: 2572
5. Kirby AJ (1983) The anomeric effect and related stereoelectronic effects at oxygen. Springer, Berlin Heidelberg New York
6. Thatcher GRJ (ed) (1993) The anomeric effect and associated stereoelectronic effects. American Chemical Society, Washington, D.C
7. Graczyk PP, Mikolajczyk M (1994) In: Eliel EL, Wilen SH (eds) Topics in stereochemistry. Wiley, New York, p 159
8. Angyal SJ (1968) Aust J Chem 21: 2737
9. Eliel EL, Wilen SH (1994) Stereochemistry of organic compounds. Wiley, New York
10. Cramer CJ (1996) J Mol Struct (Theochem) 370: 135
11. Edward JT (1955) Chem Ind 1102
12. Anderson CB, Sepp DT (1967) J Org Chem 32: 607
13. Romers C, Altona C, Buys HR, Havinga E (1969) Top Stereochem 4: 39
14. Stoddart JF (1971) Stereochemistry of carbohydrates. Wiley-Interscience, New York
15. Berman HM, Chu SSC, Jeffrey GA (1967) Science 157: 1576
16. Wolfe S, Rauk A, Tel LM, Czismadia IG (1971) J Chem Soc B 136
17. Radom L, Hehre WJ, Pople JA (1971) J Am Chem Soc 93: 289
18. Hehre WJ, Radom L, Schleyer PvR, Pople JA (1986) Ab initio molecular orbital theory. Wiley, New York
19. Radom L, Hehre WJ, Pople JA (1972) J Am Chem Soc 94: 2371
20. Tvaroska I, Bleha T (1975) Tetrahedron Lett 16: 249
21. Tvaroska I, Bleha T (1989) Adv Carbohydr Chem Biochem 47: 45
22. Wolfe S, Whangbo M-H, Mitchell DJ (1979) Carbohydr Res 69: 1
23. Thøgersen H, Lemieux RU, Bock K, Meyer B (1982) Can J Chem 60: 44
24. Cramer CJ (1992) J Org Chem 57: 7034
25. Cramer CJ, Truhlar DG (1993) J Am Chem Soc 115: 5745
26. Barrows SE, Dulles FJ, Cramer CJ, Truhlar DG, French AD (1995) Carbohydr Res 276: 219
27. Cramer CJ, Truhlar DG, French AD (1997) Carbohydr Res 298: 1
28. Barrows SE, Storer JW, Cramer CJ, French AD, Truhlar DG (1998) J Comput Chem 19: 1111
29. Cremer D, Pople JA (1975) J Am Chem Soc 97: 1354

Theor Chem Acc (2000) 103:311–312
DOI 10.1007/s002149900098

Theoretical
Chemistry Accounts
© Springer-Verlag 2000

*Perspective*

# From Xα-scattered wave to end-of-the-century applications of density functional theory in chemistry. Perspective on "Chemical bonding of a molecular transition-metal ion in a crystalline environment"

## Johnson KH, Smith FC Jr (1972) Phys Rev B 5: 831–843

**Dennis R. Salahub**[1,2]*

[1] Département de Chimie, Université de Montréal, C.P. 6128, Succursale Centre-ville, Montréal, Québec H3C 3J7, Canada
[2] CERCA – Centre de Recherche en Calcul Appliqué, 5160, boul. Décarie, bureau 400, Montréal, Québec H3X 2H9, Canada

Received: 10 August 1999 / Accepted: 16 August 1999 / Published online: 15 December 1999

**Abstract.** The paper by Johnson and Smith is representative of the Xα-scattered wave (SW) effort of the 1970s. Despite the severe approximations that were necessary at the time, the paper shows that a "first-principles" Xα calculation yields a compelling account of the electronic structure and spectrum of permanganate. Contemporaneous semiempirical and ab initio calculations were not up to the task. The quality of the results and the prospect of treating really large systems were sufficient to attract the attention of many quantum chemists (as well as the disdain of some and the ire of others). The Xα-SW work was an important link in the chain of contributions that would bring density functional theory into chemistry.

1 am.

A faint knock on the downstairs door. Or was it?

Then the unmistakable thump of a heavy boot against the door and the crack of the door jam as it shattered.

Had his sordid past caught up with him? The interrogation would be swift and on the spot. Where did that wooden chair come from? And the bare light bulb slowly swaying above it? Whose face was that, almost invisible behind the glare?

Inquisitor: Are you now or have you ever been a member of the Xα party?

Mild-mannered respectable density functional theory practitioner (MMRDFTP): What? (Where had he heard that voice before?)

Inquisitor: Are you now or have you ever been a member of the Xα party?

MMRDFTP: I'm a Mild-mannered respectable density functional theory practitioner (MMRDFTP). What do you mean by breaking into my house in the middle of the night and hauling me out of bed like that? I was just in the middle of a great dream about an exchange–correlation functional that had the right asymptotic form and took care of dispersion seamlessly. Could have done excited states too... and eminently parallelizable. And now I've forgotten what it looked like...

Inquisitor: Answer the question or you'll be eating your teeth for breakfast. Are you now or have you ever been a member of the Xα party?

MMRDFTP: OK, OK, I did do some Xα calculations a long time ago. But only a few...

Inquisitor: Why'd ya do it?

MMRDFTP: Desperation I guess. We wanted to do transition metals and then really complex systems. The semiempirical methods could only go so far and Hartree–Fock... Well, life is short... So when we heard about the Xα stuff we had to have a close look. The great physicist John Slater was pushing Xα. Slater [1] had proposed his approximate "statistical" treatment of exchange to be able to treat larger systems. Keith Johnson in his group adapted one of the solid-state band techniques (the Korringa–Kohn–Rostoker method [2, 3] – yes, the Kohn is the DFT Kohn...) to use nonperiodic, molecular, boundary conditions, the Xα-scattered wave (SW) method. This involved some pretty drastic approximations for the one-electron potential (the Kohn-Sham [4] potential in modern language, though Slater wasn't thinking in terms of a potential that included correlation – to him, Xα was derived by introducing approximations into the Hartree–Fock equations [5]). In the Xα-SW method one built spheres around each of the atoms and took a spherical average of the potential in them. Between the spheres, for want of anything better, a constant average potential was used. If you make a

---

* *Present address*: Steacie Institute for Molecular Sciences, National Research Council of Canada, 100 Sussex Drive, Ottawa, Ontario K1A 0R6, Canada

312

picture it looks like round wells around the atoms with flat areas in-between. Hence the name muffin-tin potential. Some of the quantum chemists took great joy in making fun of it – the Rin-Tin-Tin potential... etc. It was a pretty ugly potential... Actually, some of the discussions were pretty ugly too, but I digress... The muffin-tin approach had some advantages too. You didn't have to optimize basis functions, in a sense the molecule did it for you. It was fast, infinitely fast in comparison to post-Hartree–Fock methods. And in the end, the results were good enough to solve a lot of problems, but of course not everything.

Inquisitor: OK, enough blah-blah. What kind of problems could you solve with Xα?

MMRDFTP: Like I said, transition metals and the like. Things that just seemed hopeless with other methods. No one was thinking in terms of organic chemistry or 2 kcal/mol or anything like that at the time, the way you can with modern DFT. In retrospect, neither the functionals nor the techniques for solving the Kohn–Sham equations were ready yet. That would take a few more years. The 1972 paper of Johnson and Smith [6] on permanganate is a good example. For transition metals, even seemingly simple complexes such as $MnO_4^-$ present a formidable correlation problem. Hartree–Fock calculations were just too far off the mark to be realistic. So if one needed to understand such systems through quantum chemistry, it was pretty desperate, either semiempirical or Xα. Johnson and Smith did an Xα calculation, including the effects of the surrounding crystal in a rough manner (a positively charged sphere surrounding the whole permanganate anion to stabilize it) and gave a detailed analysis of the spectrum and the orbitals. They were very much concerned with being able to treat even more complex systems, impurities in solids, enzymes and the like, just the systems where modern DFT is showing such promise.

Inquisitor: You're making it all sound too rosy. There's no real link between Xα and muffin tins and all that and real legitimate DFT calculations.

MMRDFTP: "Legitimate DFT"...interesting... Actually, there are several links, a whole chain, in terms of getting DFT into better shape and into quantum chemistry. In the Florida school there were several attempts to get rid of the worst features of the muffin-tin potential but none of them were really successful. Then Dunlap, Connolly and Sabin [7] made a great contribution. Building on previous work of Sambe and Felton [8], they formulated a linear combination of atomic orbitals-Xα method using Gaussian orbitals and fitting functions for the Coulomb and exchange–correlation terms. Dunlap's program inspired others, such as deMon and DGauss. In parallel, the so-called discrete variational method [9] spawned modern codes, such as DMol and ADF. Then when the functionals got good enough to get DFT into Gaussian and the other mainstream quantum chemistry codes it all became very respectable. So can I go back to sleep now?

Inquisitor: Not quite yet, I want to ask you a question about your use of cluster models...

## References

1. Slater JC (1951) Phys Rev 81: 385
2. Korringa J (1947) Physica 13: 392
3. Kohn W, Rostoker N (1954) Phys Rev 84: 1111
4. Kohn W, Sham LJ (1965) Phys Rev A 140: 1133
5. Slater JC (1974) The self-consistent field for molecules and solids, vol 4. McGraw-Hill, New York
6. Johnson KH, Smith FC (1972) Phys Rev B 5: 831
7. Dunlap BI, Connolly JWD, Sabin JR (1979) J Chem Phys 71: 3396
8. Sambe H, Felton RH (1975) J Chem Phys 62: 1122
9. Baerends EJ, Ellis DE, Ros P (1973) Chem Phys 2: 41

Theor Chem Acc (2000) 103:313–314
DOI 10.1007/s002149900071

Theoretical
Chemistry Accounts
© Springer-Verlag 2000

*Perspective*

# Perspective on "Self-consistent molecular Hartree–Flock–Slater calculations"

## Baerends EJ, Ellis DE, Ros P (1973) Chem Phys 2:41

**A.P.J. Jansen, R.A. van Santen**

Schuit Institute of Catalysis, TUE, ST/SKA, P.O. Box 513, 5600 MB Eindhoven, The Netherlands

Received: 2 March 1999 / Accepted: 28 June 1999 / Published online: 2 November 1999

**Abstract.** This paper discusses the title paper by Baerends, Ellis, and Ros. We show the role it has played in the development of density functional theory and the further work it has initiated.

**Key words:** Density functional theory – Discrete variational quantum chemistry – Hartree–Fock–Slater method

A remarkable development of the past decade has been the general acceptance by the chemical community of ab initio computer codes as modeling tools in a wide variety of applications. While it has been possible to predict molecular spectroscopic and geometric properties for a long time, the accurate prediction of bond energies of large systems has only become possible quite recently. As shown by the 1998 Nobel award to Kohn (shared with Pople), the contribution of density functional theory (DFT) to practical ab-initio calculations is now widely recognized. Rosén has recently highlighted the crucial role of the discrete variational Xα method (DVM) in the process that has led to the current state of affairs [1]. The use of the DVM using an expansion of molecular eigenfunctions atomic orbitals was described first in the title paper.

Slater's multiple scattering Xα (MS-Xα) around 1970 was a promising quantum chemical method [2], although studies of diatomic molecules and the water molecule clearly showed that the method had shortcomings [3]. These were mainly due to the Muffin–Tin approximation that was used for the molecular potential; in particular, the assumption that the molecular potential is constant in the interatomic region.

Baerends et al. were the first to show how to do accurate Xα calculations without scattered waves and

without Muffin–Tin potentials. This was accomplished in the time-honoured fashion of quantum chemistry with a basis-set expansion in localized functions; in particular, Slater-type orbitals. This was not only used for the Kohn–Sham orbitals, but also to fit the electronic density. This allowed the accurate calculation of the molecular potential. The other important point of the paper was that numerical methods were used to calculate integrals. These methods were based on previous work by Ellis and Painter [4–6], and have since been known as the DVM. The use of a numerical method solved the problem of the complicated integrals involving the density functional; not just the one of Xα, but also later ones such as those of the local density approximation and the generalized gradient approximation [7–11].

Initial applications focused on the electronic structure of carbonyl complexes [12] and clusters [13] as models of surfaces. A special feature of the method developed by the groups of Baerends and Ziegler [14–16] is the analysis of the bond energy, which allows decomposition in terms of the σ and π bonding between the free fragment orbitals. This enables a quantitative assessment of donating and backdonating energy contributions.

The methods described in the paper by Baerends et al. were implemented in the Amsterdam density functional program (ADF) [17]. This program has played an important role in the acceptance of DFT as an alternative method to wavefunction-based methods for doing quantum chemical calculations. It was later extended, amongst other, by Ziegler's transition-state method [15] that allowed a direct calculation of the bond energy on the basis of the change in the electron density when the bond is formed. The numerical integration methods were made more user-friendly. Nowadays only the desired accuracy has to be specified. Various improved density functionals were also incorporated. Since about 1990 there has also been a version to do DFT calculations on systems with translational symmetry (ADF-BAND) [17]. The first application was to solve the problem of the adsorption site of CO on Cu [18], which cluster models of metals incorrectly predict to be

---

*Correspondence to*: R.A. van Santen

314

a high-coordination site, whereas slab calculations yield the atop site as the preferred one. A more recent application is the calculation of the six-dimensional potential-energy surface for $H_2$ dissociation on Cu(100) [19]. ADF-BAND can be considered the counterpart of DFT codes using plane-wave basis sets.

## References

1. Rosen A (1997) Adv Quantum Chem 29: 1
2. Slater JC (1972) Adv Quantum Chem 6: 1
3. Connolly JWD, Sabin JR (1972) J Chem Phys 56: 5529
4. Ellis DE (1968) Int J Quantum chem symp 2: 35
5. Painter GS, Ellis DE (1970) Int J Quantum Chem symp 3: 801
6. Ellis DE, Painter GS (1970) Phys Rev B 2: 2887
7. Vosko SH, Wilk L, Nusair M (1980) Can J Phys 58: 1200
8. Becke AD (1988) Phys Rev A 38: 3098
9. Perdew JP (1986) Phys Rev B 33: 8822
10. Lee C, Yang W, Parr RG (1988) Phys Rev B 37: 785
11. Becke AD (1993) J Chem Phys 98: 5649
12. Baerends EJ, Ros P (1975) Mol Phys 30: 1735
13. Post D, Baerends EJ (1983) J Chem Phys 78: 5663
14. Baerends EJ, Roozendaal A (1986) In: Veillard A (ed) Quantum chemistry: the challenge of transition metal and coordination chemistry. Reidel, Dordrecht, pp 159–177
15. Ziegler T, Tschrinke V, Ursenbach C (1987) J Am Chem Soc 109: 4825
16. Li J, Schreckenbach G, Ziegler T (1995) J Am Chem Soc 117: 486
17. (a) Amsterdam density functional, ADF release 2.1. Theoretical Chemistry, Vrije Universiteit, Amsterdam; (b) Baerends EJ, Ellis DE, Ros P (1973) Chem Phys 2: 41; (c) te Velde G, Baerends EJ (1992) J Comput Phys 99: 84; (d) Fonesca Guerra C et al. (1995) METECC-95 305
18. Philipsen PHT, te Velde G, Baerends EJ (1994) Chem Phys Lett 226: 538
19. Wiesenekker G, Kroes GJ, Baerends EJ (1996) J Chem Phys 104: 7344

Theor Chem Acc (2000) 103:315–316
DOI 10.1007/s002149900091

*Perspective*

# Perspective on "MO approach to electronic spectra of radicals" Čársky P, Zahradník R (1973) Top Curr Chem 43: 1

**Petr Čársky, Rudolf Zahradník**

J. Heyrovský Institute of Physical Chemistry, Academy of Sciences of the Czech Republic, Dolejškova 3, 18223 Prague 8, Czech Republic

Received: 14 January 1999 / Accepted: 16 August 1999 / Published online: 17 January 2000
© Springer-Verlag 2000

**Abstract.** The development of open-shell molecular orbital theory provided a tool for understanding the observed electronic spectra of radicals and radical ions. This then permitted chemists to understand the origin of the color of many radical ions and to explain in detail the photochemistry and reactivity of radicals and radical ions.

**Key words:** Open–shell MO methods –
Electronic spectra of radicals

In this review we summarized our experience with the development and applications of semiempirical Pariser–Parr–Pople (PPP)-type and all-valence-electron methods to electronic spectra of radicals. After the era of PPP calculations on closed-shell molecules and the advent of semiempirical all-valence-electron methods, the electronic spectra of radicals represented a new challenge for molecular orbital (MO) theory. It was a time when progress in experimental techniques resulted in accumulation of a vast amount of data on the electronic spectra of radicals of various structural types. Compared to closed-shell molecules, the electronic spectra of some radicals exhibited peculiar features: bands in the near infrared, many transitions in the whole UV/vis region, and some bands of extraordinary intensity. Clearly, without the help of MO theory, their interpretation seemed even harder than with closed-shell molecules.

There was a rich choice of available open-shell methods in the early 1960s [1] but their applications were rare and the results met with a differing degree of success. Obviously, a systematic examination was lacking.

We decided to undertake this with the aim of formulating a generally applicable computational scheme for radicals which would be a natural extension of the PPP and semiempirical all-valence-electron methods for closed-shell molecules. We used for this purpose the self-consistent-field open-shell methods of Longuet-Higgins and Pople [2] and of Roothaan [3], we derived all expressions necessary for CI-S calculations [4, 5], and we tested the semiempirical open-shell PPP-type and INDO/S calculations systematically for various classes of radicals.

Needless to say that after 30 years or so, this type of calculation, as all other semiempirical calculations, lost much of its importance. By means of highly sophisticated ab initio methods, such as SAC-CI [6] and CASPT2 [7], it is now possible to treat rather extensive open-shell systems and with remarkable accuracy. Still, we believe that for large conjugated hydrocarbon radicals, the open-shell PPP-type approach remains a method of choice. To document this we present a figure (Fig. 1) from our review on the radical anions of $\alpha$, $\omega$-diphenylpolyenes.

## References

1. Berthier G (1964) In: Löwdin PO, Pullman B (eds) Molecular orbitals in chemistry, physics and biology. Academic Press, New York, p 57
2. Longuet-Higgins HC, Pople JA (1955) Proc Phys Soc Lond Sect A 68: 591
3. Roothaan CCJ (1960) Rev Mod Phys 32: 179
4. Zahradník R, Čársky P (1970) J Phys Chem 74: 1235
5. Zahradník R, Čársky P (1973) Prog Phys Org Chem 10: 327
6. Nakatsuji H (1997) In: Leszczynski J (ed) Computational chemistry: reviews of current trends. World Scientific, Singapore, Vol. 2, p 1
7. Andersson K, Roos BO (1995) In: Yarkony DR (ed) Modern electronic structure theory. World Scientific, Singapore, p 55
8. Hoijtink GJ, van der Meij PH (1959) Z Phys Chem (Frankfurt) 20: 1

*Correspondence to:* P. Čársky

P. Čársky and R. Zahradník

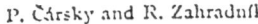

**Fig. 1.** Absorption spectra of anions of α,ω-diphenylpolyenes [8] and results of open-shell Pariser–Parr–Pople-type calculations. The *wavy lines* with *arrows* represent forbidden transitions

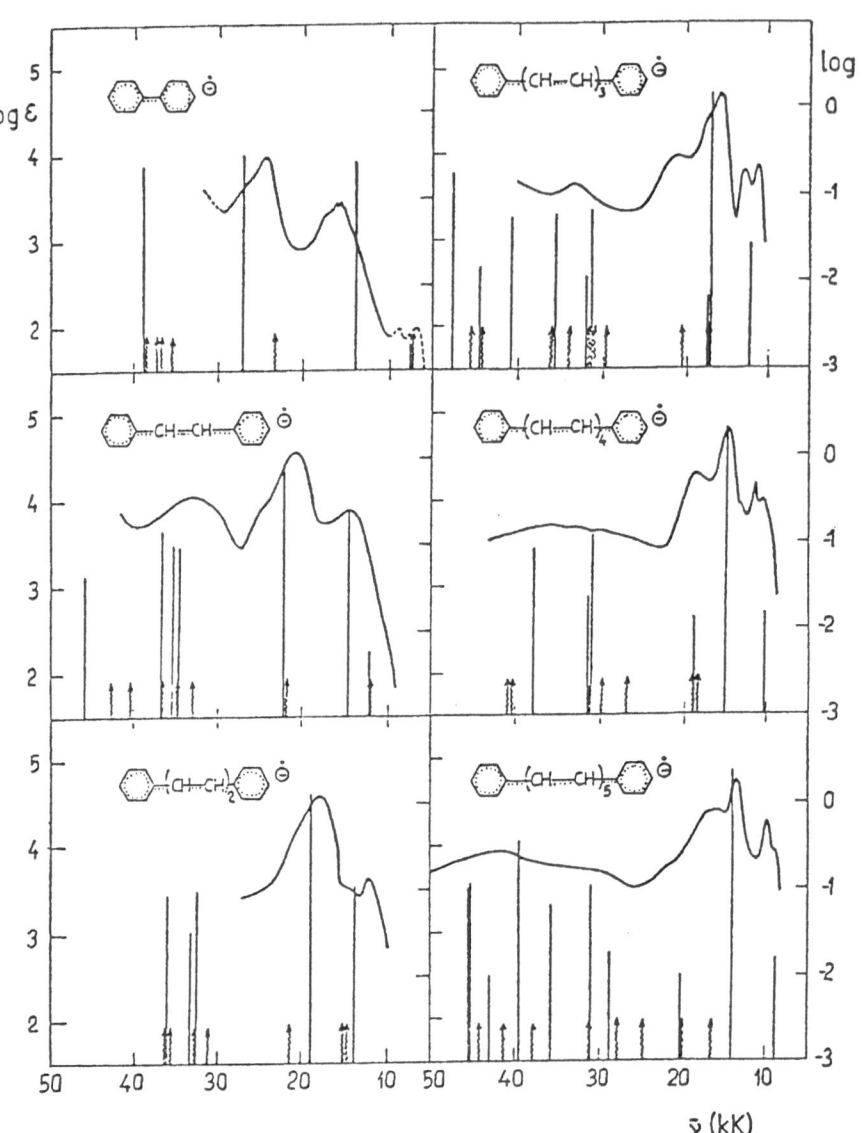

Theor Chem Acc (2000) 103:317–321
DOI 10.1007/s002149900050

Theoretical
Chemistry Accounts
© Springer-Verlag 2000

*Perspective*

# Perspective on "Group theoretical approach to the configuration interaction and perturbation theory calculations for atomic and molecular systems"

Paldus J (1974) J Chem Phys 61: 5321

**Michael A. Robb**

Department of Chemistry, King's College London, Strand, London WC2R 2LS

Received: 18 March 1999 / Accepted: 22 April 1999 / Published online: 14 July 1999

**Abstract.** This paper provides a "perspective" on the title paper of Paldus. The introduction of unitary groups into quantum chemistry has provided not only practical algorithms for many configuration-interaction-based methods still in current use, but has also provided a conceptual basis in quantum chemistry that is as important as the practical implementations.

**Key words:** Unitary groups – Configuration interaction

## 1 Introduction

A colleague recently remarked to me that a practising theoretical chemist who learned his "trade" in the 1990s has probably never heard of the unitary group approach (UGA) despite the fact that almost every quantum chemistry package in current use (excluding those devoted entirely to density functional computations) makes use, in some way, of the elegant UGA, first introduced into quantum chemistry in 1974 by Paldus. While the theory of continuous groups and the symmetric group was an essential part of the training of the theoretical chemist in the 1960s and 1970s, this subject is quite inaccessible to recently trained theoretical chemists. One of the reasons for this is that while the paper of Paldus gave a completely new insight into the configuration-interaction (CI) matrix-element problem, once this insight was achieved the applications and code development could proceed without the need to have a sound understanding of the basic mathematics (group theory) that led to this new insight in the first place.

Thus the purpose of this article is to try to describe the impact of Paldus' 1974 paper [1] (see also related accounts in Refs. [2, 3]) on the evolution of quantum chemistry at that time. The CI method itself has recently been reviewed by Shavitt [4] and the reader is referred to that paper for recent developments. We shall focus in this paper on the nature of the CI method as it existed in 1974 and the nature of the unitary group revolution introduced by Paldus that was subsequently turned into working algorithms by others [5–18] in the 10 years or so after its introduction.

## 2 The CI matrix-element problem 1957–1974

Let us begin our discussion with a simple discussion of the CI matrix-element problem. In the CI problem we seek the solution of the CI eigenvalue problem

$$\mathbf{HC} = \mathbf{EC} , \tag{1}$$

where

$$\mathbf{H} = \{\langle K|HL\rangle\} \tag{2}$$

is the representation matrix of the Hamiltonian in a basis of many-particle configuration state functions (CSF) which we denote as $\{|K\rangle\}$. The CSF, $\{|K\rangle\}$, can be chosen in a variety of ways (e.g. Slater determinants); however, in general, they are chosen to be eigenfunctions of $S^2$ and $S_z$. The central practical problem is the evaluation of the matrix $\langle K|HL\rangle$. The general formula for the matrix elements can be expressed as

$$\{\langle K|HL\rangle\} = \sum h_{ij}A_{ij}^{KL} + \frac{1}{2}\sum [ij|kl]B_{ijkl}^{KL} , \tag{3}$$

where $h_{ij}$ and $[ij|kl]$ are the usual one- and two-electron repulsion integrals (in charge-cloud notation). The $A_{ij}^{KL}$ and $B_{ijkl}^{KL}$ are numerical "vector-coupling" coefficients that depend on the nature of $|K\rangle$ and $|L\rangle$. The preceding discussion hides a problem that was the most important one in 1974. The symbol $|K\rangle$ means an orbital string (of orbitals occupied in the CSF); however, it also stands for a sequence number or index. The central problem was establishing a unique one-to-one correspondence between the two.

Equation (3) can be read in two ways. On the one hand, given $|K\rangle$ and $|L\rangle$, it gives a prescription for forming a specific matrix element $\langle K|HL\rangle$ by providing a list of integral indices $ij$ and $ijkl$ corresponding to non-zero contributions or "weights", $A_{ij}^{KL}$ and $B_{ijkl}^{KL}$, for the one- and two-electron integrals that contribute to a given matrix element. This approach is the so-called matrix-element-driven CI method that was in use in 1974 in the Boys–Reeves bonded-function CI code [19] of the Polyatom program. However, Eq. (3) can be read the other way around. One can focus on an integral $[ij|kl]$. Then Eq. (3) provides a list of matrix element labels $K, L$ to which that integral contributes. This is the direct CI approach introduced by Roos [20, 21] in 1972. In this approach, one can process the integrals in the CI eigenvalue iterations without the need to explicitly assemble the matrix $\mathbf{H}$. This direct CI approach is used in all CI-based methods today; however, Roos' method was originally formulated for the case where the self-consistent-held (SCF) result was a good starting point and the CI he used was built from all single and double replacements of the SCF orbitals. In this case the $A_{ij}^{KL}$ and $B_{ijkl}^{KL}$ are simple and can be derived using simple rules (based on the Slater formalism) found in any textbook on quantum chemistry.

As just pointed out, Roos' method was efficient but not general. One needed the methodology to carry out both full CI (all possible arrangements of electrons in the available orbitals) and a multireference CI (all possible single and double replacements from several reference configurations). Roos' method was only for single-reference CI. In the full CI and in the multireference CI case, not only does the configuration list become large, but also the indexing of the configurations becomes a major problem. For the indexing of the configurations, one needs an algorithm to associate a given string of occupied orbitals and the associated spin coupling, with a unique index $K$ in the list of CI configurations. In the early 1970s there was no general solution to the CI indexing problem. One simply generated a list of configurations in some ad hoc fashion and wrote the information to a file. (In the Boys–Reeves CI program that was implemented in the Polyatom program in the late 1960s, the input to the program was just orbital strings that were punched on cards. The program then generated all possible spin couplings, so-called canonical sets, for each orbital pattern). This configuration-list file was then processed (comparing configurations two at a time) to generate the $A_{ij}^{KL}$ and $B_{ijkl}^{KL}$ which were written to a so-called symbolic matrix-element file. This file was then merged with the numerical values of the one and two integrals in the CI eigenvalue iterations. (In the Polyatom program of the 1960s this file filled several 2400-ft magnetic tapes for a CI with a mere 2000 configurations). Thus, this symbolic matrix-element file became unmanageably large very quickly and this proved to be an insurmountable problem until the introduction of UGA methods and direct CI concepts. Further, the computation of the symbolic matrix elements was so costly that the possibility of computing matrix elements "on the fly" was out of the question. Note that every pair of configurations needed to be examined to generate the symbolic matrix elements. There was no way to tell a priori which matrix elements would be zero.

Thus the development of general CI-based methods required the solution of two major problems:

1. A genealogical or hierarchical method of generating or indexing the configuration list was essential. What was required was an algorithm that established a one-to-one correspondence between a configuration index ($K$) in a hierarchically ordered list and all the orbital and spin-coupling information about that configuration.
2. A fast and efficient method was required for generating the $A_{ij}^{KL}$ and $B_{ijkl}^{KL}$ that could be used "on the fly" so that the symbolic matrix elements did not have to be stored in a symbolic matrix-element file.

As we shall now discuss, the UGA introduced by Paldus solves both these problems.

## 3 The essential elements of UGA [1–3]

In second quantization, the numerical "vector-coupling" coefficients (the $A_{ij}^{KL}$ and $B_{ijkl}^{KL}$) appear as matrix elements of creation and annihilation operators $\mathbf{X}_{i\sigma}^{\dagger}$ and $\mathbf{X}_{j\sigma}$. The operator $\mathbf{X}_{i\sigma}^{\dagger}$ creates an electron in an orthonormal spin orbital $|i\sigma\rangle$, where $|i\sigma\rangle = |i\rangle|\sigma\rangle$, and $\sigma = \alpha$ or $\beta$. Similarly, operator $\mathbf{X}_{i\sigma}$ destroys an electron in the orthonormal spin orbital $|i\sigma\rangle$. In quantum chemistry problems in which the number of particles is conserved, the $\mathbf{X}_{i\sigma}^{\dagger}$ and $\mathbf{X}_{j\sigma}$ will always occur in pairs. The role of these operators is easily illustrated by showing their operation on a specific type of CSF, namely a Slater determinant. Thus, as an example, for the determinant

$$|\ldots\ldots i_\alpha j_\alpha \ldots\ldots k_\beta l_\beta \ldots\ldots|$$

we have simply

$$\mathbf{X}_{m\alpha}^{\dagger}\,\mathbf{X}_{i\alpha}|\ldots\ldots i_\alpha j_\alpha \ldots\ldots k_\beta l_\beta \ldots\ldots|$$
$$= |\ldots\ldots m_\alpha j_\alpha \ldots\ldots k_\beta l_\beta \ldots\ldots|\ . \tag{4}$$

A good pedagogical discussion can be found in the book by Matsen and Pauncz [22].

In this "second quantized" formalism Eq. (3) becomes

$$\langle K|HL\rangle = \sum h_{ij}\left\langle K\left|\sum_{\sigma}\mathbf{X}_{i\sigma}^{\dagger}\mathbf{X}_{j\sigma}L\right.\right\rangle$$
$$+ \sum [ij|kl]\left\langle K\left|\sum_{\sigma\gamma}\mathbf{X}_{i\sigma}^{\dagger}\mathbf{X}_{k\gamma}^{\dagger}\mathbf{X}_{l\gamma}\mathbf{X}_{j\sigma}L\right.\right\rangle\ . \tag{5}$$

Thus we can write

$$A_{ij}^{KL} = \left\langle K\left|\sum_{\sigma}\mathbf{X}_{i\sigma}^{\dagger}\mathbf{X}_{j\sigma}L\right.\right\rangle \tag{6}$$

and

$$B_{ijkl}^{KL} = \left\langle K\left|\sum_{\sigma\gamma}\mathbf{X}_{i\sigma}^{\dagger}\mathbf{X}_{k\gamma}^{\dagger}\mathbf{X}_{l\gamma}\mathbf{X}_{j\sigma}L\right.\right\rangle\ , \tag{7}$$

where the summations over $\sigma$ and $\gamma$ are over spin.

For the reader who is encountering second quantization for the first time, one can observe that application of the definition of Eq. (4) for CSF built from Slater determinants in the context of (Eq. 6) just reduces to the usual rules for matrix elements between determinants if $K$ and $L$ are determinants; however the definitions of Eqs. (6) and (7) remain true irrespective of the nature of the CSF (and become very powerful when the CSF are chosen as spin eigenfunctions).

Since the $\mathbf{X}^{\dagger}_{i\sigma}$ and $\mathbf{X}_{j\sigma}$ always occur in pairs in Eqs. (6) and (7) it becomes convenient to define a "generator"

$$E_{ij} = \sum_{\sigma} \mathbf{X}^{\dagger}_{i\sigma}\mathbf{X}_{j\sigma} \ . \tag{8}$$

The vector-coupling coefficients now take the form

$$A^{KL}_{ij} = \langle K|E_{ij}L \rangle \tag{9}$$

and

$$B^{KL}_{ijkl} = \langle K|(E_{ij}E_{kl} - \delta_{jk}E_{li})L \rangle \ . \tag{10}$$

It now remains to make the connection with unitary groups.

The unitary group $U(n)$ is the group of all $n \times n$ unitary matrices representing unitary transformations of the orthonormal orbitals $\{|i\rangle\}$. It is possible to show that the operators in Eq. (8) satisfy the commutation relations of generators of the unitary group $U(n)$.

$$[E_{ij}, E_{kl}] = \delta_{jk}E_{il} - \delta_{li}E_{kj} \ , \tag{11}$$

where the [ ] is the commutator i.e. $[A, B] = AB - BA$. Because of this fact, the matrix elements in Eq. (9) must be nothing other than group theoretical quantities (i.e. completely independent of the CI formalism). This point is important because the set of CSF remains implicit rather than explicit. There is no need to expand the CSF as linear combinations of Slater determinants. (Of course, the same fact is true for the bonded function approach of Boys and Reeves [13]). The recognition of this feature, in Paldus' 1974 paper, provides the solution of both problems in CI: the efficient evaluation of matrix elements and a genealogical classification for the CI basis functions. We now briefly discuss both points.

The completely general formulation of the representations of the unitary group was known to Paldus from the original work of Gelfand and Tsetlin [23]. Paldus' contribution was to recognise that the Gelfand and Tsetlin formalism becomes very simple for the CI problem in electronic structure theory and in his paper he presents a purely algebraic description that requires no prerequisite understanding of the details of the representation theory of continuous groups.

The basis for the classification and canonical ordering of CSF encountered in the CI method is chains of possible subgroups. When an irreducible representation (irrep) of a group is restricted to a subgroup the representation splits into a direct sum of irrep of the subgroup. If this subduction process is multiplicity-free (i.e. a given irrep of the subgroup occurs at most once in this

process) and the last group in the chain is Abelian (and thus has only one-dimensional irreps) then the chain of subgroups and irrep labels provides a unique label for the state of interest. The reader will be familiar with this process, for example, in the classification of the $l = 1$ orbital-momentum states in $O(3)$ ($L^2$) and $O(2)$ ($l_z$). For $U(n)$ this subgroup chain is very simple

$$U(n) \supset U(n-1) \supset U(n-2) \bullet \bullet \bullet U(1) \ . \tag{12}$$

The "labels" for the CSF basis functions can thus be written as a triangular Gelfand tableau

$$\begin{bmatrix} m_{1n} & m_{2n} & \bullet\bullet\bullet\bullet\bullet\bullet & m_{nn} \\ & m_{1n-1} & m_{2n-1} & \\ & & \bullet & \bullet & \\ & & & m_{11} & \end{bmatrix} \ . \tag{13}$$

Each row of the tableau gives the irrep labels for $U(n)$, $U(n-1)$ ...etc. The hierarchical ordering comes from the "betweenness conditions". The integers in any row must lie between those of the previous row. For the electronic structure problem the integers $m_{im}$ can have only three values: 2, 1 or 0. Consequently, as Paldus discovered, each row of the Gelfand tableau could be denoted by

$$\Gamma_i = 2^{a_i}1^{b_i}0^{c_i} \ , \tag{14}$$

where $a_i$ counts the number of 2s in row $i$, etc. The integers $a_i$, $b_i$ and $c_i$ thus classify the CSF in $U(i)$ and have the following simple relationship.

$$n = a + b + c \tag{15}$$

$$a = N/2 - S \tag{16}$$

$$b = 2S \ , \tag{17}$$

where $N$ is the number of electrons, $n$ is the number orbitals and $S$ is the spin.

Then one has the remarkable feature that makes the $U(n)$ method so powerful for electronic structure problems. Because of the restriction of the entries in the Gelfand tableau to 2, 1 or 0 there are only four ways of adding a lower row to a tableau and still satisfy the betweenness conditions. This leads to the genealogical representation (labelled by possible steps 0, 1, 2 or 3) of the CSF basis shown (in part) in Eq. (18) for $U(4)$ (four orbitals and four electrons) with a full CI with triplet spin multiplicity ($\Gamma_n = 2^11^20^1$).

Remarkably the sequence of step vectors [e.g. (0 1 1) for basis vector 1 or (0 3 1) for basis vector 3 ] in Eq. (18) is enough to specify all the information that is required about a CSF and to establish an algorithm for the implicit representation of the basis.

We turn briefly to the second contribution in Paldus' paper: the computation of the matrix elements themselves. Paldus derived the formulae (from the complicated formulae of Gelfand and Tsetlin [23]) for the so-called elementary generators $E_{i,i+1}$. These were both simple and, more importantly, gave only the nonzero matrix elements directly using only the step vectors as "inputs". Paldus states that the general program

320

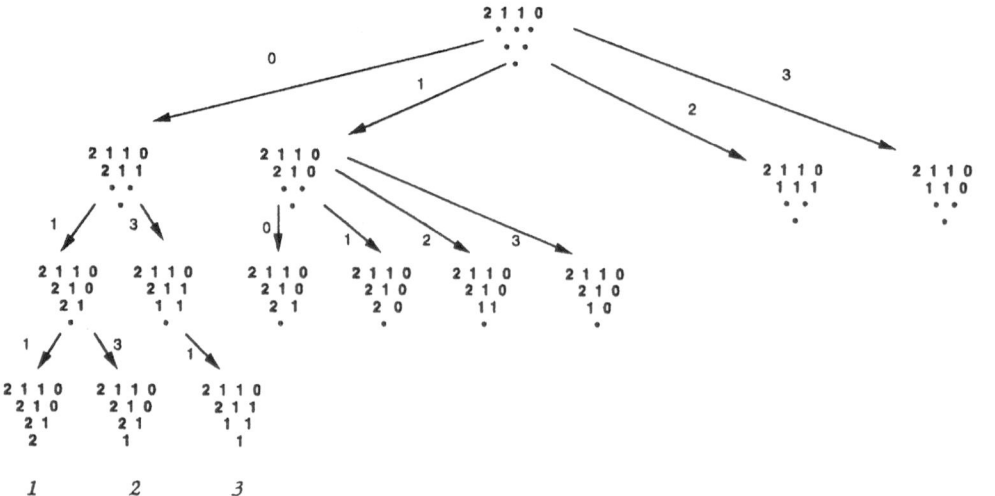

$$\tag{18}$$

required only 400 "cards". The remaining matrix elements could be found from the commutation relationships

$$[E_{ij}, E_{jl}] = E_{il} . \tag{19}$$

In addition, the two-particle matrix elements could be obtained from

$$B_{ijkl}^{KL} = \sum_M \langle K|E_{ij}M\rangle\langle M|E_{kl}L\rangle - \delta_{jk}\langle K|E_{li}L\rangle , \tag{20}$$

where the summation over $M$ was over all the basis states. In his 1974 paper, Paldus did not put these two ideas together into a working algorithm. Rather both ideas formed the basis of the practical approach that Shavitt [5] was shortly to follow.

Thus, in Paldus 1974 paper one has, in principle, the method for the hierarchical basis function generation and the computation of the matrix elements themselves. The evolution of the method to become a practical tool in quantum chemistry then proceeded rapidly. We now very briefly bring the article to its conclusion with a brief mention of these key developments by Shavitt and by Siegbahn.

## The development of UGA post 1974

Efficient subsequent general CI algorithms were based upon the subsequent graphical implementation of UGA by Shavitt [5, 6]. Shavitt derived a graphical representation [5] for the genealogical representation illustrated in Eq. (18) and showed that it could be represented efficiently on a computer via a "distinct row table". He also observed that the nonzero matrix elements corresponded to closed loops on his graph. The importance of this loop structure comes from the fact that many matrix elements have the same matrix-element loop but differed only in the "upper" and "lower" walks on the graph. This observation means that the numerical value of the matrix-element "loop" needed only to be computed once and its contribution to all the matrix elements to which it contributed could be evaluated immediately by "following" the upper and lower walks. However, the most

important breakthrough came when Shavitt recognised [6] that the evaluation of the matrix elements in the loop could be formulated in terms of sums of products of "segments" within these loops for each orbital level on the graph. This eliminated the last bottleneck which arises because of the sum over $M$ in Eq. (20). Thus a CI code could be developed that was "loop" or "shape" driven [18], where all the matrix elements could be evaluated "on the fly" without storing any formula tape.

However, while loop-driven methods provide a good approach to full CI, for multireference CI, these methods are too general. Thus the final chapter in the UGA story comes from the work of Siegbahn [7–10]. Siegbahn noticed that in multireference CI the matrix-element segments that occurred in levels in the virtual orbital space were particularly simple. He showed that one can confine the detailed analysis required for matrix formula determination to the levels that correspond to reference space orbitals. The contribution from the levels that correspond to virtual levels, which take the same form in all calculations, can be built into the structure of the computer program rather than being treated in a general way.

Thus, at the end of the story, most of the complexity of the general matrix-element problem for CI goes away. The general matrix-element problem remains only for the reference space part of multireference CI; however, Siegbahn would certainly never have discovered this fact without the general formalism introduced by Paldus and its graphical realisation by Shavitt. It is worth noting that Shavitt's graphical approach was carried out heuristically. He did not need to understand the (beautiful) theory of continuous groups. Rather it is the structure of the CI problem that emerges from Paldus work that was the important breakthrough.

## 5 Conclusion

In this perspective we have focused on the nature of the CI method as it existed in 1974 and on the nature of the unitary group revolution introduced by Paldus. This method was subsequently turned into working algorithms by others [5–18] in the 10 years or so after its introduction. Of the developments post 1974 we have only mentioned the work of Shavitt and Siegbahn because they introduced the first steps that were needed to make the method work in practice. The reader is referred to Refs. [5–18] to trace the immediately subsequent developments in more detail. Of course, there has been much development of the subject since the early 1980s and the interested reader might start by looking at Ref. 24.

The lasting importance of Paldus' 1974 paper stems from the fact that it introduced a new way of thinking into quantum chemistry problems related to the CI method.

# References

1. Paldus J (1974) J Chem Phys 61: 5321
2. Paldus J (1976) In: Eyring H, Henderson DG (eds) Theoretical chemistry, advances and perspectives. Academic Press N.Y. Vol 2, p 131
3. Paldus J (1997) In: Phariseau P, Scheire L (eds) Electrons in finite and infinite structures. Plenum Publishing, New York, p 411–429
4. Shavitt I (1998) Mol Phys 94: 3
5. Shavitt I (1997) Int J Quantum Chem Symp 11: 131
6. Shavitt I (1978) Int J Quantum Chem Symp 12: 5
7. Siegbahn Per EM (1979) J Chem Phys 70: 5391
8. Siegbahn Per EM (1980) J Chem Phys 72: 1647
9. Siegbahn Per EM (1981) In: Hinze J (ed) The unitary group for the evaluation of electronic energy matrix elements. Lecture notes in chemistry vol 22. Springer, Berlin Heidelberg New York, p 119
10. Siegbahn Per EM (1984) Chem Phys Lett 109: 417
11. Hegarty D, Robb MA (1979) Mol Phys 38: 1795
12. Handy NC (1980) Chem Phys Lett 74: 280
13. Knowles P, Handy NC (1994) Chem Phys Lett 111: 315
14. Brooks BR, Schaefer HF III (1979) J Chem Phys 70: 5092
15. Brooks BR, Laidig WD, Saxe P, Goddard JD, Schaefer HF III (1981) In: Hinze J (ed) The unitary group for the evaluation of electronic energy matrix elements. Lecture notes in chemistry, vol 22. Springer Berlin Heidelberg New York, p 158
16. Lischka H, Sheppard R, Brown FB, Shavitt I (1981) Int J Quantum Chem Symp 15: 91
17. Saunders VR, van Lenthe JH (1983) Mol Phys 48: 923
18. Saxe P, Fox DJ, Schaefer HF III, Handy NC (1992) J Chem Phys 77: 5584
19. Reeves CM (1957) PhD thesis. Cambridge University
20. Roos BO (1972) Chem Phys Lett 1: 153
21. Roos BO, Siegbahn Per EM (1977) In: Schaefer HF III (ed) Methods of electronic structure theory. Plenum, New York, p 189
22. (a) Matsen FA, Pauncz R (1986) The unitary group in quantum chemistry. Studies in physical and theoretical chemistry, vol 44, Elsevier, Amsterdam; (b) Pauncz R (1979) Spin eigen functions. Plenum, New York
23. (a) Gelfand IM, Tsetlin ML (1950) Dokl Akad Nauk SSR 71: 825; (b) eelfand IM, Tsetlin ML (1950) Dokl Akad Nauk SSR 71: 1070
24. Paldus J (1988) In: Truhlar DG (ed) Mathematical frontiers in computational chemical physics. IMA series, vol 15. Springer, Berlin Heidelberg New York, pp 262–299

Theor Chem Acc (2000) 103:322–325
DOI 10.1007/s002149900097

## Perspective

# Perspective on "Theory of self-consistent electron pairs. An iterative method for correlated many-electron wavefunctions"

**Wilfried Meyer (1976) J Chem Phys 64: 2901**

**Hans-Joachim Werner**

Institute for Theoretical Chemistry, University of Stuttgart, Pfaffenwaldring 55, D-70569 Stuttgart, Germany

Received: 13 July 1999 / Accepted: 16 September 1999 / Published online: 17 January 2000
© Springer-Verlag 2000

**Abstract.** The title paper of Wilfried Meyer is a landmark for modern electron correlation theories and their efficient implementation. It described for the first time a matrix-formulated direct configuration interaction method. This approach eliminates all coupling coefficients involving external orbitals and the need for a "formula tape". Secondly, a full integral transformation is avoided, a prerequisite for integral-direct implementations. Third, the theory is formulated in a basis of nonorthogonal virtual orbitals (or atomic orbitals), which forms the basis of current local electron correlation treatments. Meyer's paper, which was written while I was a diploma student in his group, has strongly influenced my own work on multireference configuration interaction, coupled-cluster theory, and local electron correlation methods. The current paper reviews the history of the self-consistent electron pairs method and describes its relation to current electron correlation techniques.

**Key words:** Electron correlation – Configuration interaction – Direct configuration interaction – Self-consistent electron pair theories – Local correlation

The slow convergence of the dynamical correlation energy with respect to the number of Slater determinants or configuration state functions (CSFs) has been one of the major problems of quantum chemistry since the proposal of the configuration interaction (CI) method by Boys in 1950. It was realized early that at least the first-order interacting space of the Hartree–Fock (HF) wavefunctions, which is spanned by all single and double excitations from the HF reference function into virtual orbitals, is needed to recover a substantial part of the electron correlation energy. This led to the development of electron pair theories [1, 2] in the late 1960s and early 1970s. However, even this small subspace of the full CI space grows with the fourth power of the molecular size or the number of correlated electrons (assuming that the

number of occupied and virtual orbitals grows linearly). Since, in principle, the Hamiltonian matrix must be constructed and diagonalized in the CSF basis, the applicability of the simplest and original version of the CI method was limited to very small systems. Early attempts to treat the correlation problem therefore introduced additional approximations, such as configuration selection by means of perturbational estimates of their importance. Alternatively, certain couplings were neglected, as for instance in the independent electron pair approximation (IEPA) [2]. By means of localizing the occupied orbitals and transforming the virtual orbitals to pair natural orbitals (PNOs), which are pairwise nonorthogonal, a dramatic reduction in the number of configurations could be achieved. In 1971 Meyer generalized the PNO-IEPA scheme to PNO-CI [1], which included all couplings between the different electron pairs. Moreover, he proposed the coupled-electron-pair approximation (CEPA) [1, 3], which accounts for the most important effects of higher excitations and is size-consistent. CEPA can be viewed as a good approximation to coupled-cluster theory. Using the PNO-CEPA method, Meyer, in 1973, was able to compute about 90% of the correlation energy for molecules such as $H_2O$ and $CH_4$, and this was an important landmark in electronic structure theory. Due to the optimum convergence properties of the PNO orbital basis coupled with configuration selection, this could be achieved with only a few hundred configurations. The disadvantage of the PNO-CI/CEPA method is that different orbitals are needed for each electron pair, which limits the extension to larger systems. Furthermore, the perturbative determination of the PNOs and the subsequent orbital or configuration selection leads to small discontinuities on potential-energy surfaces. Despite the tremendous success of the PNO methods it was therefore desirable to develop methods which allowed the full space of single and double excitations to be included.

The first important step in this direction was made by Roos [4] in 1972 by proposing the direct CI method. He realized that the lowest eigenvectors and eigenvalues can be computed iteratively without ever setting up the Hamiltonian matrix explicitly. Instead, it is sufficient to compute the vector $\sigma = (\mathbf{H} - E)\mathbf{c}$, where $\mathbf{c}$ is a trial vector and $\sigma$ a residual vector, from which an update to $\mathbf{c}$ can be obtained by first-order perturbation theory. The product $\mathbf{Hc}$ can be obtained directly from the two-electron integrals, the vector $\mathbf{c}$, and coupling coefficients, which depend on the formal structure of the CSFs. In the implementations by Roos and Siegbahn [5] symmetry-adapted orthonormal configuration state functions were used as $N$-electron bases. The integrals were fully transformed into the molecular orbital (MO) basis, and the construction of the $\sigma$ vector was driven by the two-electron integrals. Depending on the interaction and integral type, about 140 different nonzero coupling coefficients were distinguished, resulting in a lot of logic and overhead.

In this situation, Meyer's paper represented significant progress in several aspects. Firstly, the logic for constructing the residual vector $\sigma$ was not carried down any more to individual one- and two-electron integrals and configurations, but was reduced to deal with whole matrices of integrals and configuration coefficients. This is a generalization of what had already been proposed by Ahlrichs and Driessler [6] for the two-electron case. The integral matrices represent the half-transformed two-electron integrals with at least two occupied orbitals, while the coefficient matrices represent electron pair functions. The matrix formulation was achieved by partially giving up the normalization of the configurations: implicit in the theory is a different normalization of diagonal doubly external configurations, $\Phi_{ij}^{aa}$, and off-diagonal configurations, $\Phi_{ij}^{ab}$ ($a \neq b$). This made it possible to eliminate all coupling coefficients depending on external (virtual) orbitals. In fact, this simplification follows most naturally by constructing the CSFs so that they can be considered as tensor components of the external orbital subspaces [7]. The equations for the residual vector were given in the form of matrix operations (mainly matrix multiplications), and there are only very few simple coupling coefficients depending on the spin-coupling (singlet or triplet) of the two external electrons. Thus, all complicated logic was removed, allowing an implementation of optimum efficiency. In fact, matrix multiplications are still the fastest operations one can perform on all kinds of computers today, ranging from personal computers with the Linux operating system or RISC workstations to vector supercomputers.

Secondly, Meyer showed in this paper that a full integral transformation can be avoided. The self-consistent electron pair (SCEP) method only requires transformed integrals with at least two occupied orbitals involved. The contributions of the remaining integrals with one or zero occupied orbitals can be treated by computing for each electron pair an "external" exchange operator that can be obtained directly in the atomic orbital (AO) basis, very much like the exchange contribution to the Fock matrix in the self-consistent-field (SCF) case. In fact, in the original SCEP formulation two such operators per pair were needed for the CI with single and double excitations (CISD) case, but it

was shown soon later [8] that a single operator per pair is sufficient. This is even true for full coupled-cluster with single and double excitations (CCSD) [9]. Avoiding the full integral transformation reduces the disc space for integral storage by a factor of 2 and the processing time for the transformation from $\mathcal{O}(N^5)$ to $\mathcal{O}(mN^4)$, where $m$ is the number of correlated orbitals and $N$ the number of basis functions. At first glance this seems most significant for highly accurate calculations with large basis sets on small molecules. At the time of Meyer's paper, most researchers considered this not to be a very important advantage, since the basis sets one was able to use were relatively small anyway. Moreover, computing the external exchange operators in the AO basis rather than in the MO basis from fully transformed integrals requires additional matrix transformations, and apart from this it increases the cost formally from about $n_{\text{pair}}(N - m)^4$ to about $n_{\text{pair}}N^4$. However, from the current point of view, avoiding the full transformation is essential for modern integral-direct correlation methods [10] that allow the use of many hundreds of basis functions. If in such methods screening of the two-electron integrals in the AO basis is performed, the scaling of the computational cost with molecular size can be reduced dramatically. Using local electron correlation treatments (see later) even linear cost scaling becomes possible.

The third new and, in retrospect, very important contribution in Meyer's paper was the proposal to compute not only the external exchange operators but also the complete residual directly in the nonorthogonal AO basis. He showed that the nonorthogonality only leads to quite minor complications in the equations for the residual; namely, some additional matrix multiplications with the overlap matrix are needed. The residual in the AO basis is transformed into the MO basis only for performing the perturbational update of the configuration coefficients. At the time when the paper was written, the AO formulation probably seemed to most people an unnecessary complication. From our present viewpoint, however, it was far ahead of the time: today we know that the AO formulation is essential for exploiting the local short-range character of electron correlation and makes possible linear cost scaling with molecular size even for coupled-cluster methods.

The fourth important contribution of the paper was to show how a Brueckner CI(D) is performed efficiently. In fact, even though the paper gives all matrix elements needed for a CISD, the iterative method described by Meyer was a Brueckner theory, since the single excitations were absorbed into the reference function. He proposed two different ways for absorbing the single excitations: either by a simple first-order update scheme of the orbitals with subsequent symmetrical orthonormalization or by performing a natural orbital iteration. He also introduced projectors to the external space, similar to those used nowadays in local correlation treatments. In the SCEP framework, the projectors were used to retain strong orthogonality between the external pair functions and the reference wavefunction once the occupied orbitals were modified in the Brueckner iteration. Meyer probably favored the Brueckner iteration mainly to avoid the construction of two external exchange operators per pair in each iteration. As already mentioned, it was shown later that this is possible even with explicit inclusion of single excitations. Much later several groups showed that Brueckner theory has some important advantages, in particular in situations with symmetry-breaking.

The SCEP method was first implemented in 1976 for closed-shell reference functions by Dykstra and Meyer during a visit of Dykstra in Mainz and was first applied by Dykstra, Schaefer, and Meyer [11]. It was generalized in 1981 for multiconfiguration reference functions by Werner and Reinsch [8]. The latter method was the first implementation of an internally contracted multireference CI. The internal contraction scheme was proposed in another landmark paper of Meyer [12], which was written almost at the same time as the SCEP paper and is closely related to it. The internally contracted multireference SCEP method [8] used a similar matrix structure as Meyer's original SCEP, but was formulated in an orthogonal orbital basis and included all single and double excitations from the multiconfiguration SCF reference function. In 1988, a new implementation of the internally contracted multireference CI was completed [13, 14], which is still of unmatched efficiency. The internal contraction scheme is also widely used in multireference perturbation theory (MRPT2, CASPT2/3) [15–17], since the internally contracted configurations span exactly the first-order interaction space of a multiconfiguration reference wavefunction [12]. The SCEP theory was further generalized in 1984 by Pulay, Saebø, and Meyer [18] to the coupled-cluster doubles case [18] and later to closed and open-shell CCSD cases in our group [9, 19]. In Ref. [18] it was shown for the first time that the equations can be simplified by generating the configurations by spin-coupled pair excitation operators, $\hat{E}_{ai}\hat{E}_{bj}$. This leads not only to unnormalized but also to nonorthogonal configurations. Nevertheless, the number of matrix multiplications is reduced and for closed-shell reference functions the coupling coefficients are entirely eliminated (except for simple factors such as 2 in some terms).

As previously mentioned, the formulation of the theory with nonorthogonal correlation orbitals forms the basis for local electron correlation theories, as proposed by Pulay in 1983 [20] and first implemented for MP4(SDQ), CISD, and CEPA by Saebø and Pulay [21, 22]. A generalization for full CCSD was published in 1996 by Hampel and Werner [23]. In these methods the virtual space is spanned by a redundant set of projected AOs, which are orthogonalized on the occupied orbitals. Alternatively, AOs could be used directly as proposed by Meyer. The advantage of using AOs (or projected AOs) to span the virtual space is the fact that these are intrinsically localized. This allows the restriction of excitations from localized occupied orbitals to relatively small subspaces (domains) of virtual functions. Each pair is described by a different domain, but the size of each pair domain is independent of the molecular size. This immediately reduces the $\mathcal{O}(N^4)$ dependence of the number of configurations and variational parameters to $\mathcal{O}(N^2)$. Furthermore, by neglecting or approximating the very small contributions of distant electron pairs, linear scaling, $\mathcal{O}(N)$, of the number of electron pairs and of the configurations is obtained, but nevertheless 98–99.5% of the canonical correlation energy is recovered. Linear scaling of the computational cost with molecular size has recently been achieved in our group for local MP2 [24], and this new program has made it possible to perform overnight LMP2 calculations for molecules with about 300 valence electrons and over 1500 basis functions on a single low-cost personal computer. A local coupled-cluster program with linear-scaling behavior is presently under development.

AO formulations are currently also being investigated by other groups with the aim of reducing the computational cost for large molecules. Ayala and Scuseria [25] have presented a linear-scaling MP2 method based on the Laplace transformation formalism of Häser and Almlöf [26]. More closely related to the SCEP formalism of Meyer is a recent tensor formulation of coupled-cluster theory by Head-Gordon et al. [27]. As far as the treatment of the virtual space is concerned, this is basically the same as what was described in Meyer's SCEP paper more than 20 years ago. (see also Ref. [7]); however, Head-Gordon et al. go a step further by also using a redundant set of nonorthogonal (projected) AOs to span the occupied space. This seems very wasteful at first glance, but as Head-Gordon et al. demonstrated for a number of test cases, a relatively small number of selected configurations is sufficient to recover most of the correlation energy [28]. In fact, in a very recent paper by Scuseria and Ayala [29] it is demonstrated for a related method that the number of configurations (or coupled-cluster amplitudes) needed to obtain a fixed fraction of the canonical correlation energy scales linearly with molecular size, as is also the case in local correlation methods with orthogonal localized orbitals.

During the last decade, linear-scaling density functional theory (DFT) methods have played a crucial role in extending the applicability of electronic structure methods to very large molecular systems; however, the problem remains that there is no possibility to systematically approach the exact result using DFT. Many attempts are presently being made to improve the available functionals, but this is mainly achieved by fitting DFT results to experimental data. DFT can therefore be viewed as a semiempirical theory. Accurate wavefunction-based local electron correlation methods are currently catching up with DFT in applicability to larger systems. It is my view that Meyer's 1976 paper forms the basis for these very exciting new developments. The SCEP theory was clearly far ahead of its time.

## References

1. Meyer W (1971) Int J Quantum Chem Symp 5: 341
2. Kutzelnigg W (1977) In: Schaefer HF III (ed) Methods of electronic structure theory. Plenum, New York, pp 129–188, and references therein
3. Meyer W (1973) J Chem Phys 58: 1017
4. Roos BO (1972) Chem Phys Lett 15: 153
5. Roos BO, Siegbahn PEM (1977) In: Schaefer HF III (ed) Methods of electronic structure theory. Plenum, New York, pp 277–318

6. Ahlrichs R, Driessler F (1975) Theor Chim Acta 36: 275
7. Meyer W, Ahlrichs R, Dykstra CE (1984) In: Dykstra CE (ed) Advanced theories and computational approaches to the electronic structure of molecules. Reidel, Dordrecht, pp 19–38
8. Werner H-J, Reinsch EA (1982) J Chem Phys 76: 3144
9. Hampel C, Peterson KA, Werner H-J (1992) Chem Phys Lett 190: 1
10. Schütz M, Lindh R, Werner H-J (1999) Mol Phys 96: 719
11. Dykstra C, Schaefer HF, Meyer W (1976) J Chem Phys 65: 2740
12. Meyer W (1977) In: Schaefer HF III (ed) Methods of electronic structure theory. Plenum, New York, pp 413–446
13. Werner H-J, Knowles PJ (1988) J Chem Phys 89: 5803
14. Knowles PJ, Werner H-J (1988) Chem Phys Lett 145: 514
15. Roos BO, Linse P, Siegbahn PEM, Blomberg MRA (1982) Chem Phys 66: 197
16. Wolinski K, Pulay P (1989) J Chem Phys 90: 3647
17. Andersson K, Malmqvist P-Å, Roos BO, Sadlej AJ, Wolinski K (1990) J Phys Chem 94: 5483
18. Pulay P, Saebø S, Meyer W (1984) J Chem Phys 81: 1901
19. Knowles PJ, Hampel C, Werner H-J (1993) J Chem Phys 99: 5219
20. Pulay P (1983) Chem Phys Lett 100: 151
21. Saebø S, Pulay P (1987) J Chem Phys 86: 914
22. Saebø S, Pulay P (1988) J Chem Phys 88: 1884
23. Hampel C, Werner H-J (1996) J Chem Phys 104: 6286
24. Schütz M, Hetzer G, Werner H-J (1999) J Chem Phys (in press)
25. Ayala PY, Scuseria GE (1999) J Chem Phys 110: 3660
26. Häser M, Almlöf J (1992) J Chem Phys 96: 489
27. Head-Gordon M, Maslen PE, White CA (1998) J Chem Phys 108: 616
28. Maslen PE, Head-Gordon M (1998) J Chem Phys 109: 7093
29. Scuseria GE, Ayala PY (1999) J Chem Phys 111: 8330

Theor Chem Acc (2000) 103:326–327
DOI 10.1007/s002149900032

Theoretical
Chemistry Accounts
© Springer-Verlag 2000

## Perspective

# Perspective on "Quantum mechanical reactive scattering for three-dimensional atom plus diatom systems. II. Accurate cross sections for H + H₂"

Schatz GC, Kuppermann A (1976) J Chem Phys 65: 4668–4692

**David C. Clary**

Department of Chemistry, University College London, 20 Gordon Street, London WC1H 0AJ, UK

Received: 9 February 1999 / Accepted: 22 February 1999 / Published online: 28 June 1999

**Abstract.** An overview is given of the impact of the title paper on the field of quantum reactive scattering.

**Key words:** Quantum reactive scattering – Chemical reactions

In quantum reactive scattering the aim is to solve the Schrödinger equation for the nuclei involved in a chemical reaction. If the potential-energy surface or surfaces used in such calculations are accurate, and if the scattering calculations are performed with no approximations, then these computations should yield a variety of highly useful and reliable results. These include differential and integral cross sections selected in initial and final quantum states, and reaction rate constants. Application of the Born–Oppenheimer approximation separates the treatment of a reaction into solutions of two Schrödinger equations: first for the electrons at fixed nuclear positions, and then for the nuclei. Methods for calculating the potential-energy surface have been improved over many years but it was not until the 1976 paper by Schatz and Kuppermann on the simplest chemical reaction,

$$H + H_2 \rightarrow H_2 + H \ ,$$

that converged cross sections were reported from a quantum scattering calculation on a chemical reaction in three dimensions. This paper, therefore, set the scene for the modern theory of chemical reaction dynamics.

Prior to 1976 there had been several quantum scattering calculations on nonreactive molecular collisions [1] and on chemical reactions constrained to move in one dimension [2]. In addition there had been many classical trajectory calculations on chemical reactions in three dimensions [3] and a small number of quantum scattering calculations that modeled the three-dimensional

$H + H_2$ reaction with approximations or with incomplete convergence [4–6].

The difficulty in performing quantum scattering calculations on chemical reactions in three dimensions is in obtaining a solution for the scattering wavefunction that is continuous from reactants to products and is also a function of all the energetically available ro–vibrational quantum states. In addition, to compute converged integral or differential cross sections to compare with the results of molecular beam experiments it is necessary to repeat the calculations for a range of total angular momenta, and the computations also have to be done for several different energies to average over the cross sections to obtain rate constants.

The $H + H_2 \rightarrow H_2 + H$ reaction problem was solved by Schatz and Kuppermann by using natural collision coordinates [7] that exploited the symmetry of the reaction to ensure the continuity of the scattering wavefunction between the three different arrangement channels of the reaction [8]. They also exploited the use of body-fixed coordinate systems [9] and a close-coupling expansion of the time-independent wavefunction that was first developed and applied to nonreactive scattering problems [10]. The potential-energy surface used in the scattering computations was a semiempirical one due to Porter and Karplus [11] as an accurate potential-energy surface based on high-quality ab initio computations was not available at that time. Thus, despite the convergence of the three-dimensional quantum scattering calculations, the cross sections and rate constants obtained were not expected to give very accurate comparisons with experiment; however, the results obtained served as benchmarks for the field of reactive scattering calculations and did much to stimulate the development of new methods of quantum reactive scattering and more accurate computations of potential-energy surfaces.

The Schatz-Kuppermann paper also produced new insight into chemical reactions that derived from the detail and reliability of their results. For example, a surprisingly strong sensitivity of the reaction cross section on the $j$ and $m_j$ rotational states of the reactants and products was discovered. In addition, the benchmark results demonstrated the importance of tunneling at low temperatures and provided the first rigorous test of the widely used quasiclassical trajectory [3], reduced dimensionality [12] and transition-state [13] methods for a three-dimensional reaction. The paper also gave the first rigorous quantum scattering calculation of a differential cross section for a reaction and the predicted strong backward peaking of the angular distributions agreed with experiment [14].

It was not long before other calculations followed on from the pioneering work of Schatz and Kuppermann. Their work stimulated the ab initio computation of a highly accurate potential-energy surface for the $H + H_2 \rightarrow H_2 + H$ reaction [15] that enabled reactive cross sections of a very high accuracy to be calculated [16]. However, despite intensive work by several groups, it took some time to go beyond $H + H_2$, even to the $D + H_2 \rightarrow HD + H$ reaction. The natural collision coordinate technique used by Schatz and Kuppermann proved difficult to extend to nonsymmetric reactions in three dimensions and, eventually, different coordinate systems and new methods were developed. A hyperspherical coordinate method [17], which treats the size of the reactive system as a collision coordinate, proved to be more general and has been applied to reactions such as $F + H_2 \rightarrow HF + H$ and $H + O_2 \rightarrow OH + O$ [18], and to four-atom reactions such as $OH + H_2 \rightarrow H_2O + H$ [19]. Variational methods, involving expansion of the wavefunction in asymptotic ro–vibrational states, were also successfully applied to several atom–diatom reactions [20]. In addition, powerful wavepacket techniques that solve the time-dependent Schrödinger equation were developed and applied to several three-atom and four-atom reactions [21]. Also, an absorbing-potential method was formulated that enabled the total reactive flux into particular arrangement channels to be calculated [22]. Furthermore, new quantum methods were developed to calculate rigorously the cumulative reaction probability (i.e. reaction probabilities summed over all reactant and product states) by considering only the region close to the transition-state geometry of a reaction [23].

Although these more recent methods are not based directly on the algorithm of Schatz and Kuppermann, the psychological impact of their paper was enormous as it showed for the first time that a chemical reaction in three dimensions could be treated to convergence using quantum reactive scattering theory. Several books, reviews, and collections of papers on this subject have subsequently been published [24–28], an international conference is held regularly on this topic[1], and quantum reactive scattering was a main component of a recent Faraday Discussion on chemical reaction theory [29]. Quantum reactive scattering is now the technique of choice for comparison with experiments on the detailed dynamics of chemical reactions. The method also enables rigorous rate constants to be calculated that have useful applications in areas such as combustion, atmospheric and astrophysical chemistry.

## References

1. Allison AC, Dalgarno A (1967) Proc Phys Soc Lond 30: 609
2. Diestler DJ (1971) J Chem Phys 54: 4547
3. Karplus M, Porter RN, Sharma RD (1965) J Chem Phys 43: 3259
4. Choi BH, Tang KT (1974) J Chem Phys 61: 2462
5. Wolken G, Karplus M (1974) J Chem Phys 60: 351
6. Elkowitz AB, Wyatt RE (1975) J Chem Phys 62: 2504
7. Marcus RA (1966) J Chem Phys 45: 4493
8. Schatz GC, Kuppermann A (1976) J Chem Phys 65: 4642
9. Pack RT (1974) J Chem Phys 60: 633
10. Gordon RG (1969) J Chem Phys 51: 14
11. Porter RN, Karplus M (1964) J Chem Phys 40: 1105
12. Kuppermann A, Schatz GC, Baer M (1974) J Chem Phys 61: 4362
13. (a) Eyring H (1935) J Chem Phys 3: 107; (b) Truhlar DG, Kuppermann A (1972) J Chem Phys 56: 2232
14. Geddes J, Krause HF, Fite WL (1972) J Chem Phys 56: 3298
15. Siegbahn P, Liu B (1978) J Chem Phys 68: 2457
16. Walker RB, Stechel EB, Light JC (1978) J Chem Phys 69: 2922
17. Kuppermann A (1975) Chem Phys Lett 32: 374
18. (a) Kress JD, Pack RT, Parker GA (1990) Chem Phys Lett 170: 306; (b) Pack RT, Butcher EA, Parker GA (1995) J Chem Phys 102: 5998
19. Pogrebnya SK, Echave J, Clary DC (1997) J Chem Phys 107: 8975
20. (a) Haug K, Schwenke DW, Truhlar DG, Zhang Y, Zhang JZH, Kouri DJ (1987) J Chem Phys 87: 1892; (b) Zhang JZH, Chu SI, Miller WH (1988) J Chem Phys 88: 6233
21. Zhang DH, Zhang JZH (1994) J Chem Phys 100: 2967
22. Szichman H, Baer M (1994) J Chem Phys 101: 2081
23. Manthe U, Seideman T, Miller WH (1993) 99: 10078
24. Clary DC (ed) (1976) The theory of chemical reaction dynamics. Reidel, Dordrecht
25. Bowman JM (ed) (1994) Advances in molecular vibrations and collision dynamics. Vol. 2 JAI Press, London
26. Zhang JZH (1998) Theory and application of quantum molecular dynamics. World Scientific, Singapore
27. (a) Manolopoulos DE, Clary DC (1989) Annu Rep C Roy Soc Chem 95; (b) Miller WH (1990) Annu Rev Phys Chem 41: 245; (c) Schatz GC (1996) J Phys Chem 100: 12839; (d) Clary DC (1998) Science 279: 1879
28. Quantum theory of chemical reaction dynamics, special issue, (1997) Faraday Trans 93: 673–1016
29. Chemical reaction theory (1998) Faraday Discuss 110

[1] The Conference on the Quantum Theory of Chemical Reaction Dynamics has been held in Cambridge, UK (1990), Harvard, USA (1994), Nottingham, UK (1995), Telluride, USA (1997) and Perugia, Italy (1999)

Theor Chem Acc (2000) 103:328–329
DOI 10.1007/s002149900077

**Theoretical
Chemistry Accounts**
© Springer-Verlag 2000

*Perspective*

# Perspective on "Theoretical studies of enzymic reactions: dielectric, electrostatic and steric stabilization of the carbonium ion in the reaction of lysozyme"

## Warshel A, Levitt M (1976) J Mol Biol 103:227–249

Jiali Gao

Department of Chemistry, State University of New York, Buffalo, NY 14260, USA

Received: 14 April 1999 / Accepted: 22 July 1999 / Published online: 9 September 1999

**Abstract.** This paper provides an overview of the title paper by Warshel and Levitt. Its important contributions and influence to present-day studies of enzymatic processes are discussed.

**Key words:** Enzymatic reactions – Electrostatic effects – Hybrid quantum mechanical and molecular mechanical methods

This paper by Warshel and Levitt was one of the first describing "realistic" molecular modeling of enzymatic reactions, and paved out the way of exploring one of the central questions in molecular biology, the origin of the catalytic power of enzymes, through computer simulations. Transition-state stabilization by electrostatic interactions for the reaction of lysozyme was supported, whereas steric factors, such as the ground-state strain due to substrate–enzyme bonding, were found to be insignificant in catalysis. Although the rationale of the catalytic cleavage of a glycosidic bond by lysozyme was interesting, the study turns out to be even more fundamental because electrostatic effects are now recognized to be one of the most important factors in enzyme catalysis [1]. The paper also introduced several theoretical methods for studies of other aspects of molecular biology. The following are the new ideas that were introduced in this work.

1. Hybrid quantum mechanical (QM)/molecular mechanical (MM) methods. An important contribution of this work was the introduction of a hybrid QM/MM method that can be used to study the mechanism and energetics of enzymatic reactions. Recognizing the difficulty of treating the entire enzyme–substrate complex quantum mechanically, Warshel and Levitt proposed partitioning the whole enzyme–substrate system into a "QM" region directly involved in the reaction, and a "classical" region consisting of the rest of the enzyme residues and surrounding solvent. An important feature of this model is that microscopic dielectric effects are included in the QM calculation of enzymatic reactions by computing the electrostatic field due to the dipoles induced by polarizing the protein atoms and the dipoles induced by orienting the solvent water molecules. Consequently, along with the classical force field, all important factors that may contribute to the potential-energy surface of enzymatic reactions are adequately described. The effectiveness of the hybrid QM/MM model is evidenced by numerous recent enhancements of the methodology and applications to studies of reactions in solutions and in enzymes [2–5].

2. Computer modeling of enzymatic reactions. Another important contribution of this paper is to demonstrate the capability of computer modeling in the understanding of enzyme actions. This along with molecular dynamics simulations of proteins reported in the following year [6] provided the opportunity to explore the free-energy surface and dynamic effects in enzymatic reactions. Although extensive biochemical and structural studies, including X-ray crystallography had at that time provided useful information about the mechanisms of many enzymatic reactions, and numerous factors that may contribute to enzyme catalysis had been proposed, their specific contributions and relative importance were not known. In the mid-1970s, it was also far less clear how to use theoretical methods to treat a system as large as the whole enzyme–substrate complex using QM methods. In addition, it was not obvious whether or not it was essential to include the complete system including the surrounding solvent in these calculations since previous theoretical studies had mainly used cluster models in the gas phase. Modeling of the entire system with the newly developed hybrid QM/MM approach allowed Warshel and Levitt to examine the actual chemical process of bond-forming and bond-breaking in the enzyme active site, and to explore various factors influencing the enzyme's catalytic power. This led to the recognition that

electrostatic effects are the most important factors in enzyme catalysis and, in subsequent studies, that a pre-organized dipole orientation can more effectively stabilize the transition state than water does [8].

3. Electrostatic energy of macromolecules. The consistent treatment of electrostatic interactions that was introduced in the hybrid QM/MM method in principle provided a procedure to accurately determine electrostatic energies in proteins. Because the method is a microscopic model, the difficult task of assigning a dielectric constant to the enzyme active site was completely circumvented [7]. Furthermore, the authors immediately recognized that the $pK_a$ of ionizable groups in proteins can be estimated using computer simulations. The ability to adequately determine electrostatic energies allowed the authors for the first time to estimate the electrostatic stabilization of the carbonium ion intermediate in the reaction of lysozyme. It was found that electrostatic effects are the most important catalytic factor in lysozyme, whereas the effect of steric strain in the ground state is not important in the enzyme action.

The impact of the 1976 paper on computational studies of enzymatic reactions and on the QM treatment of large molecular systems has been enormous. Although this was not the first computer application aimed at elucidating the relationship between the structure of an enzyme and its activity, the Warshel and Levitt study did provide an early glimpse of biological supercomputing, signaling the arrival of a new era in which an unprecedented amount of information and a detailed understanding of biological processes can be obtained from computer simulations. The finding that lysozyme and perhaps most other enzymes work by an electrostatic effect has had a profound influence on the thinking about enzyme catalysis. The proposal will likely continue to be tested and validated.

Technically, the representation of the polarization of the solvent water by a grid of Langevin dipoles was used later by Warshel to develop a self-consistent-reaction-field model for the treatment of aqueous solvation. In addition, the development of consistent, polarizable intermolecular potentials for proteins, some of which are analogous to the approach used in the 1976 study, is currently being actively pursued by many research groups. In view of the importance of electrostatic interactions in macromolecular systems, more effective and accurate methods will certainly continue to emerge in the twenty-first century, taking advantage of the ever-increasing power of computers.

### References

1. Fersht A (1999) Structure and mechanism in protein science. Freeman, New York
2. Singh UC, Kollman PA (1986) J Comput Chem 7: 718
3. Field MJ, Bash PA, Karplus M (1990) J Comput Chem 11: 700
4. Gao J (1992) Science 258: 631
5. Gao J (1996) Acc Chem Res 29: 298
6. McCammon JA, Gelin BR, Karplus M (1977) Nature 267: 585
7. Sharp KA, Honig B (1990) Annu Rev Biophys Biophys Chem 19: 301
8. Warshel A (1978) Proc Natl Acad Sci USA 75: 5250

Theor Chem Acc (2000) 103:330–331
DOI 10.1007/s002149900037

Theoretical
Chemistry Accounts
© Springer-Verlag 2000

*Perspective*

# Perspective on "Theoretical interpretation of 1-2 asymmetric induction. The importance of antiperiplanarity"

## Anh NT, Eisenstein O (1977) Nouv J Chim 1: 61–70

**K.N. Houk**

Department of Chemistry and Biochemistry, University of California at Los Angeles, 405 Hilgard Avenue,
Los Angeles, CA 90095-1569, USA

Received: 8 March 1999 / Accepted: 29 March 1999 / Published online: 28 June 1999

**Abstract.** This paper describes how Anh and Eisenstein's publication in 1977 solved a classic problem in organic stereochemistry and simultaneously provided the impetus and a model for the use of quantum mechanics to understand a wide range of organic chemical phenomena in a style accessible to experimental chemists.

**Key words:** Stereoselectivity – Transition state – Ab initio quantum mechanics

In the century-old development of synthetic organic chemistry, a myriad of methods have been discovered and developed for the formation of new bonds between carbons and for the manipulation of functionality. To do so with control of the three-dimensional arrangement of atoms about the transformed bond is a central goal of organic synthesis. Control of stereoselectivity continues to be a challenging goal of synthesis and one of increasing practical significance as regulations controlling the use of stereoisomers as pharmaceuticals are created. Quantum mechanics has a surprisingly large role in this development, and the origins of this can be traced to the Anh–Eisenstein paper described here.

The 1950s saw the beginning of empirical generalizations about the stereochemistry of nucleophilic additions to chiral carbonyl compounds. The reaction in Fig. 1 is an example of such a process.

In the 1950s, Cram and Cornforth proposed models to rationalize results of this kind, invoking ideas about steric effects and their influence on transition-state con-

formations [1]. In the 1960s, Karabatsos and Felkin proposed alternatives to the Cram model, shown in Fig. 2. About the same time, mathematical formulations of stereoselectivity issues were formulated by Ruch and Ugi (1969) and by Salem (1973).

Anh and Eisenstein used quantum mechanics to determine which of these models most accurately reflected the actual transition state of the reaction. These were large systems for quantum mechanics calculations of the time, and many conformations had to be studied. STO-3G calculations on assumed model geometries were used. The tests reported in the paper are demonstrated in Fig. 3 (reprinted from Fig. 1 of the original paper). The figure shows the energies of different rotamers of the transition-state model for attack of hydride on the right-hand (solid line) or left-hand (dashed line) side of the carbonyl group. The dotted line near the bottom of the figure is the energy of rotamers of the aldehyde reactant.

For this case and many others, the Felkin model was established unequivocally to be energetically preferred, and a variety of details and interpretations were provided. In addition to the role of torsional and steric effects, Anh and Eisenstein found that the most electronegative substituent–that with the lowest-lying $\sigma_{CX}^*$ LUMO–would take the conformation anti to the attacking nucleophile. This model and its rationale are now universally known as the Felkin–Anh model. A variety of refinements and extensions occurred in following decades, such as the location of actual transition

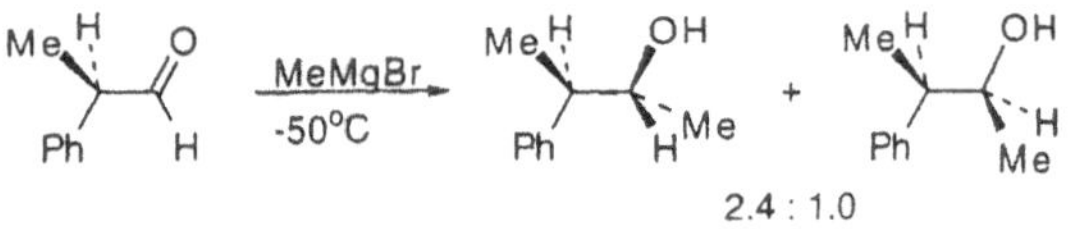

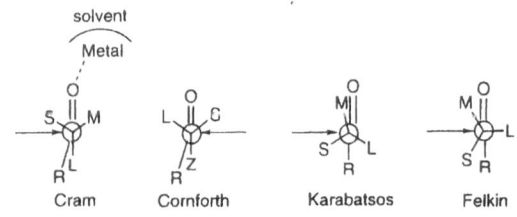

**Fig. 1.** Example of a stereoselective nucleophilic addition

2.4 : 1.0

**Fig. 2.** Cram, Cornforth, Karabatsos, and Felkin models for nucleophilic addition stereoselectivity (*S, M, L, Z* – small, medium, large, electronegative)

**Fig. 3.** An example of the results in Felkin and Anh's 1977 paper

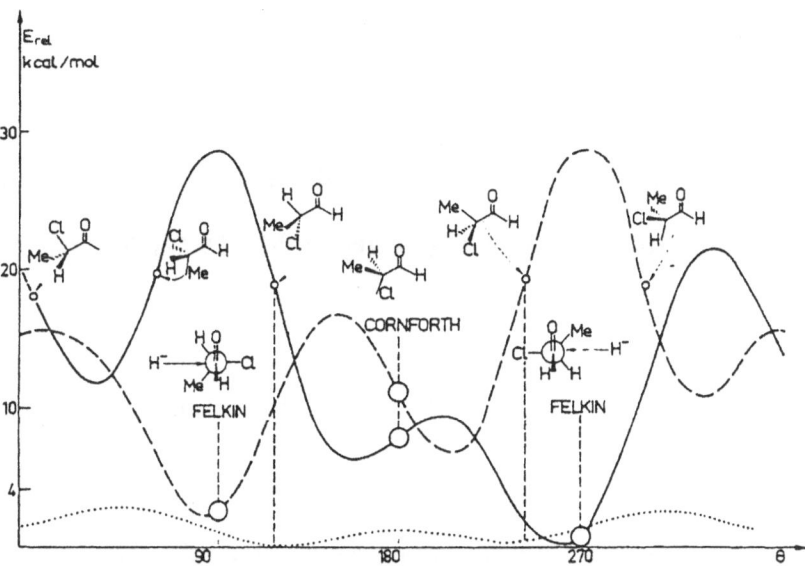

states for model reactions of this type with larger basis sets, organometallic reagents, and full optimizations [2], as well as extensions to radical and electrophilic reactions [3]. Alternative explanations have been proposed [4], but alternatives to the Felkin–Anh model have not been generally accepted [5].

The publication of a quantum mechanical investigation of this reaction in 1977 was important not only because it solved an important general problem in stereoselectivity, but because it demonstrated the power of quantum mechanical calculations to solve important problems on real organic systems. This style of reasoning, and the tests of stereoselectivity postulates through quantum mechanics, have set the style for theoretical studies relevant to synthetic organic chemistry research.

## References

1. (a) Anh NT, Eisenstein O (1977) Nouv J Chim 1: 61–70, and references cited therein; (b) Eliel EL, Wilen SH, Mander LN (1994) Stereochemistry of organic compounds. Wiley, New York, pp 875–886
2. Kaufman E, Schleyer PVR, Houk KN, Wu Y-D (1985) J Am Chem Soc 107: 5560–5563
3. Paddon-Row MN, Rondan NG, Houk KN (1982) J Am Chem Soc 104: 7162–7166
4. Cieplak AS (1994) In: Bürgi H-D, Dunitz JD (eds) Structure correlation, vol I VCH, Weinheim, pp 205–302
5. (a) Wu Y-D, Houk KN, Paddon-Row MN (1992) Angew Chem Int Ed Engl 31: 1019–1021; (b) Coxon JM, Houk KN, Luibrand RT (1995) J Org Chem 60. 418–427

Theor Chem Acc (2000) 103:332–334
DOI 10.1007/s002149900025

Theoretical
Chemistry Accounts
© Springer-Verlag 2000

*Perspective*

# Perspective on "Dynamics of folded proteins"

## McCammon JA, Gelin BR, Karplus M (1977) Nature 267: 585–590

**David A. Case**

Department of Molecular Biology, The Scripps Research Institute, La Jolla, CA 92037, USA

Received: 24 February 1999 / Accepted: 12 March 1999 / Published online: 28 June 1999

**Abstract.** This paper provides an overview of the title paper, discussing its background and significance, some details of the results, and subsequent developments that were stimulated by the work.

**Key words:** Molecular dynamics – Proteins – Historical perspective

## 1 Background

One of the key advances in computational chemistry during the last quarter of the twentieth century was the development of simulation methods to study motions of atoms and molecules in condensed-phase environments. The paper reviewed here is noteworthy for its pioneering application of molecular dynamics techniques to proteins, and (along with the 1976 Warshel and Levitt paper on lysozyme, discussed elsewhere in this issue) can be considered to herald an increased interest among the computational chemistry community on the problems dealing with the structures and dynamics of biological macromolecules.

In molecular dynamics simulations, one uses numerical integration techniques to obtain approximate solutions to Newton's equations of motion:

$$ma \equiv m(\mathrm{d}^2x/\mathrm{d}t^2) = F = -(\partial V/\partial x) \ . \tag{1}$$

Here $m$ is an atomic mass, $a$ is the acceleration (the second derivative of position $x$ with respect to time), and $F$ is the force (expressed as the derivative of a potential $V$ with respect to position). Under the assumptions usually used, the effect of electronic motion is folded into the potential-energy function $V(x)$, so that the dynamic variables become the position of each atomic nucleus as a function of time.

The use of classical trajectory techniques to study atomic and molecular collisions in the gas phase can be traced back to the 1930s, and became an established technique in the 1950s and 1960s. Some of this background is contained in the perspectives in this issue on the 1959 paper by Ford and Wheeler [1], the 1970 paper by Miller [2], and the 1965 paper by Karplus, Porter, and Sharma [3]. The first two of these deal in part with the approximation in Eq. (1) that uses classical ideas, rather than quantum dynamics, to describe nuclear motion; this greatly simplifies the computational analysis, and in most cases should lead to relatively small errors. The third paper contains a lucid description of the use of classical trajectories to analyze elementary gas-phase collisions; this background was very much "in the air" in the Karplus group, where similar calculations continued well into the 1970s [4, 5].

A second thread in computational chemistry that made it possible to carry out dynamical calculations on proteins was the development of suitable potential-energy functions (or "force fields") that describe energies as a function of configuration. The fundamental ideas trace back to ideas long used by vibrational spectroscopists to interpret infrared spectra [6]. The use of computers in conformational analysis of floppier molecules (i.e. to locate likely structures that are local minima on the potential-energy surface) had begun in earnest in the 1960s [7–12], and by the early 1970s many applications to peptide and protein systems had been reported. The work of the Lifson group in developing the consistent force field (CFF) was particularly influential to graduate student Bruce Gelin, whose principal dissertation project involved the development of a similar suite of codes to handle the "bookkeeping" involved in computing and using potential-energy functions in polypeptides, and the calibration of parameters for a force field with fewer terms than that of the CFF. In this model, the energy of a polypeptide is expressed in terms of the contributions of each local bond, angle and dihedral distortion terms (comprising a simplified version of traditional vibrational force fields), and longer-range, "nonbonded" contributions that represented steric and electrostatic interactions:

$$E_{MM} = \sum_{bonds} K_r(r - r_{eq})^2 + \sum_{angles} K_\theta(\theta - \theta_{eq})^2$$
$$+ \sum_{dihedrals} \frac{V_n}{2}[1 + \cos(n\phi - \gamma)]$$
$$+ \sum_{i<j} \left[ \frac{A_{ij}}{R_{ij}^{12}} + \frac{B_{ij}}{R_{ij}^6} + \frac{q_i q_j}{\varepsilon R_{ij}} \right] . \qquad (2)$$

This description is in many ways a quite simple one, lacking local anharmonic terms and cross-terms connecting bonds and angles, for example. Still, even for a small protein in vacuum there are many thousands of terms to be evaluated at each step, and many force constants ($K_r$, $A_{ij}$, $q_i$, etc.) to be estimated. In much early work, the value of the dielectric parameter $\varepsilon$ was taken to be the interatomic distance $R$ in angstroms, both to partially mimic solvent screening and to avoid expensive square root calculations. Gelin and postdoctoral researcher Andy McCammon (who had interests in both atomic and hydrodynamic models of proteins) worked together to construct a workable computer program to perform protein trajectories and to analyze the results, primarily in terms of time-correlation functions.

A third background thread for the application of molecular dynamics techniques to proteins came from the computational physics community on simulations of simple liquids. Here the perspectives in this issue on the 1964 paper by Rahman (on liquid argon) and on the 1971 paper by Rahman and Stillinger (on liquid water) may be helpful. The results of simulations and of other statistical models for dense liquids had led to a dynamical picture of diffusional motion whose character is dominated by relatively "hard" collisions among atoms which is augmented, but not fundamentally changed, by much "softer" (more slowly varying) attractive or cohesive forces that serve to hold the liquid together. The reasonable success of very simple potentials (such as hard-sphere models) in explaining the dynamical and equilibrium structural properties of simple liquids [13, 14] fueled the hope that many interesting features of biomolecular dynamics could be determined even with quite crude potential-energy functions, as long as the size scale and general characteristics of the rapidly varying parts of the potential were approximately correct. This belief had to contend with a common feeling (expressed at Harvard and elsewhere) that the number of adjustable parameters in force fields representing molecules as complex as proteins was so great as to severely compromise the usefulness of such calculations.

## 2 Results

The simulations reported in the paper dealt with bovine pancreatic trypsin inhibitor (BPTI), a small (58 amino acid) protease inhibitor whose structure is stabilized by three disulfide crosslinks. The simulation began with a set of X-ray coordinates and zero velocities, and ran 100 steps of numerical integration of Newton's equations of motion, with each time step being about 1 fs. Since the X-ray coordinates did not correspond to a minimum of the potential-energy surface, during the dynamics some of the potential energy was converted to kinetic energy, so that after 100 steps the mean internal temperature was 140 K. At this point, velocities were scaled by a factor of 1.5 (corresponding to scaling the kinetic energy by a factor of 2.25) and 250 additional equilibration steps were carried out, leading to an internal temperature of 285 K. Following this, 9,000 additional "production" steps were computed and analyzed, during which the average internal temperature rose to about 295 K.

The analysis of the results concentrated on the time-averaged structure and on correlation functions that illustrate the time-dependence of internal motions. A key conclusion was that the internal motion was fluidlike at ordinary temperatures, i.e. that "the dynamics of atomic displacements are dominated by collisions with neighboring atoms, at least on the picosecond time scale." Hence, it is argued, "...many of the dynamical properties (though not necessarily the correct average structure) can be obtained from any potential function which includes the forces that depend strongly upon distance (covalent and hydrogen bonds, nonbonded repulsions) and provides sufficient attractive interactions to preserve the compact structure of the native system."

## 3 Discussion

A striking feature (for most readers at the end of the twentieth century) is the short time scale of the simulation. This reflects both the speed of contemporary computers, and of the state of conformational analysis at the time, which typically analyzed calculations consisting of only a few hundreds of steps of energy minimization. The authors recognized that velocity equilibration was not complete even after 9 ps, and it was clear that significant computer resources would be needed for systematic studies of protein dynamics. For this reason, also, this initial paper is no longer consulted for quantitative details about fluctuations, although most of the results presented stand up well compared to later studies. Within 2 years, results from a 100-ps simulation were available [15], and the time scale and size of simulations continued to grow. Advances made in both algorithms and computers over the next two decades may be highlighted by comparison to a recent molecular dynamics simulation [16] of a system with about 20 times as many atoms as in the initial BPTI simulation, for a time period of 1 μs, $10^5$ times longer than the initial 10-ps time period. Even today, however, the size and time scales of affordable simulations is an important limitation to obtaining reliable answers for many interesting questions.

The field of macromolecular simulations has grown so rapidly that one no longer looks to the old primary literature for useful overviews, but rather turns to textbooks and monographs [17–21]. For those interested in some historical perspective, though, a few additional papers that represent the thinking in the Karplus group at the time are worth consulting. These include a comparison of hard-collision models with vibrational theories in understanding protein internal dynamics [22]; the inclusion of solvent effects into the simulations [23]; and analyses of oxygen binding to hemoglobin and myoglobin [24, 25]. A Scientific American article by Karplus and McCammon, although published much later, retains much of the "flavor" of the early investigations [26]. (It should be recognized, however, that the computer-generated graphics used to illustrate this article were not available in 1977; more typical of the time were line-printer plots that were later hand-traced for publication.)

The title *Nature* paper was accompanied in the same issue by a "News and Views" commentary by Barry Robson [27], who correctly noted that "the potential for future application of molecular dynamics and related techniques seems enormous", and that a key impact of molecular dynamics studies would be to counter an overly rigid view of protein structure arising from the fact that crystallographic results are typically presented as a single (average) structure. It is perhaps sobering, however, to consider Robson's suggested application for the new technique, that of using computations to address a controversy between the conventional double-helix model of DNA and a then-current "zipper" model. The difficulties to be encountered in moving from qualitative insights about the nature of protein fluctuations to reasonably reliable estimates of the relative energies of different conformations have been underestimated at one time or another by almost everyone in the field. Recent analyses of the conformational energetics of DNA, using computers many orders of magnitude more powerful than those available in 1977, illustrate problems with accuracy of potentials, treatment of electrostatics, and sampling problems that still frustrate straightforward efforts to "resolve" biochemical problems by recourse to computation [28, 29].

The introduction of molecular dynamics techniques into biomolecular simulation certainly qualifies as a key development in computational chemistry, one that required considerable foresight in tackling what were obviously very difficult problems. Martin Karplus [30] would later write, "The conceptual changes resulting from the early studies make one marvel at how much of great interest could be learned with so little – such poor potentials, such small systems, so little computer time. This, of course, is one of the great benefits of taking the initial, somewhat faltering steps in a new field where the questions are qualitative rather than quantative and any insights, even if crude, are better than none at all." In the beginning, emphasis was on picosecond motions in proteins that could be directly simulated, but it soon became clear (at least in principle) how to use the ability of molecular dynamics simulations to sample a Boltzmann distribution to make connections both to equilibrium thermodynamics [31] and to "activated" kinetic events that take place on slower time scales [32]. A few years later it was also realized that dynamical simulated annealing (i.e. a high-temperature simulation followed by slow cooling to low temperatures) can be a robust method for searching for low-energy conformers that is far superior to energy minimization techniques. The use of molecular-dynamics-based simulated annealing as part of an optimization strategy is now

nearly universal in crystallographic and NMR structure refinement [33]. Overall, biomolecular molecular dynamics simulation is now no longer a specialized task for computational chemists, but an increasingly reliable tool that is a (nearly) routine adjunct to experimental studies for those interested in the general field of structural biology.

*Acknowledgements.* It has been most enjoyable to remember the many wonderful interactions I have had, both personal and scientific, with the authors of the title paper.

# References

1. Ford KW, Wheeler JA (1959) Ann Phys (NY) 7: 287
2. Miller WH (1970) J Chem Phys 53: 1949
3. Karplus M, Porter RN, Sharma RD (1965) J Chem Phys 43: 3259
4. Brumer P, Karplus M (1973) Discuss Faraday Soc 55: 80
5. Wang ISY, Karplus M (1973) J Am Chem Soc 95: 8160
6. Wilson EB, Decius JC, Cross PC (1980) Molecular vibrations. Dover, New York, pp 312–346
7. Hendrickson JB (1961) J Am Chem Soc 83: 4537
8. Hendrickson JB (1964) J Am Chem Soc 86: 4854
9. Lifson S, Warshel A (1968) J Chem Phys 49: 5116
10. Levitt M, Lifson S (1969) J Mol Biol 46: 269
11. Scheraga HA (1971) Chem Rev 71: 195
12. Allinger NL (1976) Adv Phys Org Chem 13: 1
13. Weeks JD, Chandler D, Anderson HC (1971) J Chem Phys 54: 5237
14. Barker JA, Henderson D (1972) Annu Rev Phys Chem 23: 439
15. Karplus M, McCammon JA (1979) Nature 277: 578
16. Duan Y, Kollman PA (1998) Science 282: 740
17. Allen MP, Tildesley DJ (1987) Computer simulation of liquids. Clarendon, Oxford
18. Frenkel D, Smit B (1996) Understanding molecular simulation: from algorithms to applications. Academic Press, San Diego
19. Brooks CL III, Karplus M, Pettitt BM (1988) Adv Chem Phys 71: 1
20. van Gunsteren WF, Weiner PK, Wilkinson AJ (eds) (1993) Computer simulations of biomolecular systems, vol 2. ESCOM, Leiden
21. van Gunsteren WF, Weiner PK, Wilkinson AJ (eds) (1997) Computer simulations of biomolecular systems, vol 3. ESCOM, Leiden
22. McCammon JA, Wolynes PG, Karplus M (1979) Biochemistry 18: 928
23. Rossky PJ, Karplus M (1979) J Am Chem Soc 101: 1913
24. Gelin BR, Karplus M (1977) Proc Natl Acad Sci USA 74: 801
25. Case DA, Karplus M (1979) J Mol Biol 132: 343
26. Karplus M, McCammon JA (1986) Sci Am 254: 42
27. Robson B (1977) Nature 267: 577
28. Srinivasan J, Cheatham TE III, Kollman P, Case DA (1998) J Am Chem Soc 120: 9401
29. Jayaram B, Sprous D, Young MA, Beveridge D (1998) J Am Chem Soc 120: 10629
30. Karplus M (1993) In: van Gunsteren WF, Weiner PK, Wilkinson AJ (eds) Computer simulation of biomolecular systems: theoretical and experimental applications. vol 2. ESCOM, Leiden, pp v–xvii
31. Pearlman DA, Rao BG (1998) In: Schleyer P von R, Allinger NL, Clark T, Gasteiger J, Kollman PA, Schaefer HF III (eds) Encyclopedia of computational chemistry. Wiley, Chichester, pp 1036–1061
32. Northrup SH, Pear MR, Lee C-Y, McCammon JA, Karplus M (1982) Proc Natl Acad Sci USA 79: 4035
33. Brünger AT, Karplus M (1991) Acc Chem Res 24: 54

Theor Chem Acc (2000) 103:335–336
DOI 10.1007/s002149900059

**Theoretical
Chemistry Accounts**
© Springer-Verlag 2000

*Perspective*

# Perspective on "Statistical mechanics of isomerization dynamics in liquids and the transition state approximation"

## Chandler D (1978) J Chem Phys 68:2959

**B.J. Berne**

Department of Chemistry, Columbia University, New York, NY 10027, USA

Received: 11 March 1999 / Accepted: 19 March 1999 / Published online: 4 October 1999

**Abstract.** This paper provides a perspective on the use of the reactive flux formalism for calculating rate constants in condensed phase systems through molecular dynamics simulations. This approach makes possible the computation of rate constants even for systems with very high energy barriers.

**Key words:** Reactive flux – Molecular dynamics – Rate constants – Monte Carlo

## 1 Introduction

This article provides a discussion of the title paper by David Chandler. Here the projection operator formalism of statistical mechanics is used to derive the chemical rate equations for unimolecular reactions in classical systems. What falls out of this formalism is a microscopic expression for the the sum of the forward and backward rate constants that involves an equilibrium ensemble average of the instantaneous flux over the saddle point of the reaction weighted by a function that is unity if the reacting system is in any of its product configurations and zero if it is in any of its reactive configurations at time $t$. This quantity $\kappa(t)$, called the reactive flux, has several important limits. In the limit $t \to 0+$ the reactive flux becomes the transition-state rate constant. As $t$ increases from zero $\kappa(t)$ quickly decays to a plateau value from which it very slowly decays as an exponential with the true rate constant for the reaction. If the energy barrier for the reaction is high compared to the thermal energy, the true rate constant can be found from the plateau value of the reactive flux. Since the transition-state rate constant is an approximation based on the assumption that any trajectory that has crossed the saddle point will not recross it, the true rate constant will always be smaller than the transition-state rate constant. The beauty of this reactive flux expression is that it allows one to compute chemical rate constants for

reactions in liquids as well as in gases using molecular dynamics simulations.

For reactions involving large activation energy barriers, barrier crossing is an infrequent event. The barrier divides phase space into two regions such that most trajectories starting in one region stay in that region for very long times before making a transition to the other region. Since the crossings are infrequent, it is very difficult to determine the rate constants for barrier crossing from simulation. The reactive flux expression provides a rigorous method for accelerating the calculation of rate constants for activated barrier crossing. According to the reactive flux one can calculate the rate constant by studying trajectories originating at the barrier maximum. This allows one to propagate only initially activated trajectories and thus avoids the need to wait for an initially unactivated trajectory to get activated. This method was anticipated by the pioneering work of Keck [1, 2], Anderson [3] and Bennett [4]. By generalizing and extending the time correlation function approach of Yamamoto [5], Chandler derived the reactive flux in the form in which it is now used in simulations. He then suggested that calculations of the plateau value of the reactive flux from molecular dynamics trajectories provide a rapid method for calculating the true rate constant of the reaction.

There are two parts to a reactive flux calculation. First, one must sample initial states at the barrier maximum and then one must propagate these initial states using molecular dynamics. The sampling of the initial states is done with Monte Carlo methods or with constrained molecular dynamics. The molecular dynamics propagation can usually be carried out for a short time until the plateau value is reached. In practice the efficiency of this method depends on finding a good dividing surface in phase space, for only with a good dividing surface will the reactive flux decay to its plateau value quickly. An additional problem is met when the rate constant is very small. The reactive flux method is equivalent to computing the difference between two

336

large numbers. When this difference is small there is considerable error. In that case various approximation tricks have been devised such as the absorbing boundary method [6].

The reactive flux method is also useful in calculating rate constants in quantum systems. The path integral formulation of the reactive flux together with the use of the centroid distribution function has proved very useful for the calculation of quantum transition-state rate constants [7]. In addition new methods, such as the Meyer–Miller method [8] for semiclassical dynamics, have been used to calculate the flux–flux correlation function and the reactive flux.

The reactive flux method is widely used to compute rate constants. Several reviews [9, 10] illustrate its utility. Because of this utility, the title paper represents an important milestone in theoretical chemical kinetics.

## References

1. Keck JC (1967) Adv Chem Phys 13: 85
2. Keck JC (1962) Discuss Faraday Soc 33: 173
3. Anderson JB (1973) J Chem Phys 58: 4684
4. Bennett CH (1977) In: Christofferson RE (ed) Algorithms for chemical computation. American Chemical Society, Washington, D.C., pp 63
5. Yamamoto T (1960) J Chem Phys 33: 281
6. Straub JE, Berne BJ (1985) J Chem Phys 83: 1138
7. Voth GA, Chandler D, Miller WH (1989) J Chem Phys 91: 7749
8. Meyer HD, Miller WH (1979) J Chem Phys 70: 3214
9. Hanggi P, Talkner P, Borkovec M (1990) Rev Mod Phys 62: 251
10. Berne BJ, Borkovec M, Straub JE (1988) J Phys Chem 92: 3711

Theor Chem Acc (2000) 103:337–339
DOI 10.1007/s002149900047

Theoretical
Chemistry Accounts
© Springer-Verlag 2000

*Perspective*

# Perspective on "The energetics of enzymatic reactions"

## Warshel A (1978) Proc Natl Acad Sci USA 75: 5250

Arieh Warshel

Department of Chemistry, University of Southern California, Los Angeles, CA 90089, USA

Received: 26 February 1999 / Accepted: 14 April 1999 / Published online: 14 July 1999

**Abstract.** The origin of the catalytic power of enzymes has been one of the most important open problems in molecular biology. Our early computer modeling studies [Warshel A, Levitt M (1976) *J Mol Biol* 103: 227 indicated that electrostatic effects give the largest contributions to enzyme catalysis; however, it was not clear how enzymes can provide more electrostatic stabilization to their transition states than water does. This fundamental problem has been solved by the title paper. The paper pointed out that in reactions in water the solvent must pay significant electrostatic energy for orienting its permanent dipoles toward the transition states. It was then demonstrated that in enzymes the active site dipoles are already partially preoriented in the optimum direction and so much less electrostatic energy is lost in the reorganization process. It was further demonstrated that ion pairs and related transition states are less stable in water than in preoriented dipolar environments in general and in the active sites of real enzymes in particular. Thus, it was concluded that enzymes stabilize their transition states by preoriented dipoles and that the catalytic energy is already stored in the preorientation of these dipoles during the folding process rather than in the enzyme substrate interaction.

**Key words:** Preorganized sites – Catalysts – Electrostatic energies – Folding energy

Enzymatic reactions are involved in most biological processes; thus, there is a major practical and fundamental interest in elucidating the origin of the catalytic power of enzymes. Despite enormous progress in biochemical and structural studies [1] we still do not understand what energy contributions make enzymes so efficient.

Many prominent proposals have been put forward to explain the action of enzymes (for reviews see Refs. [1–3]). Unfortunately, it is hard to use the available experimental information to determine in a unique way which proposal is correct (see discussion in Refs. [3, 4]). It is possible, however, to use energy considerations and computer simulations to exclude different proposals [4]. For example, many of the early proposals involve ground-state destabilization rather than transition-state (TS) stabilization and this appears to be inconsistent with conceptual studies and mutation experiments [3].

Among the few proposals that can account for TS stabilization, the proposal of electrostatic stabilization [2, 5, 6] of the TS charges is probably the most reasonable proposal. This proposal has been supported by early microscopic calculations of the catalytic reaction of lysozyme [2]. Further support has been obtained from additional theoretical studies and analysis of mutation experiments.[4, 7, 8]; however, the proposal of a large electrostatic stabilization was hard to rationalize, i.e. this proposal requires that the enzyme stabilize the TS charges of the substrate more than water does (the reference reaction occurs in water). On the other hand, simulation studies have indicated that the average electrostatic interactions between the protein dipoles and the TS charges are similar in magnitude to, rather than larger than, the corresponding interactions in water. Furthermore, enzyme active sites do not provide more polar groups (e.g. hydrogen bonds) than a typical first solvation shell in water. In view of these considerations it was hard to see how enzymes could stabilize their TSs more than water does.

The 1978 paper [9] identified and solved the above-mentioned fundamental problem. The paper demonstrated that preorganized dipoles can stabilize ("solvate") ion pairs and other charge distributions more than water does. The reason for this remarkable effect is that in water about half of the energy gain from charge–dipole (solute–solvent) interactions is spent on changing the dipole–dipole (solvent–solvent) interactions. Thus the free energy of solvation is given by

$$\Delta G_{sol}^{w} \cong \Delta G_{Q\mu}^{w} - \Delta G_{\mu\mu}^{w} \cong \Delta G_{Q\mu}^{w} - \tfrac{1}{2}\Delta G_{Q\mu}^{w} = \tfrac{1}{2}\Delta G_{Q\mu}^{w} \; , \tag{1}$$

where $Q\mu$ and $\mu\mu$ designate charge–dipole and dipole–dipole, respectively. In proteins, however, the active site dipoles associated with polar groups (e.g. hydrogen bonds and C=O dipoles), the internal water molecules, and ionized residues are already partially oriented toward the TS charges. Thus, the loss in $\Delta G_{\mu\mu}$ is smaller than in water, and less free energy is spent on orienting the dipoles of the protein toward the TS of the substrate. The free-energy term $\Delta G_{\mu\mu}$ is closely related to the so-called "reorganization energy" of the given reaction [4, 10]. For example, in water we have to break water–water interactions to form good hydrogen bonds to the TS. In the enzyme, on the other hand, the hydrogen bonds are already partially oriented toward the TS charges [4, 11].

The idea that enzymes use preorganized dipoles should not be confused with proposals which are based on oversimplified macroscopic considerations, i.e. the leading term of the activation energy of charge-transfer reactions is the Marcus expression [10]

$$\Delta g^{\neq} = (\Delta G^0 + \lambda)^2 / 4\lambda + w \; , \qquad (2)$$

where $\Delta G^0$ is the free energy of the reaction, $\lambda$ is the solvent reorganization energy, and $w$ is the work of bringing the reactants to the optimum interaction distance. With this expression in mind, one may ask what factors can reduce $\Delta g^{\neq}$. A seemingly obvious suggestion is that enzymes reduce $\Delta g^{\neq}$ by having low dielectric (and presumably nonpolar) active sites where $\lambda$ would be reduced. The problem with such an oversimplified macroscopic view is that low dielectric and relatively nonpolar active sites would increase $\Delta G^0$ (relative to $\Delta G^0$ in water) for reactions that involve the formation of ionic products (or intermediates) from neutral reactants. In cases of reaction with ionic reactants the low dielectric will increase the $w$ term. Both cases will result in an overall anticatalytic effect (i.e. an increase in $\Delta g^{\neq}$ relative to the corresponding $\Delta g^{\neq}$ in water), i.e. $\Delta g^{\neq}$ will have to reflect the desolvation effect associated with bringing charges from water to nonpolar regions (see discussion in Refs. [12–14]). Thus, what is missing in the above model is the idea that enzymes reduce both $\lambda$ and $\Delta G^0$ by a preorganized polar (rather than nonpolar) environment.

The fact that the enzymes are polar and that this might be important for catalysis was realized by Krishtalik [15], who unfortunately described enzymes by using macroscopic models of a structureless sphere with low dielectric constant. This was done without noting (at least in early works [15]) that such a model leads to anticatalytic effects. Subsequent attempts [16] to include the effect of the field from the protein polar part resulted in a nonquantitative model which could not be used to explore the origin of the catalytic power of enzymes.[1] In

fact, representing the protein consistently as a partially fixed polar environment, while using continuum models with a well-defined dielectric constant, is a challenge that has not been met even today (see discussion in Ref. 17). On the other hand, our early microscopic studies [2] have overcome the continuum problems and traps by avoiding the concept of a dielectric constant altogether. This has allowed us to demonstrate that enzyme catalysis is indeed due to preoriented dipoles [9].

The idea that the active site dipoles are preoriented raises the question of the source of the energy for this catalytic effect. The 1978 paper proposed that the preorganization free energy is already invested in the folding process and therefore should not be invested during the reaction.[2] The prediction that the folding energy is used to preorient the enzyme dipoles is supported by the finding that mutations which increase the activation barrier, $\Delta g_{cat}^{\neq}$, also increase the protein stability [18]. Thus, the protein ordered structure that was optimized by evolution for catalysis cannot be optimized simultaneously for stability.

If enzymes really use a preoriented polar environment to stabilize the TS, then we understand why it was so difficult to elucidate and quantify the origin of enzyme catalysis. First, the catalysis reflects the fact that the reference reaction in water involves a large investment of reorganization energy. As argued above, a significant part of this energy is not invested in the enzymatic reaction. Thus, those who did not consider the reference reaction in water overlooked the major catalytic effect. Second, the catalytic energy appears to be stored in the enzyme itself and not in the enzyme–substrate interaction. Realizing this point is important since the tendency is to search for an especially strong interaction between the enzyme and the substrate rather than to look for the energetics stored in the enzyme; however, when we find, for example, a large catalytic contribution from a given hydrogen bond donor we have to realize that this contribution is due to the preorientation of the donor group. Similarly, the popular entropy proposal [20] attributes the catalytic effect to the loss of entropy of assembling the substrate fragments upon moving them to the active site. This entropic effect is usually quite small and the catalytic effect appears to be associated with fixing the environment (rather than the substrate).

The somewhat complex concept of a preorganized active site is starting to gain wider recognition [21, 22] and will probably emerge as one of the most important factors in enzyme catalysis. The quantitative establishment of the importance of this effect will probably continue into the next century and will involve more calculations of reorganization free energy of the type described in Refs. [12, 23] and studies of the effect of mutations on folding and catalysis [18]; however, the 1978 paper was the first to demonstrate in a semiquan-

---

[1] Krishtalik's attempt to include the field of the protein polar part in a spherical model of the protein [16] has not reproduced any catalytic effects. All the intraglobular field in chymotrypsin reported in Ref. [16] is extremely small except for the field from Asp102, which is a part of the reactant system rather than a source for a field on this system. Hence the resultant large Asp$^-$ImH$^+$ ion-pair stabilization in a low dielectric medium is an artifact of neglecting the Born energy of transferring the ions from water to the hypothetical low dielectric protein [19]

[2] This preorganization energy is not the $\lambda$ of Eq. (2), which is related to the actual reaction coordinate, but the free energy of orienting the dipoles to stabilize the TS charges. Thus, for example, an enzyme can evolve to mainly reduce $\Delta G^0$ by its fixed dipoles and in this case it will have reduced preorganization energy without a reduction in $\lambda$

titative way the catalytic effect of preorganized enzyme dipoles and to provide a rationale to the result of our early electrostatic calculations [2].

*Acknowledgement.* This work was supported by NIH grant GM24492.

## References

1. Fersht AR (1985) Enzyme structure and mechanism. Freeman, New York
2. Warshel A, Levitt M (1976) J Mol Biol 103: 227
3. Warshel A (1998) J Biol Chem 273: 27035
4. Warshel A (1991) Computer modeling of chemical reactions in enzymes and solutions. Wiley, New York
5. Vernon CA (1967) Proc R Soc Lond Ser B 167: 389
6. Perutz MF (1978) Science 201: 1187
7. Warshel A, Naray-Szabo G (eds) (1997) Computational approaches to biochemical reactivity. Kluwer, Dordrecht
8. Soman K, Yang AS, Honig B, Fletterick R (1989) Biochemistry 28: 9918
9. Warshel A (1978) Proc Natl Acad Sci USA 75: 5250
10. Marcus RA (1956) J Chem Phys 24: 966
11. Warshel A, Sussman F, Hwang JK (1988) J Mol Biol 201: 139
12. Yadav A, Jackson RM, Holbrook JJ, Warshel A (1991) J Am Chem Soc 113: 4800
13. Warshel A, Åqvist J, Creighton S (1989) Proc Natl Acad Sci USA 86: 5820
14. Warshel A, Florián J (1998) Proc Natl Acad Sci 95: 5950
15. Krishtalik LI (1980) J Theor Biol 86: 757
16. Krishtalik LI, Topolev VV (1984) Mol Biol (Moscow) 18: 721
17. Muegge I, Qi PX, Wand AJ, Chu ZT, Warshel A (1997) J Phys Chem B 101: 825
18. Shoichet BK, Baase WA, Kuroki R, Matthews BW (1995) Proc Natl Acad Sci USA 92: 452
19. Warshel A, Russell ST, Churg AK (1984) Proc Natl Acad Sci USA 81: 4785
20. Jencks WP (1986) Catalysis in chemistry and enzymology. Dover, New York
21. Rao SN, Singh UC, Bash PA, Kollman PA (1987) Nature 328: 551
22. Cannon W, Benkovic S (1998) J Biol Chem 273: 26257
23. Åqvist J, Fothergill M (1996) J Biol Chem 271: 10010

Theor Chem Acc (2000) 103:340–342
DOI 10.1007/s002149900056

Theoretical
Chemistry Accounts
© Springer-Verlag 2000

*Perspective*

# The origin of the pseudopotential density functional method. Perspective on "Microscopic theory of phase transformation and lattice dynamics of Si"

## Yin MT, Cohen ML (1980) Phys Rev Lett 45:1004–1007

**James R. Chelikowsky**

Department of Chemical Engineering and Material Science, University of Minnesota
and Minnesota Supercomputing Institute, Minneapolis, MN 55455, USA

Received: 17 February 1999 / Accepted: 7 May 1999 / Published online: 9 September 1999

**Abstract.** This paper provides an overview of the title paper by Yin and Cohen. I will briefly review some of the background for this work, provide some details of the calculations and discuss how this paper has influenced the field. In particular, this paper led to the development of the first realistic calculations for the structural energies of solids. It was the origin of the pseudopotential density functional method applied to the solid state.

**Key words:** Pseudopotentials – Density functional theory – Structural energies

Prior to 1975 or so, the words "ab initio" did not exist in the scientific vocabulary for methods describing the electronic structure of the solid state. At that time, a number of very powerful and successful methods had been developed to describe the electronic structure of solids, but these methods did not pretend to be first principles or "ab initio" methods. The foremost example of electronic structure methods at that time was the empirical pseudopotential method (EPM) [1]. The EPM was based on the Phillips–Kleinman cancelation theorem [2], which justified the replacement of the strong, all-electron potential with a weak pseudopotential [1]. The pseudopotential replicated only the chemically active valence electron states. Physically, the cancelation theorem is based on the orthogonality requirement of the valence states to the core states [1]. This requirement results in a repulsive part of the pseudopotential which cancels the strongly attractive part of the core potential and excludes the valence states from the core region. Because of this property, simple bases such as plane waves can be used efficiently with pseudopotentials.

Plane waves also take advantage of the periodicity of the crystal by expanding the wave functions and potentials in three-dimensional Fourier series. Since the pseudopotential converges so rapidly in Fourier space only a few Fourier expansion coefficients, called form factors, are required to describe the potential. The EPM uses the form factors as parameters, fixed by experiment, to describe the potential.

This simple, but elegant, approach resulted in the first realistic description of the electronic structure and optical properties of semiconductors. The EPM also yielded the first accurate "picture" of the covalent bond in solids [3]. It demonstrated conclusively that a one-electron (band picture) of solids was correct and could be used to interpret spectroscopic results. As such, it helped create the field of optical spectroscopy in solids. But, however successful the EPM was, there were issues outside its applicability: structural energies. Extensions of the EPM were contemplated, but it was widely believed at the time that no first principles or ab initio theory could be expected to describe the solid state with sufficient accuracy to obtain any useful or predictive information.

In the late 1970s, Marvin Cohen and his student Ming-Tang Yin began to explore the possibility of using density functional theory, in particular, using the local density approximation (LDA) [4] and pseudopotentials for computing the electronic structure of solids. The LDA is a powerful tool as it maps the all-electron problem onto a one-electron problem; however, in terms of realistic calculations it was largely untested for structural energies at the time of Yin and Cohen's paper. Many people, particularly in the quantum chemistry community, felt the LDA was not a serious contender with more "rigorous" techniques. Unlike "standard" quantum chemistry approaches which use the exact Hamiltonian, the LDA ground-state energy can reside below the true experimental limit. Consequently, the LDA works only if one compares two reference states and any inherent LDA errors cancel. This "uncontrolled" cancellation of errors was thought to be serious. In contrast, other methods such as configuration interaction allow one to improve the ground-state energy in a systematic manner.

However, from a computational point of view, the LDA is much easier to implement and, relative to configuration interaction methods, is much less computationally intensive. Yin and Cohen, along with others, had used the LDA to construct pseudopotentials from first principles [5]. These ab initio pseudopotentials were very helpful in describing the electronic structure of surfaces, molecules, defects, etc., but they were not viewed as tools for computing the total energy of a solid.

In the title paper, Yin and Cohen, evaluated the total energy of various silicon polymorphs using density functional theory with the LDA and pseudopotentials: an approach we can characterize as a pseudopotential density functional method (PDFM). In particular, they evaluated

$$E_{\text{total}} = \sum_i \epsilon_i - \frac{1}{2} \int \rho(r) V_H(r) \mathrm{d}^3 r$$
$$+ \int \rho(r) \{ \epsilon_{\text{xc}}[\rho] - V_{\text{xc}}[\rho] \} \mathrm{d}^3 r + E_{\text{ion-ion}} , \quad (1)$$

where the sum is over the occupied eigenvalues, $\epsilon_i$, $\rho(r)$ is the valence pseudocharge density, $V_H$ is the Hartree or Coulomb potential, $\epsilon_{\text{xc}}[\rho]$ is the exchange–correlation energy density, $V_{\text{xc}}$ is the exchange–correlation potential, and $E_{\text{ion-ion}}$ is the ion core repulsion energies. Evaluating the total energy for solids is a difficult task as one must be careful in handling various divergent Coulomb terms [6]. The eigenvalues, $\epsilon_i$, and the pseudocharge density, $\rho$, were evaluated in the same way as in the EPM, i.e., a plane-wave basis was used resulting in an eigenvalue problem. Typically fewer than 100 plane waves were used in such calculations. Today, a personal computer could easily handle the computational load for this problem.

It is important to recognize the power of the pseudopotential in this formalism. Without pseudopotentials, one has to solve the Kohn–Sham equations for all the electrons in the system. This is a very difficult exercise as the energy and length scales for the core and valence electrons' eigenvalues and eigenstate differ by orders of magnitude. (Only years after the Yin and Cohen paper were all-electron structural energy calculations implemented.)

Using the PDFM, Yin and Cohen looked at various real and hypothetical crystalline forms of silicon including close-packed structures such as the face-centered-cubic form and open structures such as the diamond and hexagonal diamond structures [7]. They evaluated the total energy for each structural phase and plotted the energies as a function of crystal volume. In doing so, they created a "phase diagram" for silicon at zero temperature as shown in Fig. 1.

Several key points can be made about the results of Yin and Cohen. They correctly found the diamond structure to be the ground-state phase relative to the other structures. The predicted lattice constant was

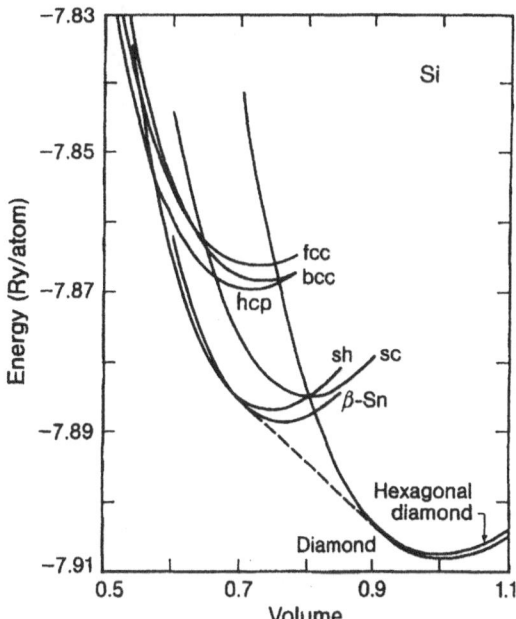

**Fig. 1.** The total electronic energy versus the atomic volume for several Si structures: diamond, hexagonal diamond, white tin ($\beta$-Sn), hexagonal close-packed (hcp) body-centered cubic (bcc), and face-centered cubic (fcc). The volume is normalized by the experimental value. The *dashed line* is the common tangent of the energy versus volume curves and gives the transition pressure between the diamond and white tin structures (see the title paper)

within 1–2% of experiment and the bulk modulus was within about 5% of experiment. The pressure to transform the diamond structure to the white tin structure ($\beta$-Sn) was also predicted to within 20%. It should also be noted that the title paper examined other structural properties such as those associated with the Grüneisen parameters and vibrational modes. Later, Cohen went on to predict a new superconducting phase of silicon on the basis of this work [8].

The impact of this paper cannot be underestimated. For the first time, it was demonstrated that accurate structural energies could be calculated and used to make predictions about the solid state. This work led to an enormous interest in utilizing pseudopotentials and density functional theory for a wide variety of systems including liquids, clusters, defects, etc. [9]. It also laid the foundation for ab initio molecular dynamics [10]. Since structural energies could be evaluated accurately, it was apparent that interatomic forces could also be calculated within this formalism and could be used in molecular dynamics simulations. It is safe to assert that this paper established the PDFM as the method of choice for the elucidation of the electronic structure of condensed matter.

### References

1. Cohen ML, Chelikowsky JR (1989) Electronic structure and optical properties of semiconductors, 2nd edn. Springer, Berlin Heidelberg New York
2. Phillips JC, Kleinman L (1959) Phys Rev 116: 287
3. Cohen ML (1973) Science 179: 1189

342

4. (a) Hohenberg P, Kohn W (1964) Phys Rev B 136: 864; (b) Kohn W, Sham LJ (1965) Phys Rev A 140: 1133
5. Chelikowsky JR, Cohen ML (1992) In: Landsberg (ed) Handbook on semiconductors, vol 1. Elsevier, Amsterdam, p 59
6. Ihm J, Zunger A, Cohen ML (1979) J Phys C 12: 4409
7. Yin MT, Cohen ML (1980) Phys Rev Lett 45: 1004
8. Chang KJ, Dacorogna MM, Cohen ML, Mignot JM, Chouteau G, Martinez G (1985) Phys Rev Lett 54: 2375
9. Chelikowsky JR, Louie SG (eds) (1996) Quantum theory of real materials. Kluwer, Dordrecht
10. Car R, Parrinello M (1985) Phys Rev Lett 55: 2471

Theor Chem Acc (2000) 103:343–345
DOI 10.1007/s002149900013

Theoretical
Chemistry Accounts
© Springer-Verlag 2000

*Perspective*

# Perspective on "Electrostatic interactions of a solute with a continuum. A direct utilization of ab initio molecular potentials for the prevision of solvent effects"

## Miertus S, Scrocco E, Tomasi J (1981) Chem Phys 55: 117

F. Javier Luque[1], Josep Maria López[1], Modesto Orozco[2]

[1] Departament de Fisicoquímica, Facultat de Famàcia, Universitat de Barcelona, Av. Diagonal s/n, E-08028 Barcelona, Spain
[2] Departament de Bioquímica i Biologia Molecular, Facultat de Química, Universitat de Barcelona,
Martí i Franquès 1, E-08028 Barcelona, Spain

Received: 29 January 1999 / Accepted: 18 February 1999 / Published online: 21 June 1999

**Abstract.** This paper provides an overview of the title paper by Miertus, Scrocco and Tomasi, including the impact that it has had on the theoretical description of solvation by means of continuum models.

**Key words:** Continuum models – Polarizable continuum method – Solvation

Current continuum solvation models provide not only qualitative insights into a wide variety of chemical phenomena in condensed phases, but in most cases are the best compromise between computational efficiency and reliability for the estimation of solvent effects. Achievement of the present status has resulted from tremendous research effort made by different groups around the world.

The initial attempts to account theoretically for solvent effects were made in the 1930s [1], but the suitability of these models for understanding chemical events in condensed phases was limited by their intrinsic simplicity. Thus, it was not until the 1970s, when continuum models were implemented within the quantum mechanical framework [2], that an accurate theoretical representation of solvent effects became possible. The last decade of this century has witnessed the spectacular growth of this new area of research [3]. It is expected that in the next century continuum methods will be the most used approach for the study of solvent effects in chemical systems.

The contribution made by Miertus, Scrocco and Tomasi (MST) and published in their 1981 *Chem Phys* paper [4] is one of those pioneering works in the development of implicit solvation methods. The relevance of this contribution in modeling and predicting chemical processes in solution can be appreciated from inspection of the citations received by this paper in the scientific literature (Figs. 1, 2). Figure 1 shows that MST's paper had achieved nearly 600 citations by 1997. More remarkably, the number of citations has been increasing in the last few years, showing that this a paper which will have a large impact on future research. The citation profile for the 1980s shows a slow increase in the number of citations, while the increase is very sharply during the 1990s. The low slope of the plot in the period until 1990 is due to the fact that most of the work performed with the polarizable continuum method (PCM) developed at Pisa was necessarily aimed at testing and refining the basic physical and mathematical assumptions of the model. The fast growth in the number of citations shown in the second half of Fig. 1 can be explained by two different factors. First, the availability of very efficient versions of the algorithm, which include treatments for computing the dielectric reaction field and also non-electrostatic effects. Second, the formidable increase in computer power and the implementation of the algorithm in widely used quantum chemistry programs. This has facilitated the application of the PCM to the study of an increasingly large number of systems and processes in diverse chemical fields.

The increasing acceptance of the PCM by the chemistry community is reflected in Fig. 2, which shows the distribution by chemistry areas of the publications referencing MST's original article in 1988 and a decade later. Inspection of Fig. 2 reveals a number of features about the evolution of the PCM. In 1988 nearly 60% of the studies were published in journals of quantum and computational chemistry, and nearly 30% in journals of physical chemistry. A decade later the contribution of quantum and computational studies had sensibly diminished to around 25%, and that of physical chemistry had increased to nearly 40%. More importantly, new areas of applications have clearly emerged, especially in

*Correspondence to*: F.J. Luque or M. Orozco

344

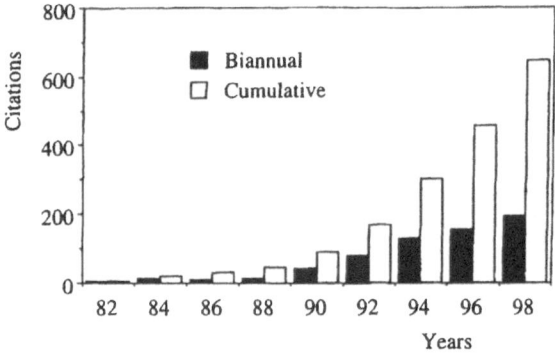

**Fig. 1.** Profile of citations in the scientific literature of "Electrostatic interactions of a solute with a continuum. A direct utilization of ab initio molecular potentials for the prevision of solvent effects" by Miertus, Scrocco and Tomasi (1981) *Chem Phys* 55: 117

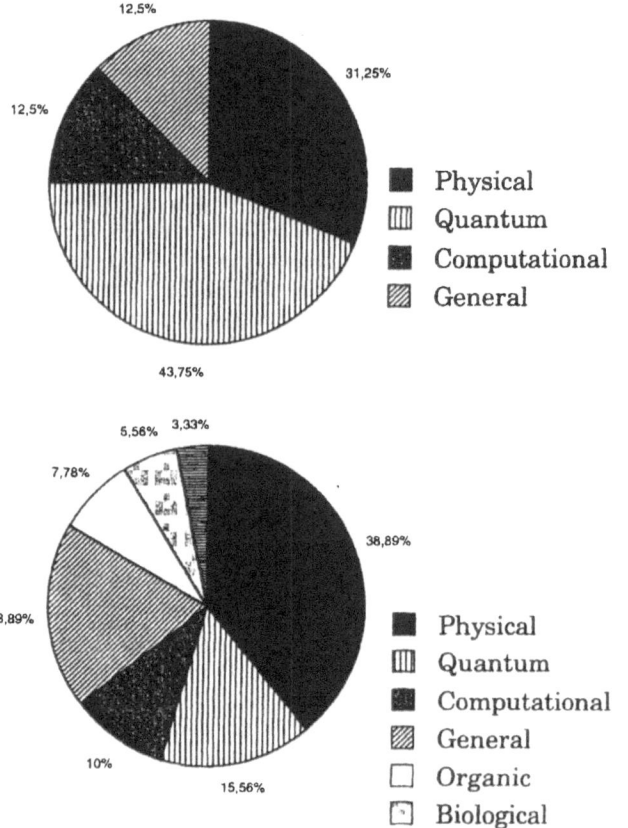

**Fig. 2.** Distribution by chemistry areas of the publications referencing *Chem Phys* 55: 117 (1981) in 1988 (top) and 1998 (bottom). Chemistry areas are defined according to the main scope of the journal where the citation was made

organic chemistry and biochemistry–medicinal chemistry. Overall, this can interpreted as the result of the progressive impact of implicit continuum models in the research conducted in very diverse areas of chemistry.

Let us now examine the significance of MST's contribution at the time of the publication of their paper and in the following years. The article in *Chem Phys* reported the basic features of the PCM for solving the

electrostatic problem in a homogeneous continuum dielectric medium. This was accomplished by reformulating the Poisson equation in terms of a boundary-element problem related to the discretization of the apparent charge density accounting for the solvent reaction field on the solute cavity surface. Compared to other methods proposed to treat the dielectric response in continuum calculations, the approach adopted by MST is characterized by three different features. First, whereas simpler spherical or ellipsoidal cavities facilitated solving the Poisson equation, MST utilized a cavity adapted to the molecular shape, which was built up from a set of interlocking spheres centered on the nuclei, allowing a more realistic treatment of the solute/solvent interface. Second, the discretized set of apparent surface charges was determined from the total electrostatic potential computed at selected points (tesserae) on the cavity. As an observable, the computation of the electrostatic potential [5] could be rigorously performed without any approximation, whereas other methods at that time utilized a truncated expansion of the solute charge distribution. Finally, once the set of apparent charges is known, their interaction with the solute is treated by adding a perturbation operator to the gas-phase solute Hamiltonian. Owing to the mutual dependence between the solute wavefunction and the apparent charges on the cavity, MST discussed a self-consistent procedure to achieve self-consistency.

The boundary-element solution adopted by MST made it necessary to solve two computational problems inherent to the physical model in the PCM. First, since the solvent reaction field originates from the total electrostatic potential, including both solute and solvent contributions, the self-polarization between the solvent apparent charges also had to be considered in determining the solvent reaction field. This question was addressed by introducing a second iterative procedure coupled to the self-consistent procedure mentioned previously. The second problem is related to the escape of tails of the solute electron density from the solute cavity. This lost electron density throws the solvent response originating from the continuous electron distribution and the discrete set of nuclei out of balance, leading to a completely biased description of the reaction field. It was then necessary to introduce a charge compensation factor to balance properly the polarization due to nuclei and electrons.

The original PCM has been further refined by the Pisa group over the years. Some of the recent modifications of different physical and mathematical features of the model are examined in the following lines. One focus of interest has been the building up of the solute/solvent interface, which has been reformulated to obtain a finer description of the cavity [6]. Another point that has deserved particular research effort has been the treatment of charge compensation, and different correction procedures have been examined [7]. The model, which was originally formulated considering exclusively the solution to the electrostatic problems, was later extended to include nonelectrostatic contributions to solvation [8]. The iterative procedure adopted to solve the problem of using the self-consistent procedure has been reformulated using other procedures, such as matrix-inversion or closure approaches [9]. Another relevant focus of interest has been the implementation of analytical expression for the free-energy derivatives [10], and their implications for geometry optimization or calculations of polarizabilities. It is also worth noting the efforts made to reformulate nonequilibrium solvation in the computational scheme of the PCM [11]. Finally, very recently many efforts have been devoted to the extension of the PCM, initially devised to deal with an isotropic homogeneous medium, to more complex systems, including anisotropic dielectrics and/or ionic solutions [12]. Even though these topics comprise just a fraction of the research developed in the framework of the PCM in the last few years, they suffice to illustrate how implicit solvation methods can be further elaborated in order to obtain a more precise description of solvation and to extend these methods to the study of chemical systems of increasing complexity.

Several groups have also made relevant contribution to the evolution of the original PCM. A related model based on conductor-like screening (COSMO) has been developed recently by Klamt and Schüürmann [13]. Likewise, another approach to the PCM has been proposed in which the cavity surface is determined in terms of an electronic isodensity surface [14]. Olivares del Valle and coworkers [15] have focused their attention on aspects such as the inclusion of correlation effects in the PCM, or on the role of nonadditive effects in solute–solvent interactions. Pascual-Ahuir et al. [16] have paid most attention to the problem of the definition of the cavity surface. The work done in Barcelona has focussed mainly on the parametrization of the PCM to treating aqueous and nonaqueous solvents, as well as the application of the PCM to the study of biochemical systems [17, 18]. Finally, we and others have made new methodological developments to allow the implementation of the PCM in molecular dynamics or in Monte Carlo calculations [19].

The list of topics mentioned above is far from being complete, and the reader is addressed to several reviews [3, 18] that offer a comprehensive view of solvation models and their applications; however, the preceding lines suffice to illustrate well the evolution and refinement of the original PCM, and how it has been reformulated and improved as new chemical challenges have been considered. Nowadays, the PCM-based versions are powerful tools for understanding a variety of chemical phenomena in condensed phases, and the accuracy achieved in determining solvent-related properties permits quantitative predictions to be formulated in some cases. These encouraging features, in conjunction with the exploration of new methodological advances, allows us to envisage that the method will be applied to an increasingly large number of chemical fields, covering more complex systems and chemical processes in condensed phases.

## References

1. (a) Born M (1920) Z Phys 1: 45; (b) Bell RP (1937) Trans Faraday Soc 27: 797; (c) Onsager L (1936) J Am Chem Soc 58: 1486; (d) Kirkwood J (1935) J Chem Phys 3: 300
2. (a) Rinaldi D, Rivail JL (1973) Theor Chim Acta 32: 57; (b) Rivail JL, Rinaldi D (1976) Chem Phys 18: 233; (c) Tapia O, Goscinski O (1975) Mol Phys 29: 1653; (d) Constancie R, Tapia O (1978) Theor Chim Acta 48: 75; (e) Klopamn G (1967) Chem Phys Lett 1: 200
3. (a) Tomasi J, Persico M (1994) Chem Rev 94: 2027; (b) Rivail JL, Rinaldi D (1995) In: Leszczynski J (ed) Computational chemistry. Review of current trends. World Scientific, New York, pp 139–74; (c) Tapia O, Bertrán J (eds) (1996) Solvent effects and chemical reactivity. Kluwer, Dordrecht; (d) Cramer CJ, Truhlar DG (1999) Chem Rev (in press)
4. Miertus S, Scrocco E, Tomasi J (1981) Chem Phys 55: 117
5. Scrocco E, Tomasi J (1973) Top Curr Chem 42: 95
6. (a) Pomelli CS, Tomasi J (1998) Theor Chem Acc 99: 34; (b) Pomelli CS, Tomasi J (1999) J Comput Chem (in press); (c) Barone V, Cossi M, Tomasi J (1997) J Chem Phys 107: 3210
7. (a) Mennucci B, Tomasi J (1997) J Chem Phys 106: 5151; (b) Cossi M, Mennucci B, Pitarch J, Tomasi J (1998) J Comput Chem 8: 833
8. (a) Floris FM, Selmi M, Tani A, Tomasi J (1997) J Chem Phys 107: 6353; (b) Floris FM, Tani A, Tomasi J (1993) Chem Phys 169: 11
9. (a) Coitiño EL, Tomasi J, Cammi R (1995) J Comput Chem 16: 20; (b) Cammi R, Tomasi J (1995) J Comput Chem 16: 1449
10. (a) Cammi R, Cossi M, Tomasi J (1996) J Chem Phys 104: 4611; (b) Cammi R, Cossi M, Mennucci B, Pomelli CS, Tomasi J (1996) Int J Quantum Chem 60: 1165; (c) Cossi M, Tomasi J, Cammi R (1995) Int J Quantum Chem Symp 29: 695
11. (a) Cammi R, Tomasi J (1995) Int J Quantum Chem Symp 29: 465; (b) Aguilar MA, Olivares del Valle FJ, Tomasi J (1993) J Chem Phys 98: 7375
12. (a) Cances E, Mennucci B, Tomasi J (1997) J Chem Phys 107: 3032; (b) Mennucci B, Cances E, Tomasi J (1997) J Phys Chem B 101: 10506; (c) Mennucci B, Cossi M, Tomasi J (1995) J Chem Phys 102: 6837, (d) Mennucci B, Cossi M, Tomasi J (1996) J Phys Chem 100: 1807
13. Klamt A, Schüürmann G (1993) J Chem Soc Perkin Trans 2: 799
14. Foresman JB, Keith TA, Wiberg KB, Snoonian J, Frisch MJ (1996) J Phys Chem 100: 16098
15. (a) Olivares del Valle FJ, Aguilar MA (1992) J Comput Chem 13: 115; (b) Contador JC, Aguilar MA, Olivares del Valle FJ (1997) Chem Phys 214: 113
16. Pascual-Ahuir JL, Silla E, Tuñon I (1994) J Comput Chem 15: 1127
17. (a) Orozco M, Hernández B, Luque FJ (1998) J Phys Chem B 102: 5228; (b) Colominas C, Luque FJ, Orozco M (1996) J Am Chem Soc 118: 6811; (c) Orozco M, Lopez JM, Colominas C, Alhambra C, Busquets MA, Luque FJ (1996) Int J Quantum Chem 60: 1179
18. Orozco M, Alhambra C, Barril X, Lopez JM, Busquets MA, Luque FJ (1996) J Mol Model 2: 1
19. (a) Colominas C, Luque FJ, Orozco M (1999) J Comput Chem (in press); (b) Varnek AA, Wipff G, Glebov AS, Feil D (1995) J Comput Chem 16: 1

Theor Chem Acc (2000) 103:346–348
DOI 10.1007/s002149900021

Theoretical
Chemistry Accounts
© Springer-Verlag 2000

*Perspective*

# Perspective on "Density-functional theory for fractional particle number: derivative discontinuities of the energy"

Perdew JP, Parr RG, Levy M, Balduz JL Jr

**Yingkai Zhang, Weitao Yang**

Department of Chemistry, Duke University, Durham, NC 27708, USA

Received: 26 February 1999 / Accepted: 16 March 1999 / Published online: 21 June 1999

**Abstract.** This paper provides an overview of the title paper by Perdew, Parr, Levy and Balduz [*Phys Rev Lett* 49:1691 (1982)]. The title paper extended density functional theory to fractional electron number by an ensemble approach and proved that the energy is a series of straight lines interpolating its values at integer numbers of electrons. It also established that the highest-occupied exact Kohn–Sham orbital energy is the negative of the ionization energy, and showed that the exchange-correlation potential jumps by a constant as the number of electrons increases by an integer. These results are fundamental and continue to inspire developments in density functional theory.

**Key words:** Density functional theory – Exchange-correlation functional – Fractional number of electrons – Self interaction error – Derivative discontinuity

The original density functional theory (DFT), based on Hohenberg–Kohn theorems [1], Kohn–Sham equations [2] and the Levy constrained search formulation [3], is a rigorous approach for determining the ground-state density and ground-state energy for any $N$-electron system. Here the electron number

$$N = \int d^3 r \rho(\mathbf{r}) \tag{1}$$

is an integer. The electronic ground-state energy of the system is a functional of the electron density $\rho$,

$$E[\rho] = F[\rho] + V_{\mathrm{ne}}[\rho] , \tag{2}$$

where $F[\rho]$ is the universal density functional and $V_{\mathrm{ne}}[\rho]$ is the electron-nuclear interaction. $F[\rho] = T_{\mathrm{s}}[\rho] + J[\rho] + E_{\mathrm{xc}}[\rho]$, where $T_{\mathrm{s}}[\rho]$ is the kinetic energy of the Kohn–Sham non-interacting reference system with the same density $\rho$, $J[\rho] = \frac{1}{2} \int \frac{\rho(\mathbf{r})\rho(\mathbf{r}')}{|\mathbf{r}-\mathbf{r}'|} d\mathbf{r}\, d\mathbf{r}'$ is the classical

*Correspondence to*: W. Yang

electron–electron repulsion energy, and $E_{\mathrm{xc}}[\rho]$ is the exchange-correlation energy functional.

Since DFT uses the electron density as the basic variable, instead of the wave function in the conventional quantum theory of electronic structure, not only would it be advantageous to have the energy functional and derivatives defined for densities with fractional number of electrons, but it is also necessary to treat systems with a fractional number of electrons. To show this necessity, one only needs to consider the dissociation of $H_2^+$. In the exact quantum-mechanical theory, at the dissociation limit, $H(a)-H(b)^+$ and its nuclear permutation $H(a)^+ - H(b)$ are two degenerate states. Any linear combination of these two states is also a ground state of the dissociation, including the state $H^{0.5+} - H^{0.5+}$. At this limit, we have two independent systems each with fractional numbers of electrons. As in Fig. 1, which shows results, obtained from a typical ab initio calculation, we can see that the one-electron system becomes two half-electron fragments when $H_2^+$ becomes dissociated. The foregoing argument also holds true for the dissociation of any homonuclear diatomic molecular ion $A_2^+$.

In the title paper by Perdew, Parr, Levy and Balduz (PPLB) [4], DFT was extended to a fractional number of electrons based on the zero-temperature grand canonical ensemble theory. Note that the ensemble description of electronic structure problems goes back to Gyftopoulos and Hatsopoulos [5]. PPLB [4] showed that at zero temperature the ground state of a system with a noninteger numbers of electrons is an ensemble of two pure states with integer numbers of electrons and $E_{N+q}^{\mathrm{g}}$, the ground-state energy of a $(N+q)$-electron system, is a linear combination of $E_N^{\mathrm{g}}$ and $E_{N+1}^{\mathrm{g}}$, the ground-state energies of the corresponding $N$- and $(N+1)$-electron systems; namely

$$E_{N+q}^{\mathrm{g}} = (1-q)E_N^{\mathrm{g}} + qE_{N+1}^{\mathrm{g}} , \tag{3}$$

where $0 < q < 1$. The energy functional was also formulated in exactly the same form as in Eq. (2), with the

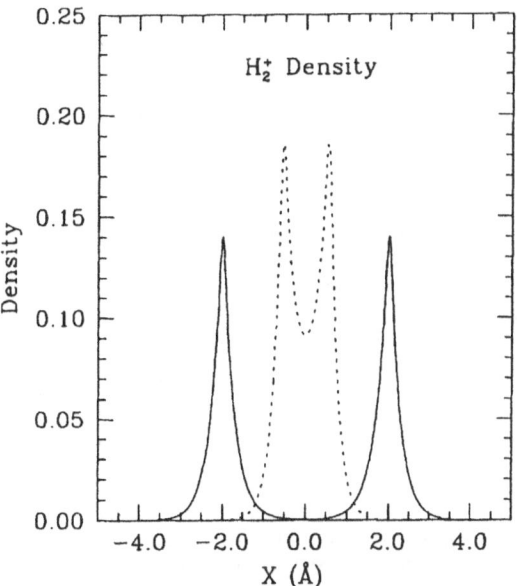

**Fig. 1.** Hartree–Fock electron density for $H_2^+$ along the bond axis with the bond middle point at zero. *Solid line* for $H_2^+$ with bond length 4.0 Å and dashed line for $H_2^+$ with bond length 1.1 Å

universal density functional defined for a fractional number of electrons by the Levy constrained search [3]

$$F[\rho_{N+q}] = \min_{\hat{\Gamma} \to \rho_{N+q}} \text{Tr}[\hat{\Gamma}(\hat{T} + \hat{V}_{ee})] \ , \qquad (4)$$

where $\hat{T}$ and $\hat{V}_{ee}$ are the kinetic and electron–electron repulsion operators, and $\hat{\Gamma}$ is any many-electron density operator that delivers the density $\rho_{N+q}$ with fractional numbers of electrons. The type of $\hat{\Gamma}$ allowed is a statistical mixture of $N$- and $(N+1)$-electron systems. The electron density entering the definition of the energy functional is also an ensemble sum

$$\rho_{N+q} = (1-q)\rho_N + q\rho_{N+1} \ . \qquad (5)$$

These are fundamental results in DFT. The density functional extended for a fractional number of electrons based on the ensemble approach gives the correct description for fractional electron number systems in the dissociation limit of $H_2^+$. A great challenge remains to construct an energy functional $E[\rho]$ that would have the correct behavior for fractional electron number systems, as described in Eq. (3).

Based on PPLB's work, Perdew [6] generalized the sum rule for the exchange-correlation hole to the fractional number of electrons, and Perdew and Levy [7] reached the conclusion that the local density approximation (LDA) and the generalized gradient approximation (GGA) energies were too low for systems with fractional numbers of electrons by analyzing this generalized sum rule.

Functionals with electron self-interaction excluded do very well in approximating the exact relation of Eq. (3) [7]. Based on the Eqs. (3) and (5), Zhang and Yang [8] have recently derived a new scaling relation necessary for the exchange-correlation functional to be self-interaction-error free.

$$E_{xc}[q\rho_1] = q^2 E_{xc}[\rho_1] \ , \qquad (6)$$

for $0 < q < 1$; however, all widely used GGA and hybrid GGA/exact-exchange functionals [9–12] fail to obey this relation. This failure leads to the consequence that even when the self-interaction error of a functional for systems with integer numbers of electrons is quite small, this error will significantly increase for fragments with noninteger numbers of electrons and so the resulting total energy can be too negative. This is demonstrated to be the main reason for the widely used functionals' failure to describe the dissociative behavior of some radicals correctly, such as $H_2^+, He_2^+, CO^-$; and it has been pointed out that the large self-interaction error for fractional numbers of electrons also accounts for the difficulty of approximate DFT to describe transition states of some chemical reactions and some charge-transfer complexes [8].

PPLB [4] have also given the physical meaning of the highest-occupied Kohn–Sham orbital energy. The Kohn–Sham orbital energies usually have no physical meanings; however, PPLB showed that for the exact density functional, the highest-occupied Kohn–Sham eigenvalue for all electron numbers $M$ between the integers $N-1$ and $N$ is the negative of the exact ionization energy of the $N$-electron system, i.e.,

$$\varepsilon_{max}^M = -I_N, \quad (N-1 < M < N) \qquad (7)$$

$$\varepsilon_{max}^M = -A_N, \quad (N < M < N+1) \ , \qquad (8)$$

where $I_N$ and $A_N$ are the ionization energy and electron affinity of the $N$-electron system. Recently, there have been some debates on this subject [7, 13, 14] and two new independent proofs have been provided [7, 15]. From an exact solution of a two-electron problem, Hooke's atom, this ionization energy theorem is confirmed [7].

Another major result from the PPLB paper [4] is the derivative discontinuities of the energy and functional, i.e., as $M$ increases by an integer $N$, the chemical potential and the Kohn–Sham potential both jump by a constant. The discontinuity of the chemical potential resolved a paradox of the chemical potential equalization principle [4, 16, 17]. The discontinuity of the exact Kohn–Sham potential, which recently has also been "exactly" constructed by Harbola [18], represents an important feature of the ground-state energy functional and is crucial for describing the band gap in an insulator or semiconductor [17, 19, 20]. Continuum approximations to exchange-correlation functionals, including all current widely used LDA and GGA functionals, fail to produce the correct derivative discontinuity. This failure accounts for the difficulty of LDA and GGA to describe the band gap correctly. Recently this property of derivative discontinuity has been used by Tozer and Handy [21] to design new exchange-correlation functionals.

In summary, the title paper is a seminal contribution that extended DFT to a fractional number of electrons. All the theorems, relations and properties for the exact DFT mentioned above, which are either presented in the PPLB paper or were developed later based on it, are not only very interesting in theory, but have also become more and more important from the practical viewpoint, especially in the development of a new generation of the exchange-correlation functional. Some well-known difficulties for the widely used functionals, such as the dissociate behavior of some radicals, the reaction barrier of some reactions and the band gap, are found to originate from the failure to satisfy these exact relations. This kind of difficulty for approximate functionals cannot be solved in the present framework of GGAs. We believe that the future breakthrough to further extend the applicability of density functional calculations needs a novel approach to construct the exchange-correlation functional that has the correct behavior for a fractional electron number, as shown by this title paper.

*Acknowledgement.* Financial support from the National Science Foundation is gratefully acknowledged.

## References

1. Hohenberg P, Kohn W (1964) Phys Rev B 136: 864
2. Kohn W, Sham L (1965) Phys Rev A 140: 1133
3. Levy M (1979) Proc Natl Acad Sci USA 76: 6062
4. Perdew JP, Parr RG, Levy M, Balduz Jr JL (1982) Phys Rev Lett 49: 1691
5. Gyftopoulos EP, Hatsopoulos GN (1968) Proc Natl Acad Sci (USA) 68: 786
6. Perdew JP (1985) In: Dreizler RM (ed) Density functional methods in physics. Plenum, New York, pp 265–308
7. Perdew JP, Levy M (1997) Phys Rev B 56: 16021
8. Zhang Y, Yang W (1998) J Chem Phys 109: 2604
9. Becke AD (1988) Phys Rev A 38: 3098
10. Lee C, Yang W, Parr RG (1988) Phys Rev B 37: 785
11. Becke AD (1993) J Chem Phys 98: 5648
12. Perdew JP, Burke K, Ernzerhof M (1996) Phys Rev Lett 77: 865
13. Kleinman L (1997) Phys Rev B 56: 12042
14. Kleinman L (1997) Phys Rev B 56: 16029
15. Casida ME (1999) Phys Rev B 59: 4694
16. Parr R, Yang W (1989) Density-functional theory of atoms and molecules. Oxford University Press, New York
17. Dreizler R, Gross E (1990) Density-functional theory. Springer Berlin Heidelberg New York
18. Harbola MK (1998) Phys Rev A 57: 4253
19. Perdew JP, Levy M (1983) Phys Rev Lett 51: 1884
20. Sham LJ, Schlueter M (1983) Phys Rev Lett 51: 1888
21. Tozer DJ, Handy NC (1998) J Chem Phys 108: 2545

Theor Chem Acc (2000) 103:349–352
DOI 10.1007/s002149900072

Theoretical
Chemistry Accounts
© Springer-Verlag 2000

*Perspective*

# Perspective on "Principles for a direct SCF approach to LCAO-MO ab initio calculations"

## Almlöf J, Faegri K Jr, Korsell K (1982) J Comput Chem 3:385–399

**Donald G. Truhlar**

Department of Chemistry and Supercomputer Institute, University of Minnesota, Minneapolis, MN 55455-0431, USA

Received: 24 February 1999 / Accepted: 1 July 1999 / Published online: 4 October 1999

**Abstract.** The direct self-consistent-field (SCF) method recalculates all two-electron integrals each time they are needed in an SCF calculation. This perspective article discusses how the original paper on direct SCF by Almlöf et al. developed the principles by which this could be made efficient and thereby provided an example of the semantic approach to computational chemistry in which algorithm development and coding are not compartmentalized.

**Key words:** Computational chemistry – Direct self-consistent field method – Molecular orbital – Quantum mechanics – Two-electron integrals

Modern computational chemistry has come a long way. In the early days the advances often came by applying computers syntactically to algorithms that provided literal translations of theories developed without regard to computing platforms; however, such an approach would now be very old-fashioned. A modern computation has several distinct elements:

1. Fundamental equations. The underlying fundamental equations, sometimes called the first principles, are well known: Hamilton's equations, the Maxwell equations, the Liouville equation, and the Schrödinger equation were all well known by the 1920s and well integrated by the 1930s [1]. Feynman's path integral approach to quantum mechanics and statistical mechanics [2] was well appreciated by the 1960s.

2. Theory. Usually we do not solve the fundamental equations directly. We use a theory, for example, Hartree–Fock theory [3], Møller-Plesset perturbation theory [4], coupled-cluster theory [5], Kohn's [6, 7], Newton's [8], or Schlessinger's [9] variational principle for scattering amplitudes, the quasiclassical trajectory method [10], the trajectory surface hopping method [11], classical S-matrix theory [12], the close-coupling approximation

[13–16], transition-state theory [17], variational-transition-state theory [18], self-consistent reaction-field theory [19], and so forth. Some of these theories date back to the 1930s, and new ones are continually being developed.

3. Algorithms. Most theories consist of differential or integral equations, sometimes integrodifferential equations, and these may be linear or nonlinear and are almost always multidimensional. Usually we cannot solve them by the classical, analytical methods of mathematics. Instead we reduce them to numerical algorithms, typically involving interpolation, extrapolation, quadrature, linear algebra, iteration, Monte Carlo sampling, and other general techniques. Prior to the days of automatic computing machines, this was the last step. For example, some of the early solutions to the Hartree–Fock equations were carried out using pencil and paper by Hartree's father, a retired sea captain, but he did not submit any articles about computational strategies to a scientific journal. Now, however, step 4 is a sine qua non, and the process of creating the program often entails as much research as developing the algorithm.

4. Software. To get numbers, one must convert the algorithm to a working computer program. In modern computational chemistry there are many opportunities for improved theories and improved algorithms, but more and more it is becoming clear that there is room for systematic, scientific progress in step 4 as well as in steps 2 and 3. In fact, treating step 4 as an afterthought no longer constitutes state-of-the-art work.

One of the first consequences of the increasing prominence of step 4 was to provide feedback to step 3 and promote the introduction of new algorithms. For example, the Monte Carlo method [20] of integration, unlike Gaussian quadrature or the Runge–Kutta algorithm, is hardly conceivable without computers. Other examples are the Cannon algorithm [21] or Strassen's algorithm [22, 23] for matrix multiplication or for the matrix multiplication steps in matrix inversion carried out by a matrix-times-vector formulation [22, 24]. In comparing the Monte Carlo, Cannon, and Strassen algorithms to their traditional counterparts, one is inevitably forced to consider the details of a given computer architecture, especially its ability to perform computational steps in parallel or in pipelines and also to consider any communication bottlenecks that may arise in trying to take advantage of such parallelism. This has led to the emergence of parallel computation as a well-recognized field of study, with its own internal rules and dynamics. Parallel computation is a subfield of the broader emerging field of scientific computation.

At what point did the field of computational chemistry emerge as a distinct subfield of scientific computation? I would say that this occurred gradually, but certain milestones can be recognized. These milestones occurred whenever chemists recognized a new tradeoff (such as computation speed versus communication speed) that needed to be considered in applying state-of-the-art computer hardware to solve more challenging chemical problems. Usually (but not always) the challenge comes from increasing the size of the system.

When one takes a step back to gain a broader view, one finds that chemists, more than other computational scientists, have been stymied by memory bottlenecks. These were severe already in the 1960s, and by the 1980s it was very clear that many large-system calculations that were affordable in terms of computer time were not doable simply for lack of storage space to hold all the intermediate results. This was true especially for electronic structure calculations, where the aggravating quantities were two-electron integrals. As a student one quickly learned that the number of two-electron integrals in a calculation with $N$ orbitals (this number scales as $N^4$ for large $N$). Then, if you knew how many integrals fitted on a 2400 foot tape, you could calculate how many tapes you needed. I have selected the paper by Almlöf, Faegri, and Korsell [25] for this perspective because it provided the first systematic attack on the memory bottleneck in computational chemistry and thereby, in my opinion, opened many people's eyes to the possibilities for progress if one rethinks the way one marshals the available computer resources to attack a computational problem. As one pursues this kind of thinking, the boundaries between theory, algorithms, and software actually begin to blur.

The desirability of blurring the boundary was made especially clear in the early days of vector computing by Kascic [26], who popularized the semantic approach to vectorization, which was the first form of parallel computing to benefit from systematic development. As Kascic pointed out, there are two ways to arrive at a vectorized algorithm. In the syntactic approach, we convert the physical problem into an algorithm (usually conceptualized as a scalar algorithm in those days), then we vectorize the algorithm. In the semantic approach, we proceed from the physical problem to the vectorized algorithm. It is like the difference between thinking in English and then translating into a foreign language or thinking directly in a foreign language. This merging of the design of step 3 (algorithm) and step 4 (computer code) is now well established for parallel computing. The paper by Almlöf et al. provides an analogous example of

the benefits of redesigning the algorithm with the hardware capabilities in mind, but here the issue is the balance of computer speed with memory capacity, whereas in much parallel computation design the issue is balancing computer speed with communication speed. These issues are of course related since, if one is willing to tolerate slower access to data, one can store more of it. The capacity of the arithmetic registers is very low, level-2 cache is faster but is still very limited, high-speed memory is slower but larger, disk has a big latency cost but is even larger, and tape libraries still provide the largest, slowest storage capacity.

The method proposed by Almlöf et al. is called the direct self-consistent-field (SCF) method. It is the basis of virtually every large SCF-type calculation (Hartree–Fock or density functional theory) that is run today. It was motivated by the unsymmetrical rate of advance in central processor (CPU) technology and storage technologies. Originally CPU time was expensive, so each two-electron integral was saved and reused every time it was needed. Eventually though, as system size increased, these integrals had to be "out of core," typically on disks, and the time and effort to store them and the load that their retrieval placed on input/output capacity became the real bottleneck. The proposed solution is simply to recalculate the integrals every time they are needed rather than to store them and recycle them. The obvious drawback is that one increases the computer time, and the key advances were to ameliorate this problem. There were five elements in this:

1. As soon as one decides not to store and retrieve the integrals, one recognizes that one need not compute, store, retrieve, and use them in an order that makes retrieval efficient; rather one can restructure the algorithm for better CPU efficiency [27]. In particular one can switch to an integral-driven order of events. When an integral is calculated it should be used to the maximum possible extent.

2. In some implementations, direct methods place a greater premium on the use of symmetry [28] to identify integrals with permuted indices that are identical to each other and rearranging the algorithm to take advantage of that. (In other implementations the direct method diminishes rather than increases the importance of symmetry.)

3. Direct methods place a renewed emphasis on decreasing the number of SCF cycles by improving the iterative strategy [29].

4. Emphasis on the new integral bottleneck made one realize that not all integrals need be calculated after all [30]. One can devise efficient upper-bound estimates, and these can be used to prescreen the integrals. If the inexpensive estimate of the upper bound indicates that the integral will be negligible, the more expensive evaluation of the integral itself is omitted. Eventually a considerable amount of sophistication can be built into the prescreening process [31, 32].

5. Sometimes one does not need an integral, not because it is small, but because it will only be multiplied by small numbers in the rest of the calculation. In SCF calculations, the integrals are multiplied by density matrix elements; thus the magnitudes of the corresponding

density matrix elements are also used in deciding whether to calculate a particular integral. In particular one calculates a bound on the error in a Fock matrix element due to not calculating an integral [25]. Density-weighted integral estimates reduce the complexity of integral evaluation from $O(N^4)$ to $O(N^2)$ [33]. At an even higher level of sophistication one recognizes that only those integrals are required that are related to significant changes in the density matrix from one iteration to the next.

All these kinds of considerations get amplified as one proceeds to use direct methods in a wider context, for example, for energy gradients and Hessians [34], for linear scaling algorithms [35–38], or for relativistic effects [39]. Direct SCF methods are very well suited for parallel computation [40–42]. In addition the whole process of rearranging the algorithms makes one rethink the semantics of the problem in a very fruitful and stimulating way; such benefits would be missed if one simply syntactically translated the traditional algorithm into computer code.

Two other interesting algorithmic consequences of direct methods for quantum chemistry may be mentioned. First, viewing electron repulsion integrals as only intermediates on the way to the final results of interest (Fock matrix, energy, forces) has had snowballing consequences in electronic structure theory. The work of White and co-workers on fast multipole methods [43, 44] and J-matrix engines [45] are examples of more drastically redefining the intermediates. Second, optimizing the algorithm to the computer architecture can be usefully generalized as follows: given particular amounts of memory and disk, with given access rates, design the algorithm with the shortest time-to-solution. This has been pursued in so-called semidirect methods for second-order Møller–Plesset theory [46].

The ideas behind direct SCF calculations are also echoed in various ways in more far-afield areas of research. A very literal analog comes in basis-set approaches to scattering calculations [6–9] where we have explored [47] the tradeoff between recalculating integrals versus writing them to disk. A less obvious analog occurs in an optimized quadrature scheme [48] we developed. The widespread use of Gaussian quadrature formulas owes to their general efficiency for a wide variety of integrands. In some cases though there are tremendous cost savings by using more specialized quadrature formulas derived specifically for problem-specific functions that occur in a large number of integrals in a specific calculation. Modern computers can easily store large numbers of function-specific quadrature weights. If this allows one to decrease the number of quadrature points by a factor of 2 or 3 per dimension it can make not doable problems doable [49], for a cost of a few thousand storage locations. This would have been a considerable cost on the computer I used as a graduate student (total storage = 32K words), but is hardly worth a second thought now.

I hope this perspective conveys some of the benefits that can accrue from taking a more holistic view of the computational process and letting hardware considerations become more strongly coupled to algorithm design. This is the critical step in developing computational chemistry into a master discipline that integrates theoretical chemistry with scientific computation. I believe that the paper by Almlöf, Faegri, and Korsell provides a classic example of this kind of development.

*Acknowledgements.* I am grateful to Matt Challacombe, Martin Head-Gordon, Knut Faegri, Eric Schwegler, David Schwenke, and Peter Taylor for comments on the manuscript.

## References

1. Tolman RC (1938) The principles of statistical mechanics. Oxford University Press, New York
2. Feynmann RP (1972) Statistical mechanics. Addison-Wesley, Reading
3. Fock V (1930) Z Phys 62: 795
4. Møller C, Plesset MS (1934) Phys Rev 46: 618
5. Bartlett RJ, Dykstra CE, Paldus J (1984) In: Dykstra C (ed) Advanced theories and computational approaches to the electronic structure of molecules, Reidel, Dordrecht, p 127
6. Kohn W (1948) Phys Rev 74: 1763
7. Truhlar DG, Abdallah J Jr, Smith RL (1974) Adv Chem Phys 25: 211
8. Newton RG (1966) Scattering theory of waves and particles, McGraw-Hill, New York
9. Schlessinger L (1968) Phys Rev 167: 1411
10. Karplus M, Porter RN, Sharma RD (1965) J Chem Phys 43: 3259
11. Tully JC (1998) In: Thompson DL (ed) Modern methods for multidimensional dynamics computations in chemistry. World Scientific, Singapore, p 34
12. Miller WH (1974) Adv Chem Phys 25: 69
13. Massey HSW, Smith RA (1933) Proc R Soc London Ser A142: 142
14. Wheeler JA (1937) Phys Rev 52: 1083
15. Lane NF, Geltman S (1967) Phys Rev 160: 53
16. Allison AC, Dalgarno A (1967) Proc Phys Soc 90: 609
17. Eyring H (1935) J Chem Phys 3: 107
18. Truhlar DG, Garrett BC (1984) Annu Rev Phys Chem 35: 159
19. Rinaldi D, Rivail JL (1973) Theor Chim Acta 32: 57
20. Siepmann JI (1999) Adv Chem Phys 105: 1
21. Cannon L (1969) Ph D thesis. Montana State University, Bozeman
22. Strassen V (1969) Numer Math 13: 354
23. Bailey D (1988) SIAM J Sci Stat Comput 9: 603
24. Press WH, Teukolsky SA, Vetterling WT, Flannery BP (1992) Numerical recipes in FORTRAN, 2nd edn. Cambridge University Press, Cambridge, p 95
25. Almlöf J, Faegri K Jr, Korsell K (1982) J Comput Chem 3: 385
26. Kascic MJ Jr (1983) Semantic and syntactic vectorization: whence cometh intelligence in supercomputing? Summer Computer Simulation Conference, Vancouver
27. Almlöf J, Faegri K Jr (1990) In: Carbo R, Klobukowski M (eds) Self-consistent field: theory and applications. Elsevier, Amsterdam, p 195
28. Almlöf JE (1997) Theor Chem Acc 97: 10
29. Pulay P (1980) Chem Phys Lett 73: 393
30. Almlöf J, Taylor PR (1984) In: Dykstra C (ed) Advanced theories and computational approaches to the electronic structure of molecules. Reidel, Dordrecht, p 107
31. Häser M, Ahlrichs R (1989) J Comput Chem 10: 104
32. Ruud K, Jonsson D, Norman P, Agren H, Save T, Jensen HJA, Dahle P, Helgaker T (1998) J Chem Phys 108: 7973
33. Strout DL, Scuseria GE (1995) J Chem Phys 102: 8448
34. Helgaker T, Jørgensen P (1992) In: Wilson S, Diercksen GHF (eds) Methods in computational molecular physics. Reidel, Dordrecht, p 353
35. Challacombe M, Schwegler E, Almlöf J (1996) J Chem Phys 104: 4685

352

36. Yang W, Perez-Jorda JM (1998) In: Schleyer PvR, Allinger, NL, Clark T, Gasteiger J, Kollman PA, Schaefer HF III, Schreiner PR (eds) Encyclopedia of computational chemistry. Wiley, Chichester, p 1496
37. White CA, Johnson BG, Gill PMW, Head-Gordon M (1994) J Chem Phys 230: 8
38. Ochenfeld C, White CA, Head-Gordon M (1998) J Chem Phys 109: 1663
39. Laerdahl JK, Save T, Faegri K Jr (1997) Theor Chem Acc 97: 177
40. Lüthi HP, Almlöf J (1993) Theor Chim Acta 84: 289
41. Lüthi HP, Mertz JE, Feynmann MW, Almlöf JE (1992) J Comput Chem 113: 160
42. Petterson LGM, Faxen T (1993) Theor Chim Acta 85: 345
43. White CA, Johnson BG, Gill PMW, Head-Gordon M (1994) Chem Phys Lett 230: 8
44. White CA, Johnson BG, Gill PMW, Head-Gordon M (1996) Chem Phys Lett 253: 268
45. White CA, Head-Gordon M (1996) J Chem Phys 104: 2620
46. Frisch MJ, Head-Gordon M, Pople JA (1990) Chem Phys Lett 166: 281
47. Schwenke DW, Mladenovic M, Zhao M, Truhlar DG, Sun Y, Kouri DJ (1989) In: Laganà A (ed) Supercomputer algorithms for reactivity, dynamics, and kinetics of small molecules. Kluwer, Dordrecht, p 191
48. Schwenke DW, Truhlar DG (1984) Comput Phys Commun 34: 57
49. Schwenke DW, Truhlar DG (1985) In: Numrich RW (ed) Supercomputer applications. Plenum, New York, p 215

Theor Chem Acc (2000) 103:353–360
DOI 10.1007/s002149900093

Theoretical
Chemistry Accounts
© Springer-Verlag 2000

*Perspective*

# Perspective on "Density functional approach to the frontier-electron theory of chemical reactivity"

## Parr RG, Yang W (1984) J Am Chem Soc 106: 4049–4050

**Paul W. Ayers[1], Mel Levy[2]**

[1] Department of Chemistry, The University of North Carolina at Chapel Hill, Chapel Hill, NC 27599, USA
[2] Department of Chemistry and Quantum Theory Group, Tulane University, New Orleans, LA 70118, USA

Received: 3 June 1999 / Accepted: 16 September 1999 / Published online: 15 December 1999

**Abstract.** The Fukui function, $f(\vec{r})$, was proposed as a tool for deducing the relative reactivity of different positions in a molecule by Parr and Yang in 1984. Herein we sketch the theory of the Fukui function, with special emphasis on its logical motivation, interpretation, qualitative characteristics, and practical computation. We conclude with some words about the Fukui function's extensions, limitations, and importance.

**Key words:** Fukui function – Site reactivity – Chemical potential – Hardness-softness

## 1 Introduction

Chemistry is the study of how atoms and molecules act when exposed to external stimuli (other electronic systems, lasers, electrodes, Bunsen burners, etc.). Theoretical chemistry is tasked with both the explanation of existing chemical data and the prediction of new results. Even the former of these tasks is formidable; perusal of a chemical catalog or one of the many handbooks of chemistry will convince one that separate theoretical treatment of each cataloged molecule and chemical reaction is neither possible nor desirable. What we desire are "primitive patterns of understanding" [1] – general principles that both impart order to the chemical storehouse of knowledge and help to predict which molecules and chemical reactions are the most likely to achieve some desired goal.

One such principle is provided by the Fukui function, proposed by Parr and Yang in 1984 as a tool for understanding and predicting the relative reactivity of

different sites in a molecule [2]. This paper provides a perspective on the Fukui function through the lens of 15 years of subsequent developments.

## 2 Theoretical background

We start by asking the following question. Given a molecule, A, how does it react with another molecule, B? When addressing this problem, it is convenient to work with the three-dimensional ground-state electron density, $\rho(\vec{r})$. With the use of $\rho(\vec{r})$, the total energy is given through

$$E_{v_0}[\rho] = F[\rho] + \int \rho(\vec{r})v_0(\vec{r})\mathrm{d}\vec{r} \ , \tag{1}$$

where $F$ is either the original Hohenberg–Kohn functional for nondegenerate ground states or its later extension to degenerate ground states. $F$ is the sum of the kinetic energy functional, $T$, and the electron–electron repulsion energy functional, $V_{ee}$ [3, 4].

The electron density not only determines the total energy, but also how the total energy changes as the number of electrons, $N$, and the external potential, $v_0(\vec{r})$, change. To see this, start with the expression for the first-order change in the energy [5]:

$$\mathrm{d}E[N, v_0(\vec{r})] = \left(\frac{\partial E}{\partial N}\right)_{v_0(\vec{r})} \mathrm{d}N + \int \left(\frac{\delta E}{\delta v_0(\vec{r})}\right)_N \delta v_0(\vec{r})\mathrm{d}\vec{r} \ . \tag{2}$$

Using Eq. (1) to derive

$$\left(\frac{\delta E}{\delta v_0(\vec{r})}\right)_N = \rho(\vec{r}) \ , \tag{3}$$

defining the chemical potential through

$$\mu \equiv \left(\frac{\partial E}{\partial N}\right)_{v_0(\vec{r})} \ , \tag{4}$$

*Correspondence to*: P. W. Ayers and M. Levy

and substituting Eqs. (3) and (4) into Eq. (2) gives the working equation [3]

$$dE[N, v_0(\vec{r})] = \mu \, dN + \int \rho(\vec{r}) \delta v_0(\vec{r}) d\vec{r} \ . \tag{5}$$

It has been shown by Perdew et al. [6] that $\mu$ is discontinuous at integer values of $N$.

The identification of $(\partial E/\partial N)_{v_0(\vec{r})}$ as an electronic chemical potential gives several insights into reactivity. First of all, $\mu$ can be regarded as measuring the escaping tendency of electrons from a system; electrons flow from places with high chemical potential to places with low chemical potential until $\mu$ is constant throughout the molecule [3]. So $\mu$ is related to the negative of the electronegativity [3, 7]. (Indeed, Mulliken's electronegativity scale is simply the finite difference approximation to $-\mu$ [7, 8].) The principle of electronegativity equalization follows from the fact that $\mu$ is a global property of the molecule [7, 9–12]. The second term of Eq. (5) is just the Hellmann-Feynman [13] expression in density functional theory (DFT).

To get more detailed information about reactivity we must consider the second-order change in the energy due to changes in electron number and external potential. These effects are carried in the first-order changes of the chemical potential and the density [14]:

$$d\mu[N, v_0(\vec{r})] = \left(\frac{\partial \mu}{\partial N}\right)_{v_0(\vec{r})} dN + \int \left(\frac{\delta \mu}{\delta v_0(\vec{r})}\right)_N \delta v_0(\vec{r}) d\vec{r} \tag{6}$$

$$\delta\rho[N, v_0(\vec{r}); \vec{r}] = \left(\frac{\partial \rho(\vec{r})}{\partial N}\right)_{v_0(\vec{r})} dN$$
$$+ \int \left(\frac{\delta \rho(\vec{r})}{\delta v_0(\vec{r}')}\right)_N \delta v_0(\vec{r}') d\vec{r}' \ . \tag{7}$$

The first term in Eq. (6) has been identified as the absolute hardness, $\eta$ [15]:

$$\eta \equiv \left(\frac{\partial \mu}{\partial N}\right)_{v_0(\vec{r})} = \left(\frac{\partial^2 E}{\partial N^2}\right)_{v_0(\vec{r})} \ . \tag{8}$$

This identification has paved the way for enhanced understanding of the hard/soft acid/base principle [15–18] and the maximum hardness principle [19–23]. The second term in Eq. (7) is called the linear response function [3, 24, 25]

$$\omega(\vec{r}, \vec{r}') \equiv \left(\frac{\delta \rho(\vec{r})}{\delta v_0(\vec{r}')}\right)_N \ ; \tag{9}$$

$\omega(\vec{r}, \vec{r}')$ measures the way the density changes when the external potential changes.

The Maxwell relation for Eq. (5) is used to define the Fukui function [2]:

$$f(\vec{r}) \equiv \left(\frac{\partial \rho(\vec{r})}{\partial N}\right)_{v_0(\vec{r})} = \left(\frac{\delta \mu}{\delta v_0(\vec{r})}\right)_N \ . \tag{10}$$

Substitution of Eqs. (8)–(10) simplifies Eqs. (6) and (7), yielding the working equations

$$d\mu[N, v_0(\vec{r})] = \eta \, dN + \int f(\vec{r}) \delta v_0(\vec{r}) d\vec{r} \tag{11}$$

$$\delta\rho[N, v_0(\vec{r}); \vec{r}] = f(\vec{r}) dN + \int \omega(\vec{r}, \vec{r}') \delta v_0(\vec{r}') d\vec{r}' \ . \tag{12}$$

## 3 Interpreting the Fukui function

We return to a consideration of the chemical reaction

$$A + B \rightarrow A - B \ . \tag{13}$$

Equations (5), (11), and (12) do much to explain this reaction. For instance, consideration of the chemical potentials $\mu_A$ and $\mu_B$ and the hardnesses $\eta_A$ and $\eta_B$ not only allows one to predict whether or not reaction Eq. (13) will occur, but if the reaction occurs it allows one to estimate the binding energy of the product molecule, A-B [3, 12, 15, 17]. Consideration of the linear response function for the reactant molecules allows one to approximate the change in the density associated with reaction Eq. (13). However, because $\mu$ and $\eta$ are global constants they contain no information on site reactivity. It is precisely information about site reactivity that is contained in the Fukui function.

Consider the case where the chemical potential of reactant B is much higher than that of reactant A. Then, in accord with the principle of electronegativity equalization, molecule B transfers electrons to molecule A. Hence, for reaction Eq. (13), $\Delta N_A > 0$ and $\Delta N_B < 0$. Now, consider the change in $\rho_A$ due to an increase in the number of electrons, $N_A$:

$$f_A^+(\vec{r}) \equiv \left(\frac{\partial \rho_A(\vec{r})}{\partial N_A}\right)_{v_{0,A}(\vec{r})}^+ \ . \tag{14}$$

In Eq. (14) the superscript "+" on the derivative indicates that the derivative is taken from above. We conclude that molecule A readily accepts electrons into regions where $f_A^+(\vec{r})$ is large.

Now consider the change in $\rho_B$ upon loss of electrons:

$$f_B^-(\vec{r}) \equiv \left(\frac{\partial \rho_B(\vec{r})}{\partial N_B}\right)_{v_{0,B}(\vec{r})}^- \ . \tag{15}$$

We conclude that molecule B readily donates electrons from regions where $f_B^-(\vec{r})$ is large.

We expect that when the product, A-B, forms, B will have donated electrons to molecule A from those regions where $f_B^-(\vec{r})$ is large and A will have accepted electrons from molecule B into regions where $f_A^+(\vec{r})$ is large. Accordingly, we expect chemical bonds to form between atom(s) in fragment A where $f_A^+(\vec{r})$ was large and atom(s) in fragment B where $f_B^-(\vec{r})$ was large.

What happens if the chemical potentials of molecules A and B are similar, so that

$$\Delta N_A \approx \Delta N_B \approx 0 \ . \tag{16}$$

In this case we suspect that reactant A will both accept electrons from and donate electrons to reactant B. So here we need the true derivative (as opposed to the one-sided derivatives of Eqs. 14 and 15):

$$f^0(\vec{r}) \equiv \lim_{\varepsilon \to 0} \frac{\rho[N + \varepsilon, v_0(\vec{r})] - \rho[N - \varepsilon, v_0(\vec{r})]}{2\varepsilon} \; . \qquad (17)$$

Unfortunately, since $\rho$ has slope discontinuities at integer numbers of electrons [2, 6, 26–28] (see Eq. 29), the limit in Eq. (17) does not exist. But defining $f^0(\vec{r})$ as the average of $f^+(\vec{r})$ and $f^-(\vec{r})$ makes good sense, as is seen from the following analysis:

$$f^0(\vec{r}) \equiv \lim_{\varepsilon \to 0} \frac{\rho[N + \varepsilon, v_0(\vec{r})] - \rho[N - \varepsilon, v_0(\vec{r})]}{2\varepsilon}$$

$$= \lim_{\varepsilon \to 0} \left\{ \begin{array}{c} \dfrac{(\rho[N + \varepsilon, v_0(\vec{r})] - \rho[N, v_0(\vec{r})])}{2\varepsilon} \\ + \dfrac{(\rho[N, v_0(\vec{r})] - \rho[N - \varepsilon, v_0(\vec{r})])}{2\varepsilon} \end{array} \right\}$$

$$= \lim_{\varepsilon \to 0} \frac{(\rho[N + \varepsilon, v_0(\vec{r})] - \rho[N, v_0(\vec{r})])}{2\varepsilon}$$

$$+ \lim_{\varepsilon \to 0} \frac{(\rho[N, v_0(\vec{r})] - \rho[N - \varepsilon, v_0(\vec{r})])}{2\varepsilon} \qquad (18)$$

$$= \frac{f^+(\vec{r}) + f^-(\vec{r})}{2} \; .$$

So, in this case we use

$$f_A^0(\vec{r}) \equiv \frac{f_A^-(\vec{r}) + f_A^+(\vec{r})}{2} \qquad (19)$$

and

$$f_B^0(\vec{r}) \equiv \frac{f_B^-(\vec{r}) + f_B^+(\vec{r})}{2} \qquad (20)$$

to predict site reactivity. Extension of our previous result leads us to conclude that for reactions in which $\mu_A \approx \mu_B$, bonds form between atom(s) in fragment A where $f_A^0(\vec{r})$ was large and atom(s) in fragment B where $f_B^0(\vec{r})$ was large [29].[1]

In summary, $f^+(\vec{r})$ (Eq. 14) measures the reactivity of a site in a molecule relative to nucleophilic attack; $f^-(\vec{r})$ (Eq. 15) measures the reactivity of a site relative to electrophilic attack; and $f^0(\vec{r})$ (Eqs. 19, 20) measures the reactivity of a site relative to neutral (or radical) attack [2].

We note that one can obtain the same conclusions when one considers [2]

---

[1]Berkowitz shows that during chemical reaction (Eq. 13), the amount of charge transfer between the two molecular fragments A and B increases as

$$J_f \equiv \iint \frac{f_A(\vec{r}) f_B(\vec{r}')}{|\vec{r} - \vec{r}'|} \, d\vec{r} \, d\vec{r}'$$

increases. This result suggests that the Fukui potentials,

$$v_{f_{A/B}}(\vec{r}) \equiv \int \frac{f_{A/B}(\vec{r}')}{|\vec{r} - \vec{r}'|} d\vec{r}' \; ,$$

indicate favorable orientations for the reacting fragments. Since the Fukui potentials are large where $f_{A/B}$ is large, Berkowitz's conclusions agree with those in the text

$$f(\vec{r}) \equiv \left( \frac{\delta\mu}{\delta v_0(\vec{r})} \right)_N \; . \qquad (21)$$

In this formulation the key realization is that $d\mu$ represents the extent of a chemical reaction and hence that "$|d\mu|$ big is good" [2, 3]. One-sided functional derivatives are once again required since $\mu[N, v_0(\vec{r})]$ is discontinuous at integer values of N [6, 26–28]. Use of the definitions

$$\mu^{+/-} = \left( \frac{\partial E}{\partial N} \right)_{v_0(\vec{r})}^{+/-} \qquad (22)$$

$$f_A^+(\vec{r}) \equiv \left( \frac{\delta\mu_A^+}{\delta v_{0,A}(\vec{r})} \right)_{N_A} \qquad (23)$$

$$f_B^-(\vec{r}) \equiv \left( \frac{\delta\mu_B^-}{\delta v_{0,B}(\vec{r})} \right)_{N_B} \qquad (24)$$

and the "$|d\mu|$ big is good" rule leads to the recovery of our previous results [2].

## 4 Qualitative properties of the Fukui function

Before attempting to accurately compute a quantity it is useful to know something about its qualitative behavior. Here we consider some results of this type for the Fukui function.

1. Normalization. The Fukui function is normalized to unity [3]:

$$\int f(\vec{r}) d\vec{r} = 1 \; . \qquad (25)$$

In proving this result we introduce the shape function, $\sigma(\vec{r}) \equiv \rho(\vec{r})/N$; note that $\sigma$ is normalized to unity [30, 31]. Then

$$\int f(\vec{r}) d\vec{r} = \int \left( \frac{\partial\rho(\vec{r})}{\partial N} \right)_{v_0(\vec{r})} d\vec{r} = \left( \frac{\partial[\int N\sigma(\vec{r}) d\vec{r}]}{\partial N} \right)_{v_0(\vec{r})}$$

$$= \left( \frac{\partial[N \int \sigma(\vec{r}) d\vec{r}]}{\partial N} \right)_{v_0(\vec{r})} = 1 \; . \qquad (26)$$

2. Cusp condition. In a molecule, the density has cusps at each nuclear position, $\vec{R}_\alpha$. These cusps satisfy the relation

$$\lim_{|\vec{r} - \vec{R}_\alpha| \to 0} \left( \frac{(\vec{r} - \vec{R}_\alpha)}{|\vec{r} - \vec{R}_\alpha|} \frac{\vec{\nabla}_{\vec{r} - \vec{R}_\alpha} \rho(\vec{r})}{\rho(\vec{r})} \right) = -2Z_\alpha \; , \qquad (27)$$

where $Z_\alpha$ is the nuclear charge of the nucleus at the point $\vec{R}_\alpha$ and the notation indicates that the initial point of the vectors is $\vec{R}_\alpha$ [32–35]. Chattaraj et al. [36] showed that Eq. (27) implies that the Fukui function also satisfies the cusp condition

$$\lim_{|\vec{r} - \vec{R}_\alpha| \to 0} \left( \frac{(\vec{r} - \vec{R}_\alpha)}{|\vec{r} - \vec{R}_\alpha|} \frac{\vec{\nabla}_{\vec{r} - \vec{R}_\alpha} f(\vec{r})}{f(\vec{r})} \right) = -2Z_\alpha \; . \qquad (28)$$

3. Asymptotic decay. If one uses the zero-temperature limit of the grand canonical ensemble to interpolate between integer numbers of electrons, then

$$\rho[N + \varepsilon, v_0; \vec{r}] = (1 - \varepsilon)\rho[N, v_0; \vec{r}] + \varepsilon\rho[N + 1, v_0; \vec{r}] ,$$
(29)

where $\varepsilon$ is between zero and one, inclusive [6, 26–28]. Then

$$f^+(\vec{r}) \equiv \left(\frac{\partial\rho(\vec{r})}{\partial N}\right)_{v_0}^+ = \lim_{\varepsilon\to 0^+}\left\{\frac{\rho[N + \varepsilon, v_0; \vec{r}] - \rho[N, v_0; \vec{r}]}{\varepsilon}\right\}$$

$$= \rho[N + 1, v_0; \vec{r}] - \rho[N, v_0; \vec{r}]$$
(30)

$$f^-(\vec{r}) \equiv \left(\frac{\partial\rho(\vec{r})}{\partial N}\right)_{v_0}^- = \lim_{\varepsilon\to 0^+}\left\{\frac{\rho[N, v_0; \vec{r}] - \rho[N - \varepsilon, v_0; \vec{r}]}{\varepsilon}\right\}$$

$$= \rho[N, v_0; \vec{r}] - \rho[N - 1, v_0; \vec{r}] .$$
(31)

Equations (30) and (31) are remarkable since they indicate that the finite difference approximation to Eqs. (14) and (15) with $\Delta N \leq 1$ is exact (for exact DFT).

Far from any finite system, the electron density decays exponentially according to

$$\rho(\vec{r}) \xrightarrow{\text{big }|\vec{r}|} Ae^{-\sqrt{8\text{IP}}\,r} ,$$
(32)

where $A$ is a constant of proportionality and IP is the ionization potential of the system [37–44]. Due to the empirically observed convexity of $E$ versus $N$ [3, 6],

$$\text{IP}[N, v_0(\vec{r})] > \text{IP}[N + 1, v_0(\vec{r})]$$
(33)

as long as the $N + 1$st electron is bound. When the $N + 1$st electron is unbound, $\rho[N, v_0(\vec{r})] = \rho[N + 1, v_0(\vec{r})]$ and hence the Fukui function is a generalized function which is zero everywhere but which is normalized to 1 (as the $\alpha \to 0^+$ limit of the function $\frac{\alpha^3}{8\pi} \cdot e^{-\alpha r}$).

Equations (30)–(33) reveal that at long range the dominant contribution to the Fukui function is from the density with the greatest number of electrons (as the contribution from the other density decays exponentially faster). Hence

$$f^+(\vec{r}) \xrightarrow{\text{big }|\vec{r}|} A[N + 1, v_0]e^{-\sqrt{8\text{EA}}\,r}$$
(34)

and

$$f^-(\vec{r}) \xrightarrow{\text{big }|\vec{r}|} A[N, v_0]e^{-\sqrt{8\text{IP}}\,r} ,$$
(35)

where EA and IP denote the electron affinity and ionization potential of the $N$-electron system. Restating Eqs. (34) and (35) and including higher-order terms [37–44] one finds that

$$\left[\frac{\partial\ln(f^+(\vec{r}))}{\partial r}\right] \xrightarrow{\text{big }r} -\sqrt{8\,\text{EA}} + 2\left(\frac{(Z_{\text{total}} - N)}{\sqrt{2\,\text{EA}}} - 1\right)\left(\frac{1}{r}\right)$$

$$+ \left(\text{higher powers of }\frac{1}{r}\right)$$
(36)

$$\left[\frac{\partial\ln(f^-(\vec{r}))}{\partial r}\right]$$

$$\xrightarrow{\text{big }r} -\sqrt{8\,\text{IP}} + 2\left(\frac{[Z_{\text{total}} - (N - 1)]}{\sqrt{2\,\text{IP}}} - 1\right)\left(\frac{1}{r}\right)$$

$$+ \left(\text{higher powers of }\frac{1}{r}\right) ,$$
(37)

where $Z_{\text{total}}$ is the sum of all the nuclear charges.

It should be noted that Eqs. (30) and (31) also provide an alternative derivation of the cusp condition for the Fukui function. This can be seen by substituting Eqs. (30) and (31) into Eq. (28) and using Eq. (27) to simplify the result.

## 5 Computing the Fukui function

In Sect. 3 we showed that the Fukui function should be useful for predicting site reactivity. How do we calculate it? One possibility is to use Eqs. (30) and (31). The first problem with doing this is that it requires us to do ab initio calculations not only for the neutral system, but also for the cation and the anion. Having to perform three calculations is difficult enough; the fact that one of these calculations is for an anion complicates matters still further. Another problem is that Eq. (29) is valid only for the exact theory; approximate density functionals do not satisfy this condition and hence Eqs. (30) and (31) are only approximately true for DFT calculations performed with approximate density functionals. Indeed, one sometimes obtains qualitatively incorrect answers if Eq. (30) is applied when Eq. (29) is not valid [45].

A different approach is to perform a gradient expansion of the Fukui function. For an atom, Chattaraj et al. proposed the expansion

$$f(\vec{r}) = \frac{\rho(\vec{r})}{N} + \frac{\alpha}{N\rho^{\frac{2}{3}}(0)}\left\{\left[\left(\frac{\rho(0)}{\rho(\vec{r})}\right)^{\frac{2}{3}} - 1\right]\nabla^2\rho(\vec{r})\right.$$

$$\left. - \frac{2}{3}\left(\frac{\rho(0)}{\rho(\vec{r})}\right)^{\frac{2}{3}}\frac{\vec{\nabla}\rho(\vec{r})\,\vec{\nabla}\rho(\vec{r})}{\rho(\vec{r})}\right\} ,$$
(38)

where $\rho(0)$ is the value of the density at the nucleus and $\alpha$ is an empirical parameter [36, 46]. Unfortunately, expansions of this type only produce one Fukui function: $f^+(\vec{r})$ and $f^-(\vec{r})$ are not found [36, 46].

A variational method for obtaining the Fukui function has been devised [47]. Define the hardness kernel as [3, 24]

$$\eta(\vec{r}, \vec{r}') = \frac{\delta^2 E_{v_0}[\rho]}{\delta\rho(\vec{r})\delta\rho(\vec{r}')} = \frac{\delta^2 F[\rho]}{\delta\rho(\vec{r})\delta\rho(\vec{r}')} = \frac{\delta(\mu - v_0(\vec{r}))}{\delta\rho(\vec{r}')} .$$
(39)

Then minimize the functional

$$\eta[g] \equiv \int\int g(\vec{r})g(\vec{r}')\eta(\vec{r}, \vec{r}')d\vec{r}\,d\vec{r}'$$
(40)

with respect to $g(\vec{r})$ which obey the normalization constraint

$$\int g(\vec{r})d\vec{r} = 1 .$$
(41)

The Fukui function, $f(\vec{r})$, is the function which minimizes Eq. (40) and $\eta[f] = \eta$, the absolute hardness from Eq. (8) [47]. If one enforces the normalization constraint (Eq. 41) with a Lagrange multiplier, $\lambda$, one finds that $\lambda = 2\eta$ and that [47, 48]

$$\eta = \int f(\vec{r})\eta(\vec{r},\vec{r}')\mathrm{d}\vec{r}' \ . \tag{42}$$

Equation (42) is remarkable because the integration is only over one set of coordinates.

While the gradient expansion approach (Eq. 38) does not give results for the one-sided Fukui functions, $f^+(\vec{r})$ and $f^-(\vec{r})$ may be obtained from the variational approach by using the one-sided hardness kernels, $\eta^{+/-}(\vec{r},\vec{r}')$, in Eqs. (40) and (42) [47]. If we use the one-sided hardness kernels obtained from the zero-temperature limit of the grand canonical ensemble [6], then $\eta^+ = \eta^- = 0$; hence, the one-sided hardness kernels have no inverse.

It seems that this variational method should be the method of choice for generating the Fukui function. Unfortunately, accurate determination of the hardness kernel is complicated by the lack of an accurate explicit kinetic energy functional, $T[\rho]$. Nonetheless, Eq. (40) has been applied to the Hückel model [49] and Eq. (42) has been applied to the Thomas–Fermi–Dirac–Weizsäcker approximation of $F[\rho]$ [47].

Since most modern DFT calculations use the Kohn–Sham method, it seems desirable to use the information available from a solution of the Kohn–Sham equations to determine the Fukui function. To this end, Cohen and coworkers [50–52] showed that one can relate the Fukui function to the frontier orbital densities of Kohn–Sham theory through

$$f^{+/-}(\vec{r}) = \int \left[ \left( \frac{\delta v_0(\vec{r}')}{\delta v_{\mathrm{eff}}(\vec{r})} \right)_\mu \right]^{-1}$$
$$\times \sum_{\mathrm{spin}} \left| \phi^{\mathrm{LUMO/HOMO}}(\vec{r}',s) \right|^2 \mathrm{d}\vec{r}' \ . \tag{43}$$

Here $[(\delta v_0(\vec{r}')/\delta v_{\mathrm{eff}}(\vec{r}))_\mu]^{-1}$ is the inverse of the transpose of the potential–potential response function, $(\delta v_0(\vec{r})/\delta v_{\mathrm{eff}}(\vec{r}'))_\mu$, $v_{\mathrm{eff}}$ is the Kohn–Sham effective potential, and $\phi^{\mathrm{LUMO}}$ ($\phi^{\mathrm{HOMO}}$) is the lowest unoccupied (highest occupied) Kohn–Sham orbital. Fortunately, one can express $(\delta v_0(\vec{r})/\delta v_{\mathrm{eff}}(\vec{r}'))_\mu$ in terms of quantities from a Kohn–Sham DFT equation:

$$\left( \frac{\delta v_0(\vec{r})}{\delta v_{\mathrm{eff}}(\vec{r}')} \right)_\mu$$
$$= \delta(\vec{r}' - \vec{r}) - \int \left[ \left( \frac{\delta \rho(\vec{r}'')}{\delta v_{\mathrm{eff}}(\vec{r})} \right)_\mu \cdot \left( \frac{1}{|\vec{r}'' - \vec{r}'|} + \frac{\delta v_{\mathrm{xc}}(\vec{r}')}{\delta \rho(\vec{r}'')} \right) \right] \mathrm{d}\vec{r}'', \tag{44}$$

where $v_{\mathrm{xc}}$ is the exchange–correlation potential [53]. Ignoring the second term in Eq. (44), recovers the "frozen orbital" approximation of Parr and Yang [2]. Accordingly, the second term in Eq. (44) must represent the effects of orbital relaxation.

We encounter problems when we decide to compute the effects of orbital relaxation. Consider, for a moment, just the problems associated with computing $\delta v_{\mathrm{xc}}(\vec{r}')/\delta \rho(\vec{r}'')$. While $\delta v_{\mathrm{xc}}(\vec{r}')/\delta \rho(\vec{r}'')$ is indeed readily computed from a given approximation of $E_{\mathrm{xc}}$, it is

probably not accurately modeled by approximate functionals which were designed to reproduce energetic data (and not the exchange–correlation potential, much less the functional derivative thereof).

Another approach to computing the Fukui function was provided by Yang et al. [12, 54]:

$$f^+(\vec{r}) = \left| \phi_{N+1}(\vec{r}) \right|^2 + \sum_{i=1}^{N} \left( \frac{\partial |\phi_i(\vec{r})|^2}{\partial N} \right)^+_{v_0(\vec{r})} \tag{45}$$

$$f^-(\vec{r}) = \left| \phi_N(\vec{r}) \right|^2 + \sum_{i=1}^{N-1} \left( \frac{\partial |\phi_i(\vec{r})|^2}{\partial N} \right)^-_{v_0(\vec{r})} \ . \tag{46}$$

Here $N$ is the number of electrons in the system and the $\phi_i$ are Kohn–Sham spin orbitals. It is apparent that if one neglects the relaxation of the core orbitals (the frozen core approximation), Eqs. (45) and (46) reduce to

$$f^+(\vec{r}) \approx \rho_{\mathrm{LUMO}}(\vec{r}) \ , \tag{47}$$
$$f^-(\vec{r}) \approx \rho_{\mathrm{HOMO}}(\vec{r}) \ , \tag{48}$$

and hence (from Eq. 19)

$$f^0(\vec{r}) \approx \frac{\rho_{\mathrm{HOMO}}(\vec{r}) + \rho_{\mathrm{LUMO}}(\vec{r})}{2} \ . \tag{49}$$

Equations (47)–(49) are precisely the densities of the frontier molecular orbitals; hence, in a frozen core approximation the (exact) theory of the Fukui function reverts to an essential element of the approximate frontier molecular orbital theory of Fukui [55–57].

Michelak et al. [45] have computed the Fukui function using a variety of the methods previously described as well as some more sophisticated approximate approaches. Specifically, they considered the finite difference approximation to the derivative with $\Delta N = 1$ and $\Delta N = 0.01$, a modified finite difference formula in which only terms linear in orbital changes and occupation number changes are retained, the frozen core approximation (Eqs. 47, 48) and an approximation to Eq. (46).

Their approximation to Eq. (46) deserves further comment. If one differentiates the Kohn–Sham equations with respect to the particle number at constant external potential, one obtains

$$\left( -\frac{\nabla^2}{2} + v_0(\vec{r}) + \int \frac{\rho(\vec{r}')}{|\vec{r} - \vec{r}'|} \, \mathrm{d}\vec{r}' + v_{\mathrm{xc}}(\vec{r}) - \varepsilon_i \right)$$
$$\times \left( \frac{\partial \phi_i(\vec{r})}{\partial N} \right)_{v_0(\vec{r})} + \left[ \int \frac{f(\vec{r}')}{|\vec{r} - \vec{r}'|} \, \mathrm{d}\vec{r}' + \left( \frac{\partial v_{\mathrm{xc}}(\vec{r})}{\partial N} \right)_{v_0(\vec{r})} \right.$$
$$\left. - \left( \frac{\partial \varepsilon_i}{\partial N} \right)_{v_0(\vec{r})} \right] \phi_i(\vec{r}) = 0 \ , \tag{50}$$

One can then get the Fukui function by combining Eqs. (50) and (46). By omitting the derivative of the $v_{\mathrm{xc}}$ with respect to $N$ (a term which is approximated to questionable accuracy by existing functionals), one may eliminate all the integrals except those which were already calculated in the course of the Kohn–Sham calculation (which provides an advantage over methods using Eqs. 43, 44).

358

The results of Michelak et al. [45] indicate that while the frontier molecular orbital approximation is sometimes qualitatively different from the other methods of computing the Fukui function, all the other methods usually give qualitatively similar results. Since the Fukui function is used to make qualitative judgements about site reactivity, these results suggest that one may compute the effects of orbital relaxation on the Fukui function in whatever reasonable manner one finds most convenient.

# 6 Miscellany

There are numerous other applications and extensions of the Fukui function. For instance, we have not included the extension of the Fukui function formalism to conductors (where there are "bands" of occupied states rather than occupied orbitals) [58, 59]. One can also fruitfully consider "condensed Fukui functions" [12, 45, 60–64], where one affixes a Fukui "index" to each atomic center (by partitioning the molecule into regions in either real space or function space and integrating the Fukui function over that region). The relationships between the Fukui function and the grand canonical ensemble of DFT and the relationships between the Fukui function and the softness kernel and local softness are also important [24, 51, 65].

In Sect. 2 we hinted at some of the Fukui function's shortcomings. The Fukui function measures the change in the density of a system when the number of electrons changes at constant external potential. As such, the Fukui function accurately reflects the component of chemical reactivity that is conveyed through the transfer of charge between systems; however, the Fukui function explicitly ignores the effects of the approaching reagent's external potential, assuming, in effect, that the transfer of charge between reactant molecules occurs at such large distances that charge distributes itself in a reactant molecule exactly as it would were the molecule isolated. This assumption may be good when the transition state occurs very early along the reaction coordinate (equivalently, when the transition state more strongly resembles the reactants than the products) because in this case the "point of no return" which determines how the reactant molecules will bind to each other is encountered before the external potentials (hence the density, hence the Fukui function) of the reactants are significantly different from those of the isolated molecules. Of course, the Fukui potential response [66], $(\delta f(\vec{r})/\delta v_0(\vec{r}'))_N$, enters into the energy expression in the third order and includes these effects [51]. For cases in which more than one electron is transferred, the discontinuities in the Fukui function at integer $N$ cause further complications.

Similar to the previous caveat is the following observation: the Fukui function is generally computed for an isolated reactant molecule (using one of the approximations from Sect. 4). When the reactant is placed into solution, however, the external potential felt by the molecule changes, and both the ground-state density and the Fukui function can be expected to differ appreciably from their values in vacuo. Viewed in this context, it is

remarkable that the frontier molecular orbital theory does as good a job at predicting reactivity in solution as it does – especially when one considers the ability of solvents to stabilize and destabilize charges on molecules.

One can rationalize the success of the gas-phase Fukui functions in solution by noting that the chemical potential, $\mu$, is relatively insensitive to solvent effects [67]. If this insensitivity carries through to the Fukui function (which is a functional derivative of $\mu$ – see Eq. 21) then the use of gas-phase Fukui functions is placed on a more solid footing. Alternatively, consider the argument of Dewar and Storch [68]: if the solvent is pushed out of the "reacting region" as the reactant molecules approach each other then solvent effects may be small enough for the gas-phase reactivity indices to make qualitatively correct predictions. Further exploration of the gas-phase Fukui function's success in solution is needed.

To the extent that the Fukui function is a qualitative index, solvent effects and the way $f(\vec{r})$ changes with $v_0(\vec{r})$ and $N$ are unimportant whenever such effects do not change the ordering of site reactivities (as in this case the predicted site reactivity preferences of the molecule are unchanged from the gas phase). Certainly we are not always so fortunate. For instance, the site at which an ambidentate ligand binds to a substrate will generally depend both on the solvent in which the reaction is conducted and on the identity of the substrate. Inasmuch as gas-phase Fukui functions predict that one of the two possible binding sites is preferable in all circumstances, the gas-phase Fukui function fails to predict the correct results whenever circumstances conspire to make the other binding site preferable. (Contreras et al. [69] have recently considered such a case, discovering that including the effects of the Fukui response was sufficient to give the correct site reactivity predictions for acetaldehyde enolate.)

Looking towards the future, designing better ways to compute the Fukui function is still a priority. In this regard, the search for accurate, explicit, kinetic energy functionals (so that Eq. 40 or 42 represents a viable alternative to Kohn–Sham-based methods like Eqs. 45, 46) is of great importance. Of course, finding an accurate kinetic energy functional would also revolutionize DFT by providing a simple alternative to solving the Kohn–Sham equations. In the more immediate future, it is anticipated that the recent progress in computing the Fukui function from a single Kohn–Sham calculation will continue, with more and more reliable approximations being made, while concurrent progress is made in increasing the methods' computational efficiency.

As we deepen our understanding of the zero-temperature limit of the grand canonical ensemble [6, 26–28] and other possible schemes for interpolating between integer numbers of electrons we may gain new insights into the Fukui function and its associated reactivity measures.

Recently there has been some interest in the Fukui responses, which, we argued, may be important for systems in which the charge transfer between reactants only occurs once the reactants are too close together for either reactant's Fukui function to resemble its value in isolation [25, 69]. As efficient methods for computing Fukui responses become available we may see first-order corrections to the Fukui function routinely included in many calculations.

Finally, recent work by Chattaraj and coworkers [70, 71] has focussed on extending the conceptual tools of the ground-state DFT to excited states. These studies have the potential to significantly deepen our understanding of excited-state chemistry.

We close with some additional words on the significance of the Fukui function. The Fukui function shores up the theoretical foundations of frontier molecular orbital theory. Equations (43)–(46) reveal that the site reactivity indices of frontier molecular orbital theory (Eqs. 47–49) may be regarded as the frozen orbital approximation to the Fukui function. The Fukui function is the zeroth-order index for site reactivity; various functional derivatives of the Fukui function represent higher-order corrections to the zeroth-order site reactivity map provided by the Fukui function. For instance, the first-order correction to the Fukui function is

$$\delta f^{+/0/-}[N, v_0(\vec{r}); \vec{r}] = \left(\frac{\partial f^{+/0/-}(\vec{r})}{\partial N}\right)_{v_0(\vec{r})} dN$$
$$+ \int \left(\frac{\delta f^{+/0/-}(\vec{r})}{\delta v_0(\vec{r}')}\right)_N \delta v_0(\vec{r}') d\vec{r}' \, . \tag{51}$$

Hence, frontier molecular orbital theory may be considered as an approximation to the theory of the Fukui function, which is itself an approximation (since it is corrected by Eq. 51 and higher-order corrections) to the exact site reactivity map of a molecule.

Also significant is the fact that the Fukui function can be used to provide an "aufbau principle" for DFT. To see this, find $f_B^-(\vec{r})$ and subtract this from $\rho(\vec{r})$ to get the $N - 1$ electron density $\rho_{N-1}(\vec{r})$ (see Eqs. 30, 31). Continuing this process until all the electrons have been removed, one finds that

$$\rho(\vec{r}) = \sum_{n=1}^{N} f_n^-(\vec{r})$$
$$= \sum_{n=1}^{N} \int_{n-1}^{n} f_k^-(\vec{r}) dk \, , \tag{52}$$

where $f_n^-(\vec{r})$ is the Fukui function from below for the $n$-electron system. The second equality in Eq. (52) extends this "aufbau principle" to approximate density functionals (where Eqs. 30, 31 are invalid). Equation (52) has considerable potential as a conceptual tool since it allows one to build an $N$-electron density from one-electron densities just as one might build an $N$-electron wavefunction from orbitals.

## 7 Conclusion

We started this review by indicating that we should strive for general principles that explain a wide range of chemical phenomena. The use of the Fukui function provides one such principle [2, 3, 12, 72–75]. While the Fukui function may give erroneous results in situations where the first- and higher-order corrections thereto are important, it is the leading-order term in an exact theory of the site reactivity of a molecule.

The Fukui function successfully predicts relative site reactivities for most chemical systems. As such it provides a method for understanding and categorizing chemical reactions. More importantly, the Fukui function can be used to predict what the products of a given reaction will be. As computing the Fukui function becomes faster and easier, its predictive ability might be routinely used to winnow the list of potentially useful reagents, catalysts, etc. before performing the types of experiments or calculations necessary to fully characterize a chemical reaction. This predictive ability renders the Fukui function an important tool of the chemist.

*Acknowledgements.* Financial support from the National Science Foundation is greatly appreciated. The authors thank R.G. Parr, W. Yang, and M. Berkowitz for helpful comments.

## References

1. Attributed to C.A. Coulson by R. McWeeny in the preface of McWeeny R (1979) Coulson's valence. Oxford University Press, Oxford
2. Parr RG, Yang WT (1984) J Am Chem Soc 106: 4049
3. Parr RG, Yang WT (1989) Density functional theory of atoms and molecules. Oxford University Press, New York
4. Dreizler RM, Gross EKU (1990) Density functional theory. Springer, Berlin Heidelberg New York
5. Nalewajski RF, Parr RG (1982) J Chem Phys 77: 399
6. Perdew JP, Parr RG, Levy M, Balduz JL (1982) Phys Rev Lett 49: 1691
7. Parr RG, Donnelly RA, Levy M, Palke WE (1978) J Chem Phys 68: 3801
8. Mulliken RS (1934) J Chem Phys 2: 782
9. Sanderson RT (1951) Science 114: 670
10. Sanderson RT (1976) Chemical bonds and bond energy. Academic, New York
11. Parr RG, Bartolotti LJ (1982) J Am Chem Soc 104: 3801
12. Chermette H (1999) J Comput Chem 20: 129
13. Feynman RP (1939) Phys Rev 56: 340
14. Liu S, Parr RG (1997) J Chem Phys 106: 5578
15. Parr RG, Pearson RG (1983) J Am Chem Soc 105: 7512
16. Pearson RG (1963) J Am Chem Soc 85: 3533
17. Chattaraj PK, Lee H, Parr RG (1991) J Am Chem Soc 113: 1855
18. Pearson RG (1995) Inorg Chim Acta 240: 93
19. Pearson RG (1987) J Chem Educ 64: 571
20. Parr RG, Chattaraj PK (1991) J Am Chem Soc 113: 1854
21. Chattaraj PK (1996) Proc Indian Natl Sci Acad Part A 62: 513
22. Pearson RG (1993) Acc Chem Res 26: 250
23. Pearson RG (1999) J Chem Educ 76: 267
24. Berkowitz M, Parr RG (1988) J Chem Phys 88: 2554
25. Senet P (1996) J Chem Phys 105: 6471
26. Perdew JP, Levy M (1997) Phys Rev B 56: 16021
27. Chan GK-L (1999) J Chem Phys 110: 4710
28. Zhang Y, Yang WT Theor Chem Acta
29. Berkowitz M (1987) J Am Chem Soc 109: 4823
30. Parr RG, Bartolotti LJ (1983) J Phys Chem 87: 2810
31. Cedillo A (1994) Int J Quantum Chem Symp 28: 231
32. Steiner E (1963) J Chem Phys 39: 2365
33. Kato T (1957) Commun Pure Appl Math 10: 151

34. Handy NC (1996) In: Bicout D, Field M (eds) Quantum mechanical simulation methods for studying biological systems. Springer Berlin Heidelberg New York, p 1
35. Pack RT, Brown WB (1966) J Chem Phys 45: 556
36. Chattaraj PK, Cedillo A, Parr RG (1995) J Chem Phys 103: 10621
37. Morrell MM, Parr RG, Levy M (1975) J Chem Phys 62: 549
38. Levy M, Parr RG (1976) J Chem Phys 64: 2707
39. Katriel J, Davidson ER (1980) Proc Natl Acad Sci USA 77: 4403
40. Hoffmann-Ostenhof M, Hoffmann-Ostenhof T (1977) Phys Rev A 16: 1782
41. Aldrichs R, Hoffmann-Ostenhof M, Hoffmann-Ostenhof T, Morgan JD III (1981) Phys Rev A 23: 2106
42. Levy M, Perdew JP, Sahni V (1984) Phys Rev A 30: 2745
43. Ambladh CO, von Barth U (1985) Phys Rev B 31: 3231
44. Patil SH (1989) J Phys B 22: 2051
45. Michalak A, De Proft F, Geerlings P, Nalewajski RF (1999) J Phys Chem A 103: 762
46. Pacios LF, Gmez PC (1998) J Comput Chem 19: 488
47. Chattaraj PK, Cedillo A, Parr RG (1995) J Chem Phys 103: 7645
48. Ghosh SK (1990) Chem Phys Lett 172: 77
49. Cedillo A, Parr RG (1996) J Chem Phys 105: 9557
50. Cohen MH, Ganduglia-Pirovano MV, Kudrnovsky J (1994) J Chem Phys 101: 8988
51. Cohen MH, Ganduglia-Pirovano MV, Kudrnovsky J (1995) J Chem Phys 103: 3543
52. Cohen MH (1996) In: Nalewajski RF (ed) Topics in current chemistry: density functional theory IV: theory of chemical reactivity. Springer, Berlin Heidelberg New York, pp 143
53. Senet P (1997) J Chem Phys 107: 2516
54. Yang WT, Parr RG, Pucci R (1984) J Chem Phys 81: 2862
55. Fukui K, Yonezawa Y, Shingu H (1952) J Chem Phys 20: 722
56. Fukui K (1973) Theory of orientation and stereoselection. Springer, Berlin Heidelberg New York
57. Fukui K (1982) Science 217: 747
58. Yang W, Parr RG (1985) Proc Natl Acad Sci USA 82: 6723
59. Cohen MH, Ganduglia-Pirovano MV, Kudrnovsky J (1994) Phys Rev Lett 72: 3222
60. Yang W, Mortier W (1986) J Am Chem Soc 108: 5708
61. Langenaeker W, Demel K, Geerlings P (1992) J Mol Struct (THEOCHEM) 259: 317
62. Balawender R, Komorowski L (1998) J Chem Phys 109: 5203
63. De Proft F, Langenaeker W, Geerlings P (1995) Int J Quantum Chem 55: 449
64. Roy RK, Pal S, Hirao K (1999) J Chem Phys 110: 8236
65. Harbola MK, Chattaraj PK, Parr RG (1991) Isr J Chem 31: 395
66. Fuentealba P, Parr RG (1991) J Chem Phys 94: 5559
67. Pearson RG (1986) J Am Chem Soc 108: 6109
68. Dewar MJS, Storch DM (1985) Proc Natl Acad Sci USA 82: 2225
69. Contreras R, Domingo LR, Andrés J, Pérez P, Tapia O (1999) J Phys Chem A 103: 1367
70. Chattaraj PK, Sengupta S (1997) J Phys Chem A 101: 7893
71. Chattaraj PK, Poddar A (1998) J Phys Chem A 102: 9944
72. Nalewajski RF (ed) (1996) Topics in current chemistry: density functional theory IV: theory of chemical reactivity. Springer, Berlin Heidelberg New York; see also Refs [3, 12, 73–75] and references therein
73. Geerlings P, Langenaeker W, De Proft F, Baeten A (1996) In: Murray JS, Sen K (eds) Molecular electrostatic potentials: concepts and applications. Theoretical and computational chemistry, vol. 3. Elsevier, Amsterdam, pp 587
74. Geerlings P, De Proft F, Martin JML (1996) In: Seminario JM (ed) Recent developments and applications of modern density functional theory. Theoretical and computational chemistry, vol 4. Elsevier, Amsterdam, pp 773
75. Parr RG, Yang W (1995) Annu Rev Phys Chem 46: 701

Theor Chem Acc (2000) 103:361–363
DOI 10.1007/s002149900065

Theoretical
Chemistry Accounts
© Springer-Verlag 2000

*Perspective*

# Perspective on "Density functional thermochemistry. III. The role of exact exchange"

## Becke AD (1993) J Chem Phys 98:5648–52

**Krishnan Raghavachari**

Bell Laboratories, Lucent Technologies, Murray Hill, NJ 07974, USA

Received: 30 March 1999 / Accepted: 27 April 1999 / Published online: 4 October 1999

**Abstract.** Recent developments in density functional theory have transformed the entire field of quantum chemistry. This paper provides a perspective on Becke's landmark papers in 1992 and 1993 that led to the popular density functionals such as B3LYP.

**Key words:** Quantum chemistry – Density functional theory – Theoretical thermochemistry

One of the most dramatic changes in the standard theoretical model used most widely in quantum chemistry occurred in the early 1990s. Until then, ab initio quantum chemical applications [1] typically used a Hartree–Fock (HF) starting point, followed in many cases by second-order Møller–Plesset perturbation theory. For small molecules requiring more accuracy, additional calculations were performed with coupled-cluster theory, quadratic configuration interaction, or related methods. While these techniques are still used widely, a substantial majority of the papers being published today are based on applications of density functional theory (DFT) [2]. Almost universally, the researchers use a functional due to Becke, whose papers in 1992 and 1993 contributed to this remarkable transformation that changed the entire landscape of quantum chemistry.

DFT within the local density approximation (LDA) has been used extensively for decades in computational physics; however, its deficiencies, such as the substantial overestimation of molecular binding energies, were well known. Consequently, applications using LDA for molecular systems [3] were limited, particularly relative to the large number of quantum chemical studies using HF-based methods. The emergence of gradient-corrected density functionals in the late 1980s increased the accuracy of such methods significantly. Becke proposed one of the most successful gradient-corrected exchange functionals (B88) in 1988 [4]. Used in conjunction with

gradient-corrected correlation functionals such as those of Perdew (P86) [5], Lee, Yang, and Parr (LYP) [6], or Perdew and Wang (PW91) [7], density functional methods were successfully used in many applications [8]; however, these important developments did not initially have much impact on the mainstream ab initio quantum chemistry community.

It was at this juncture that Becke started a series of publications on "Density functional thermochemistry" [9–13]. The title publication on the "role of exact exchange" is the third paper in this series [11] and has had an astonishing impact on the entire field of quantum chemistry. One clear measure of the impact of this 1993 paper is that it has already accumulated more than 2200 citations in the scientific literature in less than 6 years [14]. Interestingly, most of the citations are from practitioners of quantum chemistry who have investigated a variety of chemical applications using functionals such as B3LYP with commercial software packages.

One of the most important reasons for the impact of Becke's papers was his decision to assess the DFT methods systematically on the same set of test molecules assembled by Pople and coworkers in their development of the Gaussian-$n$ methods [15–18]. The successful G1 [15, 16] and G2 [17] methods were the first two methods developed for the prediction of the thermochemistry of small molecules within chemical accuracy ($\pm 2$ kcal/mol). In order to assess the performance of these theoretical methods, Pople and coworkers also developed a set of test molecules for which accurate experimental thermochemical information is known. Becke adopted the initial G1 test set containing 55 neutral molecules as the standard test set for the development and assessment of new density functionals. This immediately provided a mechanism for an unbiased comparison of the performance of DFT methods with the state-of-the-art ab initio quantum chemical techniques for the same set of molecules.

The first two papers in the series by Becke [9, 10] explored the improvement in the predicted thermochemistry of DFT methods due to the inclusion of gradient corrections from exchange (B88) and correlation

(PW91), respectively. While the PW91 correlation had only a small effect, the inclusion of B88 exchange caused a dramatic improvement. The mean absolute deviation for the calculated dissociation energies of 55 molecules in the G1 test set decreased from 36.2 kcal/mol (LDA) to 3.7 kcal/mol (B88). For comparison, the state-of-the-art G2 method had a deviation of 1.2 kcal/mol [17]. Becke presented these results at the 7th International Congress in Quantum Chemistry at Menton, France, in July 1991. The performance of the B88 functional was even more remarkable for a method that is no more expensive computationally than HF theory. The implications for larger molecules and for the future of quantum chemistry were obvious.

Becke's initial papers immediately caught the attention of the mainstream ab initio quantum chemistry community. In particular, Pople and coworkers started to investigate gradient-corrected density functional methods. Even more importantly, they were quickly able to implement them efficiently in the Gaussian 92 computer program [19]. In fact, their implementation included a fully self-consistent solution of the Kohn–Sham equations for the available gradient-corrected density functionals. This was an improvement over Becke's original implementation, which was only post-LDA (self-consistent treatment at the LDA level followed by evaluation of the functional using LDA densities). Pople presented a preliminary assessment of the applicability of DFT methods at the Sanibel quantum chemistry symposium in 1992 [20]. The widespread availability of the Gaussian program coupled with Pople's influence resulted in a large number of quantum chemists starting to use DFT methods. Interestingly, Pople and coworkers performed most of their initial assessment [21, 22] using the BLYP functional (B88 exchange and LYP correlation) which immediately became one of the most popular density functionals.

The third paper by Becke [11] was published at this point and introduced a key theoretical concept – the role of exact exchange. In an earlier paper, Becke had used ideas from the "adiabatic connection" formula to show that a novel "half-and-half" mixing of exact exchange energy (as evaluated in HF theory) with that from LDA yielded improved results [23]. The title paper explored this idea further and concluded that exact exchange must play a role in highly accurate density functional methods. In addition, Becke introduced a semiempirical generalization to determine the extent of mixing of exact exchange based on fitting to the G1 thermochemical data. In fact, he proposed the following formula containing three different semiempirical parameters:

$$E_{xc} = E_{xc}^{LDA} + a_0 \left( E_x^{exact} - E_x^{LDA} \right) + a_x \Delta E_x^{B88} + a_c \Delta E_c^{PW91} .$$

The semiempirical coefficient $a_0$ determines the extent of replacement of electron-gas exchange with exact exchange, and $a_x$ and $a_c$ determine the optimum inclusion of gradient-correction for exchange (B88) and correlation (PW91), respectively. Becke determined the coefficients by a linear least-squares fit to 56 atomization energies (the original 55 molecules + $H_2$), 42 ionization

potentials, and 8 proton affinities in the G1 test set along with 10 first-row atomic energies to get $a_0 = 0.20$, $a_x = 0.72$, and $a_c = 0.81$. Using the three-parameter fit, Becke obtained a mean absolute deviation of 2.4 kcal/mol for the atomization energies of the molecules in the G1 test set. This was only a factor of 2 from that of G2 theory (1.2 kcal/mol) and very close to the target value of $\pm 2$ kcal/mol to reach chemical accuracy.

There are several interesting points to note about this expression proposed by Becke. First of all, the use of a small number of semiempirical parameters obtained by fitting to the G1 test set is somewhat reminiscent of the parameters ("higher-level corrections") used by Pople and coworkers in their formulations of the Gaussian-$n$ methods [15–18]. More interestingly, in today's notation, Becke's functional as defined above is represented as B3PW91, where B3 refers to the three parameters in the defining expression and PW91 refers to the correlation functional used. The functional B3LYP that is most popular in the quantum chemical literature today uses LYP instead of PW91. Becke never did propose or endorse the use of the B3LYP functional; however, the Gaussian 94 program [24] allowed the use of different combinations of exchange and correlation functionals and included an implementation of the B3LYP functional (with an easy-to-use keyword). As defined, the B3LYP functional uses the same semiempirical parameters derived by Becke (for B3PW91) without any re-optimization. Nevertheless, B3LYP appears to perform somewhat better than B3PW91 in careful assessments. For example, in a recent critical evaluation of seven different density functionals, B3LYP had the lowest mean absolute deviation from experiment for the heats of formation of 148 molecules in the G2 test set [25]. Surprisingly, refitting of the three parameters in the B3LYP functional yields only a slight improvement.

Overall, Becke's landmark series of papers were the key ingredients in causing an important transformation in quantum chemistry. The first two papers were instrumental in getting the attention of the ab initio quantum chemistry community, while the third paper provided the key idea of the role of exact exchange that led to the highly successful functionals such as B3LYP. Interestingly, the need to evaluate explicit exchange integrals has made it difficult to incorporate such functionals in the conventional software packages used by the computational physics community; however, the quantum chemistry community has embraced this approach that has led to a dramatic increase in the number of users and applications using such hybrid density functionals. The powerful DFT methods are now an essential part of the arsenal of computational quantum chemistry.

## References

1. Hehre WJ, Radom L, Pople JA, Schleyer PvR (1987) Ab initio molecular orbital theory. Wiley, New York
2. (a) Hohenberg P, Kohn W (1964) Phys Rev B 136: 864; (b) Kohn W, Sham LJ (1965) Phys Rev A 140: 1133
3. Labanowski JK, Andzelm JW (eds) (1991) Density functional methods in chemistry. Springer, Berlin Heidelberg New York
4. Becke AD (1988) Phys Rev A 38: 3098

5. (a) Perdew JP (1986) Phys Rev B 33: 8822; (b) Perdew JP (1986) Phys Rev B 34: 7406 (erratum)
6. Lee C, Yang W, Parr RG (1988) Phys Rev B 37: 785
7. Perdew JP, Wang Y (1992) Phys Rev B 45: 13244
8. (a) Ziegler T (1991) Chem Rev 91: 651; (b) Parr RG, Yang W (1995) Annu Rev Phys Chem 46: 701, and references therein
9. Becke AD (1992) J Chem Phys 96: 2155
10. Becke AD (1992) J Chem Phys 97: 9173
11. Becke AD (1993) J Chem Phys 98: 5648
12. Becke AD (1996) J Chem Phys 104: 1040
13. Becke AD (1997) J Chem Phys 107: 8554
14. Institute for Scientific Information – Citation database – 2257 citations as of March 23, 1999
15. Pople JA, Head-Gordon M, Fox DJ, Raghavachari K, Curtiss LA (1989) J Chem Phys 90: 5622
16. Curtiss LA, Jones C, Trucks GW, Raghavachari K, Pople JA (1990) J Chem Phys 93: 2537
17. Curtiss LA, Raghavachari K, Trucks GW, Pople JA (1991) J Chem Phys 94: 7221
18. Curtiss LA, Raghavachari K, Redfern PC, Rassolov A, Pople JA (1998) J Chem Phys 109: 7764
19. Frisch MJ, Trucks GW, Head-Gordon M, Gill PMW, Wong MW, Foresman JB, Johnson BG, Schlegel HB, Robb MA, Replogle ES, Gomperts R, Andres JL, Raghavachari K, Binkley JS, Gonzales C, Martin RL, Fox DJ, Defrees DJ, Baker J, Stewart JJP, Pople JA (1992) Gaussian 92. Gaussian Inc., Pittsburgh, Pa
20. Gill PMW, Johnson BG, Pople JA, Frisch MJ (1992) Int J Quantum Chem Symp 26: 319
21. Gill PMW, Johnson BG, Pople JA, Frisch MJ (1992) Chem Phys Lett 197: 499
22. Johnson BG, Gill PMW, Pople JA (1993) J Chem Phys 98: 5612
23. Becke AD (1993) J Chem Phys 98: 1372
24. Frisch MJ, Trucks GW, Schlegel HB, Gill PMW, Johnson BG, Robb MA, Cheeseman JR, Keith TA, Petersson GA, Montgomery JA, Raghavachari K, Al-Laham MA, Zakrzewski VG, Ortiz JV, Foresman JB, Cioslowski J, Stefanov BB, Nanayakkara A, Challacombe M, Peng CY, Ayala PY, Chen W, Wong MW, Andres JL, Replogle ES, Gomperts R, Martin RL, Fox DJ, Binkley JS, Defrees DJ, Baker J, Stewart JJP, Head-Gordon M, Gonzales C, Pople JA (1995) Gaussian 94. Gaussian, Inc., Pittsburgh, Pa
25. Curtiss LA, Raghavachari K, Redfern PC, Pople JA (1997) J Chem Phys 106: 1063